Nanoemulsions in Food Technology

Nanoemulsions in
Food Technology

Food Analysis & Properties
Series Editor: Leo M. L. Nollet
University College Ghent, Belgium

This CRC series **Food Analysis and Properties** is designed to provide a state-of-art coverage on topics to the understanding of physical, chemical and functional properties of foods including: (1) recent analysis techniques of a choice of food components, (2) developments and evolutions in analysis techniques related to food, (3) recent trends in analysis techniques of specific food components and/or a group of related food components.

Fingerprinting Techniques in Food Authenticity and Traceability
Edited by K.S. Siddiqi and Leo M.L. Nollet

Hyperspectral Imaging Analysis and Applications for Food Quality
Edited by Nrusingha Charan Basantia, Leo M.L. Nollet, Mohammed Kamruzzaman

Ambient Mass Spectroscopy Techniques in Food and the Environment
Edited by Leo M.L. Nollet and Basil K. Munjanja

Food Aroma Evolution: During Food Processing, Cooking and Aging
Edited by Matteo Bordiga, Leo M. L. Nollet

Mass Spectrometry Imaging in Food Analysis
Edited by Leo M. L. Nollet

Proteomics for Food Authentication
Edited by Leo M. L. Nollet, Otles, Semih

Analysis of Nanoplastics and Microplastics in Food
Edited by Leo M. L. Nollet and Khwaja Salahuddin Siddiqi

Chiral Organic Pollutants: Monitoring and Characterization in Food and the Environment
Edited by Edmond Sanganyado, Basil Munjanja, and Leo M. L. Nollet

Sequencing Technologies in Microbial Food Safety and Quality
Edited by Devarajan Thangadurai, Leo M.L. Nollet, Saher Islam, and Jeyabalan Sangeetha

Nanoemulsions in Food Technology: Development, Characterization, and Applications
Edited by Javed Ahmad and Leo M.L. Nollet

For more information, please visit the Series Page: https://www.crcpress.com/Food-Analysis–Properties/book-series/CRCFOODANPRO

Nanoemulsions in Food Technology

Development, Characterization, and Applications

Edited by
Javed Ahmad
Leo M.L. Nollet

CRC Press
Taylor & Francis Group
Boca Raton London New York

CRC Press is an imprint of the
Taylor & Francis Group, an **informa** business

First edition published 2022
by CRC Press
6000 Broken Sound Parkway NW, Suite 300, Boca Raton, FL 33487-2742

and by CRC Press
2 Park Square, Milton Park, Abingdon, Oxon, OX14 4RN

Library of Congress Cataloging-in-Publication Data
A catalog record has been requested for this book

ISBN: 978-0-367-61492-8 (hbk)
ISBN: 978-1-032-10442-3 (pbk)
ISBN: 978-1-003-12112-1 (ebk)

DOI: 10.1201/9781003121121

Typeset in Sabon
by KnowledgeWorks Global Ltd.

I would like to thank Javed from the bottom of my heart for his continued and immediate commitment to making this book a reality.

Leo M.L. Nollet

Contents

SECTION III   APPLICATIONS

Series Preface

There will always be a need to analyze food compounds and their properties. Current trends in analyzing methods include automation, increasing the speed of analyses, and miniaturization. Over the years, the unit of detection has evolved from micrograms to pictograms.

A classical pathway of analysis is sampling, sample preparation, clean up, derivatization, separation, and detection. At every step, researchers are working and developing new methodologies. A large number of papers are published every year on all facets of analysis. So, there is a need for books that gather information on one kind of analysis technique or on the analysis methods for a specific group of food components.

The scope of the CRC Series on Food Analysis & Properties aims to present a range of books edited by distinguished scientists and researchers who have significant experience in scientific pursuits and critical analysis. This series is designed to provide state-of-the-art coverage on topics such as:

1. Recent analysis techniques on a range of food components.
2. Developments and evolution in analysis techniques related to food.
3. Recent trends in analysis techniques for specific food components and/or a group of related food components.
4. The understanding of physical, chemical, and functional properties of foods.

The book *Nanoemulsions in Food Technology* is volume 16 of this series.

I am happy to be a series editor of such books for the following reasons:

- I am able to pass on my experience in editing high-quality books related to food.
- I get to know colleagues from all over the world more personally.
- I continue to learn about interesting developments in food analysis.

Much work is involved in the preparation of a book. I have been assisted and supported by a number of people, all of whom I would like to thank. I would especially like to thank the team at CRC Press/Taylor & Francis, with a special word of thanks to Steve Zollo, senior editor.

Many, many thanks to all the editors and authors of this volume and future volumes. I very much appreciate all their effort, time, and willingness to do a great job.

I dedicate this series to:

- My wife, for her patience with me (and all the time I spend on my computer).
- All patients suffering from prostate cancer; knowing what this means, I am hoping they will have some relief.

Preface

Nanoemulsion techniques are nowadays in increasing demand to meet the expectation of healthier and safer food products. Currently, researchers and people from academia and the food industry are making more efforts on the utilization of the nanoemulsion technique to encapsulate, protect, and deliver functional compounds for food science applications. This technique is of particular interest for the development of encapsulating nanosystems for the delivery of functional components because of significant advantages compared to conventional emulsification techniques. In addition, this technique is helpful in the design of delivery systems of nano dimension by utilizing low-energy emulsification methods excluding the need of any solvent, heat, or even sophisticated instruments in its production. The use of essential oils in improving food flavor or food production (particularly preservation and safety) is of significant importance. The nanoemulsion technique is really helpful in utilizing the essential oils of different biological sources for various purposes in nanoemulsified forms for food science application.

The whole book is divided into three sections.

The first section (*Fundamental Concepts and Technique Overview*) will provide a detailed discussion about the technology overview of nanoemulsions. The chapters of Section 1 mainly deal with fundamentals of the nanoemulsion technology; ingredients, and composition of food-grade nanoemulsions; methods of preparation and recent advancements in its manufacturing along with stability perspectives of this technique.

The second section (*Characterization*) of the book describes the techniques involved in nanoemulsion characterization. It mainly deals with interfacial and nanostructural characterization of nanoemulsions, different physical characterization techniques as well as various imaging and separation techniques involved in its characterization.

The last section of the book (*Applications*) is devoted to the discussion related to the applicability of nanoemulsion techniques in food sciences.

Each chapter concludes with a general reference section, which is a bibliographic guide to more advanced texts.

This book will provide in-depth information and a comprehensive discussion on a technology overview, physical and nanostructural characterization, as well as applicability of the nanoemulsion technique in food sciences. Researchers, industries, and students would be interested in this concise body of information. The contributing authors are drawn from a rich blend of experts in various areas of scientific field exploring nanoemulsion techniques for wider applications.

The editors thank all contributors for their timeless efforts to bring this book to a good end.

Javed Ahmad
Leo M.L. Nollet

About the Editors

Dr. Javed Ahmad is an Assistant Professor at the Department of Pharmaceutics College of Pharmacy, Najran University, KSA. He received his Ph.D. in Pharmaceutics from Jamia Hamdard, New Delhi, India.

His principal working experiences are nanotechnology–mediated drug delivery through different routes to improve the clinical efficacy and pharmacokinetic profile of different therapeutics for better management of various diseases.

He has received numerous awards for his scientific findings and reviewing tasks.

He is the author of nearly 100 publications and/or book chapters. He is the editor of *Bioactive Phytochemicals: Drug Discovery to Product Development* (Bentham Science Publisher, 2020).

Leo M.L. Nollet earned an MS (1973) and Ph.D. (1978) in biology from the Katholieke Universiteit Leuven, Belgium. He is an editor and associate editor of numerous books. He edited for M. Dekker, New York—now CRC Press/Taylor & Francis Publishing Group—the first, second, and third editions of *Food Analysis by HPLC* and *Handbook of Food Analysis*. The last edition is a two-volume book. Dr. Nollet also edited the *Handbook of Water Analysis* (first, second, and third editions) and *Chromatographic Analysis of the Environment*, third and fourth editions (CRC Press). With F. Toldrá, he coedited two books published in 2006, 2007, and 2017: *Advanced Technologies for Meat Processing* (CRC Press) and *Advances in Food Diagnostics* (Blackwell Publishing—now Wiley). With M. Poschl, he coedited the book *Radionuclide Concentrations in Foods and the Environment*, also published in 2006 (CRC Press). Dr. Nollet has also coedited with Y.H. Hui and other colleagues on several books: *Handbook of Food Product Manufacturing* (Wiley, 2007), *Handbook of Food Science, Technology, and Engineering* (CRC Press, 2005), *Food Biochemistry and Food Processing* (first and second editions; Blackwell Publishing—now Wiley—2006 and 2012), and the *Handbook of Fruits and Vegetable Flavors* (Wiley, 2010). In addition, he edited the *Handbook of Meat, Poultry, and Seafood Quality*, first and second editions (Blackwell Publishing—now Wiley—2007 and 2012). From 2008 to 2011, he published five volumes on animal product-related books with F. Toldrá: *Handbook of Muscle Foods Analysis, Handbook of Processed Meats and Poultry Analysis, Handbook of Seafood and Seafood Products Analysis, Handbook of Dairy Foods Analysis,* and *Handbook of Analysis of*

Edible Animal By-Products. Also, in 2011, with F. Toldrá, he coedited two volumes for CRC Press: *Safety Analysis of Foods of Animal Origin* and *Sensory Analysis of Foods of Animal Origin.* In 2012, they published the *Handbook of Analysis of Active Compounds in Functional Foods.* In a coedition with Hamir Rathore, *Handbook of Pesticides: Methods of Pesticides Residues Analysis* was marketed in 2009; *Pesticides: Evaluation of Environmental Pollution* in 2012; *Biopesticides Handbook* in 2015; and *Green Pesticides Handbook: Essential Oils for Pest Control* in 2017. Other finished book projects include *Food Allergens: Analysis, Instrumentation, and Methods* (with A. van Hengel; CRC Press, 2011) and *Analysis of Endocrine Compounds in Food* (Wiley-Blackwell, 2011). Dr. Nollet's recent projects include *Proteomics in Foods* with F. Toldrá (Springer, 2013) and *Transformation Products of Emerging Contaminants in the Environment: Analysis, Processes, Occurrence, Effects, and Risks* with D. Lambropoulou (Wiley, 2014). In the series Food Analysis & Properties, he edited (with C. Ruiz-Capillas) *Flow Injection Analysis of Food Additives* (CRC Press, 2015) and *Marine Microorganisms: Extraction and Analysis of Bioactive Compounds* (CRC Press, 2016). With A.S. Franca, he coedited *Spectroscopic Methods in Food Analysis* (CRC Press, 2017), and with Horacio Heinzen and Amadeo R. Fernandez-Alba he coedited *Multiresidue Methods for the Analysis of Pesticide Residues in Food* (CRC Press, 2017). Further volumes in the series Food Analysis & Properties are *Phenolic Compounds in Food: Characterization and Analysis* (with Janet Alejandra Gutierrez-Uribe, 2018), *Testing and Analysis of GMO-containing Foods and Feed* (with Salah E. O. Mahgoub, 2018), *Fingerprinting Techniques in Food Authentication and Traceability* (with K.S. Siddiqi, 2018), *Hyperspectral Imaging Analysis and Applications for Food Quality* (with N.C. Basantia, Leo M.L. Nollet, Mohammed Kamruzzaman, 2018), *Ambient Mass Spectroscopy Techniques in Food and the Environment* (with Basil K. Munjanja, 2019), *Food Aroma Evolution: During Food Processing, Cooking, and Aging* (with M. Bordiga, 2019), *Mass Spectrometry Imaging in Food Analysis* (2020), *Proteomics in Food Authentication* (with S. Ötleş, 2020), *Analysis of Nanoplastics and Microplastics in Food* (with K.S. Siddiqi, 2020), and *Chiral Organic Pollutants, Monitoring and Characterization in Food and the Environment* (with Edmond Sanganyado and Basil K. Munjanja, 2020).

List of Contributors

Javed Ahamad
Department of Pharmacognosy
Faculty of Pharmacy
Tishk International University
Kurdistan Region, Iraq

Javed Ahmad
Department of Pharmaceutics
College of Pharmacy
Najran University
Najran, Kingdom of Saudi Arabia

Mohammad Zaki Ahmad
Department of Pharmaceutics
College of Pharmacy
Najran University
Najran, Kingdom of Saudi Arabia

Usama Ahmad
Department of Pharmaceutics
Faculty of Pharmacy
Integral University
Lucknow, India

Ayesha Akhtar
Quality Control
Med City Pharma
Jeddah, Kingdom of Saudi Arabia

Md. Shoaib Alam
Research and Development
Jamjoom Pharmaceuticals
Jeddah, Kingdom of Saudi Arabia

Hibah Mubarak Aldawsari
Department of Pharmaceutics
Faculty of Pharmacy
King Abdulaziz University
Jeddah, Kingdom of Saudi Arabia

Fahad Khalid Aldhafiri
Public Health Department
College of Applied Medical
    Sciences
Majmaah University
Riyadh, Kingdom of Saudi Arabia

Nabil Abdulhafiz Alhakamy
Department of Pharmaceutics
Faculty of Pharmacy
King Abdulaziz University
Jeddah, Kingdom of Saudi Arabia

Faraat Ali
Department of Inspection and Enforcement
Laboratory Services
Botswana Medicines Regulatory Authority
    Gaborone, Botswana

Faisal Obaid Alotaibi
College of Pharmacy (Boys)
Al-Dawadmi Campus
Shaqra University
Riyadh, Kingdom of Saudi Arabia

Sultan Alshehri
Department of Pharmaceutics
College of Pharmacy
King Saud University
Department of Pharmaceutical
    Sciences
College of Pharmacy
Almaarefa University
Riyadh, Kingdom of Saudi Arabia

Mohammed Faiz Arshad
College of Pharmacy
   (Boys)
Al-Dawadmi Campus
Shaqra University
Riyadh, Kingdom of
   Saudi Arabia

Mohammed Aslam
Faculty of Pharmacy
Al Hawash Private
   University
Homs, Syria

Shaimaa M. Badr-Eldin
Department of Pharmaceutics
Faculty of Pharmacy
King Abdulaziz University
Jeddah, Kingdom of
   Saudi Arabia

Md. Abul Barkat
Department of Pharmaceutics
College of Pharmacy
University of Hafr Al Batin
Al Jamiah, Hafr Al Batin, Kingdom
   of Saudi Arabia

Georgeos Deeb
Faculty of Science
Damascus University
Damascus, Syria

Jamia Firdous
DS College of Pharmacy
Aligarh, India

Anuj Garg
Institute of Pharmaceutical
   Research
GLA University
Mathura, India

Sadaf Jamal Gilani
Department of Basic Health
   Sciences
Princess Nourah bint Abdulrahman
   University
Riyadh, Kingdom of Saudi Arabia

Khaled M. Hosny
Department of Pharmaceutics
Faculty of Pharmacy
King Abdulaziz University
Jeddah, Kingdom of Saudi Arabia
Department of Pharmaceutics
   and Industrial Pharmacy
Faculty of Pharmacy
Beni-Suef University
Beni-Suef, Egypt

Syed Sarim Imam
Department of Pharmaceutics
College of Pharmacy
King Saud University
Riyadh, Kingdom of Saudi Arabia

Mohammed Asadullah Jahangir
Department of Pharmaceutics
Nibha Institute of Pharmaceutical Sciences
Rajgir, India

Keerti Jain
Department of Pharmaceutics
National Institute of Pharmaceutical
   Education and Research (NIPER)
Raebareli, India

Pawan Kaushik
Institute of Pharmaceutical Sciences
Kurukshetra University
Kurukshetra, India

Sabna Kotta
Department of Pharmaceutics
Faculty of Pharmacy
King Abdulaziz University
Jeddah, Kingdom of Saudi Arabia

Deepak Kumar
Department of Pharmaceutics
National Institute of Pharmaceutical
   Education and Research (NIPER)
Raebareli, India

Wei Meng Lim
Department of Pharmaceutical Technology
School of Pharmacy
International Medical University
Kuala Lumpur, Malaysia

Shadab Md
Department of Pharmaceutics
Faculty of Pharmacy
King Abdulaziz University
Jeddah, Kingdom of Saudi Arabia

Showkat R. Mir
Department of Pharmacognosy
School of Pharmaceutical Education
& Research
Jamia Hamdard
New Delhi, India

Mohd. Aamir Mirza
Department of Pharmaceutics
School of Pharmaceutical Education
& Research
Jamia Hamdard
New Delhi, India

Abdul Aleem Mohammad
Department of Pharmaceutics
College of Pharmacy
Najran University
Najran, Kingdom of Saudi Arabia

Md. Ali Mujtaba
Department of Pharmaceutics
Faculty of Pharmacy
Northern Border University
Rafha, Kingdom of Saudi Arabia

Abdul Muheem
Department of Pharmaceutics
School of Pharmaceutical Education
& Research
Jamia Hamdard
New Delhi, India

Gulam Mustafa
College of Pharmacy
(Boys)
Al-Dawadmi Campus
Shaqra University
Riyadh, Kingdom of
Saudi Arabia

Leo M.L. Nollet
University College
Ghent, Belgium

Abdul Qadir
Department of Pharmaceutics
School of Pharmaceutical Education &
Research
Jamia Hamdard
New Delhi, India

Md. Rizwanullah
Department of Pharmaceutics
School of Pharmaceutical Education &
Research
Jamia Hamdard
New Delhi, India

Abdul Samad
Department of Pharmaceutical Chemistry
Faculty of Pharmacy
Tishk International University
Kurdistan Region, Iraq

Rasheed Ahemad Shaik
Department of Pharmacology &
Toxicology
Faculty of Pharmacy
King Abdulaziz University
Jeddah, Kingdom of Saudi Arabia

Rewati Raman Ujjwal
Department of Pharmaceutics
National Institute of Pharmaceutical
Education and Research (NIPER)
Raebareli, India

Musarrat Husain Warsi
Department of Pharmaceutics and
Industrial Pharmacy
College of Pharmacy
Taif University
Taif-Al-Haweiah, Kingdom of Saudi Arabia

Sheetal Yadav
Department of Pharmaceutics
National Institute of Pharmaceutical
Education and Research (NIPER)
Raebareli, India

Ameeduzzafar Zafar
Department of Pharmaceutics
College of Pharmacy
Jouf University
Sakaka, Al-Jouf, Kingdom of Saudi
   Arabia

Farrukh Zeeshan
Department of Pharmaceutical
   Technology
School of Pharmacy
International Medical University
Kuala Lumpur, Malaysia

Fundamental Concepts and Technique Overview

CHAPTER 1

Fundamental Aspects of Food-Grade Nanoemulsions

Javed Ahmad[1] and
Leo M. L. Nollet[2]

[1]College of Pharmacy, Najran University, Najran, Kingdom of Saudi Arabia
[2]University College, Ghent, Belgium

CONTENTS

1.1 INTRODUCTION

At present, food industries across the world are exploring the area of nanotechnology particularly nanoemulsion to encapsulate, protect, and deliver bioactive compounds for food applications. The delivery of encapsulated bioactive compounds by means of

DOI: 10.1201/9781003121121-1

3

nanoemulsions offers substantial nutritive and therapeutic advantages. This encapsulation of bioactive moiety/functional compounds in nanoemulsions can be done efficiently by utilizing a low-energy emulsification method without the need of any solvent, heat, or sophisticated instrument.

The applicability of nanoemulsions in the food industry is highly influenced by their physicochemical characteristics, mainly mean particle size, viscosity, optical clarity, and stability. There are diverse areas in the food industry where nanoemulsions play an important role, such as nutraceuticals, packaging films, natural preservatives, and many more (Acevedo Fani et al., 2015; Artiga-Artigas et al., 2017). The essential oils are being widely used in nanoemulsified forms for preserving food products and keeping their flavor and aroma intact. The nanoemulsion technique is helpful in utilizing the essential oils of different biological sources for different purposes in nanoemulsified forms for food applications.

Many bioactive molecules that possess therapeutic and/or nutritive value have bioavailability constraints due to their lipophilic nature. Such bioavailability constraints can be overcome by encapsulating these bioactive molecules in nanoemulsions. Recent studies have demonstrated that therapeutic and nutritive effects of bioactive molecules have been greatly improved by means of nanoemulsions (Cheong et al., 2018; Golfomitsou et al., 2018; Silva et al., 2018). Besides bioavailability, appearance, shelf life, color, aroma, and flavor of the food products have also been improved through encapsulation of the active/functional moiety into a nanoemulsion system (Yang et al., 2011; Goindi et al., 2016; Tian et al., 2017).

This chapter provides a comprehensive account of the nanoemulsion systems, and it particularly highlights the fundamental overview, characterization perspectives, and applicability of nanoemulsions in the food industry.

1.2 FUNDAMENTAL CONCEPTS AND TECHNIQUE OVERVIEW

This section provides a brief discussion about a nanoemulsion system. It mainly deals with the fundamentals of the nanoemulsion technique, ingredients and composition of food-grade nanoemulsions, methods of preparation, and recent advancements in its manufacturing. The fate of nanoemulsions in biological systems is also discussed along with the stability perspectives of this technique.

1.2.1 Ingredients and Composition of Nanoemulsions

A nanoemulsion is made up of two immiscible liquids of which one is aqueous in nature and the other one is oleaginous, such as oils. The immiscibility of the two phases is overcome by the addition of an emulsifier or a surfactant (Figure 1.1). The emulsions are composed of a dispersed phase and a continuous phase depending upon the type of emulsion which may be oil-in-water (o/w) type or water-in-oil (w/o). The oil phase consists of short-chain, long-chain, or medium-chain fatty acids. Non-polar essential oils, lipid substitutes, mineral oils, or several lipophilic components can be used as oil phase. The structure and nature of the oil phase selected for preparation govern the formation, stability, and functionality of the developed nanoemulsion. The structure is mainly constituted by different number of carbons or different chain length fatty acids, and it has been reported that long-chain, short-chain, and medium-chain triglycerides have different *in vivo* fate (Tadros et al., 2004; Wooster et al., 2008). Therefore, the structure of the oil phase

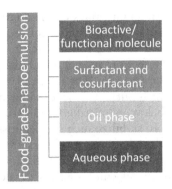

FIGURE 1.1   Basic components of food-grade nanoemulsion.

influences the absorption and thus bioavailability of the encapsulated bioactive ingredient. The physical characteristics such as viscosity, refractive index, interfacial tension, density, and phase behavior also have direct impact on the stability and preparation of a nanoemulsion (Tadros et al., 2004; Wooster et al., 2008; McClements and Rao, 2011). The long-chain fatty acids are generally preferred for their favorable attributes, such as natural origin, low cost, easy availability, and nutritional value (Witthayapanyanon et al., 2006). The medium-chain triglycerides are of synthetic or semisynthetic origin, but they offer the benefits of solubility and miscibility of bioactive components as it has been established that bioactive molecules are more soluble in medium-chain triglycerides as compared to long-chain triglycerides. And with short-chain triglycerides, the major benefit is the direct absorption of the encapsulated bioactive ingredients into systemic circulation. Therefore, the different nature and structure of glycerides influence the development of bioactive encapsulated nanoemulsions and their biological fate.

The nanoemulsions are kinetically stable, but there are stability concerns with them such as phase separations, sedimentation or creaming (which is the gravity-driven phase separation due to the density difference between the two liquids), droplet coalescence (merging of two droplets into a single larger one), and Ostwald ripening (increase in mean droplet size over time). Therefore, to prevent these instabilities, surfactants are added to these preparations (Schmitt et al., 2004). The selection of a surfactant is governed by hydrophilic-lipophilic balance (HLB) values. For o/w nanoemulsion surfactants with HLB values > 10 are used; for example, Tween 80 and sodium lauryl sulfate. The surfactants can be cationic, anionic, non-ionic, or zwitterionic in nature. More than one surfactant can also be used to stabilize the nanodispersion. When the combination of both hydrophilic and lipophilic (or cosurfactant) emulsifiers is used, a significant decrease in surface tension has been observed (Kabalnov et al., 1990; Taisne et al., 1996). The surfactants stabilize the nanoemulsion by averting particle aggregation by different mechanisms including repulsive interaction between adsorbed layers, the interfacial coverage, the steric effects, and the high dilational viscoelasticity of the interfacial layers (Boos et al., 2013; Llamas et al., 2018). The repulsive interactions are pertinent in case of ionic surfactants where even a small fraction of adsorbed molecules at the drop surface is enough to avert coalescence (Llamas et al., 2018) and keep the stability of nanoemulsions intact. The interfacial coverage is the comparative area of the interface inhabited by the surfactant, and the higher the coverage area, the lesser the coalescence, and more stable the nanoemulsion. Then steric hindrance is related to composite surface layers with large surfactant molecules at high adsorption coverage, which plays an important role in stabilizing the nanoemulsion. The high values of the dilational viscoelasticity make the liquid films between drops in nanoemulsions more stable.

Nevertheless, surfactants stabilize the formulation to a great extent, but they have been found to exert an irritant effect because of which their use is limited in foodstuffs. This has resulted in opting for w/o emulsions formulated without using surfactants (Glatter and Glatter, 2013). Such surfactant-free nanoemulsions can be prepared by cooling the continuous phase below its melting temperature (Ridel et al., 2015; Duffus et al., 2016; Chen et al., 2018). To stabilize such nanoemulsions, weighting agents, ripening retarders, thickening agents, and texture modifiers are being used. Weighting agents block the gravitational forces and reduce sedimentation and creaming. They include ester and dammar gum which is used in o/w emulsions as their density is comparable with that of the oil phase surrounding aqueous phase. Ripening retarders exert hydrophobic effect which impedes ripening. Examples include mineral oils and long-chain triacylglycerols (Schuch et al., 2014). The texture modifiers interact with the aqueous phase and increase its viscosity by thickening it. Some of the commonly used texture modifiers are biopolymers including gums, vegetable proteins, and polysaccharides (Imeson, 2011).

1.2.2 Preparation Methods of Nanoemulsions

The nanoemulsions are composed of small-size droplets with high surface area, and to generate such increased surface area, energy is required that is provided by external sources. The fundamental reason behind the formation of a nanoemulsion is the ability to significantly lower interfacial tension at the oil-water interface that requires a surfactant. The surfactants stabilize the formulation by lowering the interfacial tension. The nanoemulsion can be prepared either by high-energy or low-energy method. The mean droplet size is governed by the method of preparation, operating conditions, and surfactants used. The formation of an emulsion via high-energy method involves the breakdown of a crude emulsion into fine small-size droplets, which is achieved via adsorption of surfactants and collision of droplets. In the case of the high-energy method, high pressure is provided by a homogenizer, sonication, and/or microfluidizer to break the macro droplets of crude emulsion into smaller ones and further adsorption of the surfactant on to the interface ensuing steric stabilization of the resulting nanodispersion (Marzuki et al., 2019). The adsorption kinetics also affect the stability and droplet size of the prepared nanoemulsions (Silva et al., 2015). The high-energy method is mainly used in the food industry. In the low-energy method, a nanoemulsion is prepared by temperature and composition alteration of the o/w system. The energy is gained from the chemical potential of the constituents. The nanoemulsions are formed spontaneously at oil and water phase interface by mild mingling of the constituents. Nanoemulsion formation by low-energy methods depends on physicochemical properties such as temperature, composition, and solubility (Anton and Vandamme, 2009). The low-energy methods include phase inversion temperature (PIT), phase inversion composition (PIC), and solvent diffusion method.

1.2.3 Fate of Nanoemulsions in Biological Systems

It is essential to be aware of the biological fate of the administered encapsulated nutraceuticals in nanoemulsions. The physicochemical properties of materials change as we move toward the nanometer scale, and this will affect the pharmacokinetic profile that is the biological fate of the bioactive compound. The fate of nanoemulsions within the biological systems usually depends upon their physicochemical properties which include mean droplet diameter, surface morphology, electrical charge, and interfacial characteristics. However, many more investigations are needed to identify the release pattern of encapsulated components within the biological system (McClements and Xiao, 2012).

After administration in the body, nanoemulsions are subjected to the diverse physiological milieu as they pass through the gastrointestinal tract (GIT) (McClements et al., 2009; Singh et al., 2009). A basic understanding of all the possible environments throughout the GIT is essential to address the biological fate of a nanoemulsion (McClements et al., 2009; Singh et al., 2009).

Throughout the GIT, the pH, ionic compositions, enzyme functions, biopolymer levels, biosurfactant levels, flow/stress patterns, and the distinct GIT regions vary. All these environments can greatly cause significant changes in the physicochemical properties of a nanoemulsion. Therefore, this must be taken into consideration while designing an effective nutraceutical delivery system (Zhang and McClements, 2018).

The majority of the nutraceuticals are administered orally, and the first area that comes in contact is the mouth or oral cavity (McClements et al., 2008; McClements et al., 2009; Singh et al., 2009). It becomes necessary to understand this area, as the droplet of the formulation administered might be altered depending upon the time it remains in the oral cavity which might range from few seconds to minutes. Due to the process of mastication, i.e., chewing and mixing with saliva, both structural and compositional changes may occur in the nanoemulsion which might affect the fate of formulation. Human saliva contains multiple components. It possesses different biopolymers (such as mucin), salts (such as sodium and calcium), enzymes (such as amylase), and has a reasonably neutral pH (Vitorino et al., 2004; Amado et al., 2005; Kalantzi et al., 2006). Due to flocculation, mucin may encourage the accumulation of nanoemulsion within the oral cavity. This structural modification of nanoemulsion might alter the overall fate and efficacy of the administered nutraceutical, as shown in Figure 1.2. The composition of the bolus coming to the stomach from the mouth might play an important role during lipid digestion.

The masticated food passes to the esophagus and reaches the gastric site. The nanoemulsion is further treated in the stomach and is converted into such a form that it may be absorbed efficiently in the small intestine. Inside the stomach, the gastric fluids have about 1–3 pH, with an ionic strength of about 100 mM under fasting conditions and involve multiple digestive enzymes (such as gastric proteases, lipases) and surface-active materials (such as proteins and secretory phospholipids) (Kalantzi et al., 2006). Immediately after the food reaches the inside stomach, the pH of the stomach increases dramatically and then gradually decreases as the food moves toward the small intestine. The pH-time pattern of the stomach alters the nature of food ingested (Kalantzi et al., 2006).

Apart from the pH of the stomach, digestive enzymes also play an important role in determining the fate of nanoemulsions. Gastric pepsin and lipase are the most significant digestive enzymes found inside the stomach (Sams et al., 2016). Protein hydrolysis by pepsin can modify the fate of both free as well as entrapped lipid droplets. If these droplets are protein-coated in a nanoemulsion, then pepsin hydrolysis of the adsorbed protein layer will cause instability as the droplets are no longer controlled from accumulation. Although if the lipid globules are stuck inside a protein microgel, hydrolysis of the protein matrix will result in the release of the lipid droplets. The intensity and extent of protein hydrolysis of pepsin under gastric conditions depends on the protein type, with flexible open proteins (like casein) digesting more quickly than compact globular proteins (like innate β-lactoglobulin) (Mandalari et al., 2009).

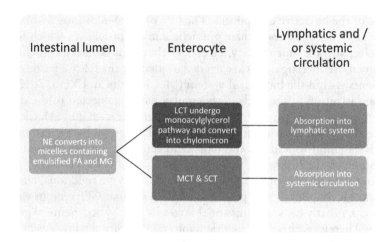

FIGURE 1.2 Fate of nanoemulsion (NE) containing fatty acids (FA) of long-chain triglycerides (LCT), medium-chain triglycerides (MCT), and short-chain triglycerides (SCT) after oral administration. MG: monoglycerides.

The stability of nanoemulsions inside the stomach is governed by many more factors. The existence of polysaccharides such as chitosan, pectin, methylcellulose, and fucoidan can stimulate the accumulation of protein-coated lipids inside the stomach and modify their stability. Due to polysaccharides' capability to encourage bridge or depletion flocculation, droplets may become unstable or they may become more stable if polysaccharides form a protective layer around them (McClements, 2015; Zhang et al., 2015). Therefore, to assess the overall fate of a nanoemulsion in GIT, the amount of polysaccharide must be carefully taken into account.

The processed food (from the stomach) which is now in the fluid form (chyme) passes across the pylorus sphincter and move into the small intestine where maximum digestion and absorption of macro- and micronutrients takes place (Singh et al., 2009). In this region, multiple digestive enzymes and salts are at work, which promotes digestion, haulage, and assimilation of nutrients. Phospholipids, bile salts, digestive enzymes and coenzymes (for example, lipases, phospholipases, colipases, amylases, proteases), bicarbonate salts, and other minerals are present in the fluids in the small intestine. Maximum lipid digestion takes place inside the duodenum, the proximal area of the small intestine. The acidic chyme combines with basic intestinal secretions and makes the complex's pH to neutral. Pancreatic enzymes work best at this pH (Zhang and McClements, 2018).

In the small intestine, there are two types of factors that alter digestion, solubilization, and absorption of lipid globules: (i) the quality of the lipid ingested and (ii) the characteristics of the intestinal fluids. Here, the focus will remain mainly on the effect of lipid digestion as they affect solubilization of free fatty acids (FFAs) and monoacylglycerol (MAGs) by enabling lipase to access the corresponding triacylglycerol (TAGs) (Zhang and McClements, 2018). In comparison, high multivalent cation levels can hinder lipid ingestion by facilitating droplet flocculation, thus plummeting lipase accessibility to the surfaces of the lipidic droplet. (McClements, 2015; Zhang and McClements, 2018). Gelation of certain forms of ionic biopolymers (such as alginate) that form hydrogels and capturing of lipid droplets is facilitated, multivalent ions thereby obstruct lipid digestion. The lipase is unable to have access to these captured lipid droplets. The fate of nanoemulsion inside the small intestine greatly depends upon the quality of food ingested and the overall health status of an individual (Tso and Crissinger, 2000).

The ultimate destination of any administered food in the GIT is the large intestine. This region can be divided into cecum, colon, and rectum. The colon has a significant role in inflaming the swallowed dietary fibers that have traveled throughout the GIT (Basit, 2005). Most of the nanoemulsion components are consumed throughout the GIT before reaching the large intestine. It is necessary to ensure that the components making up to the colon do not cause any harmful effect in the region. Surplus maltodextrin, for instance, exhibits inhibition of cellular antibacterial activities and disrupts intestinal antimicrobial protection processes in the colon. Further, unabsorbed iron may increase pathogen accumulation in infants and trigger inflammation in the intestinal (Nickerson et al., 2014; Nickerson et al., 2015).

Lipidic carrier systems and the existence of food collectively augment the oral absorption of nutraceuticals that are lipophilic in nature. The physiological process for the digestion and absorption of such lipids via the lymphatic system is very pertinent for this augmented nutraceutical bioavailability. Such lipidic carriers may enhance the lymphatic haulage of hydrophobic molecules by promoting the fabrication of the chylomicron.

The intestinal lymphatic transport system has been established as a promising nutraceutical delivery pathway. The anatomy of the mesenteric lymphatic system is unique and allows the transport of ingested bioactives bypassing the first-pass effect. Subsequent to administration of lipids and several lipophilic bioactives orally, they get distributed inside the intestinal enterocyte and are associated with secreted enterocyte lipoproteins. Such an intestinal lymphatic transport is recognized as an absorptive pathway. The bioactive compound is conjugated to chylomicron and is released into the lymphatic circulation, rather than site circulation, from the enterocyte. Thus, the metabolically active liver is avoided but inevitably reverts to the systemic circulation. This advantage enables the lymphatic system to play a key role in the bioavailability enhancement of certain bioactive compounds. For this mode of transport, the biological responses of lipid digestion and absorption are important. Lipid vehicles by stimulating the development of chylomicrons, improve the lymphatic transport of lipophilic materials. In combination with the triglyceride center of chylomicrons, lipophilic nutraceuticals join the lymphatic system. Besides lipid digestion, the physicochemical characteristics of bioactive compounds also determine its in vivo fate. The bioactive compound loading per chylomicron is affected by its partition coefficient and solubility in the triglyceride. In 1986, Charman and Stella suggested that bioactive molecules for lymphatic transport should have a logP > 5, given the discrepancy between intestinal blood vs lymph blood flow (500:1) and the reality that chylomicron makes up about 1% of lymph (Charman and Stella, 1986; Nickerson et al., 2015). Liu et al. demonstrated the significance of the physiology of lipid digestion in promoting lymphatic transportation. After administering soybean emulsion stabilized by milk fat globule membrane, 19.2% of the total vitamin D3 dose reached into the lymph, whereas administering this emulsion system with pancreatic lipase and bile salt, the amount of lymphatic transport was increased to 20.4% (Liu et al., 1995). The real impact of the dispersed phase of the vehicle on the degree of lymphatic delivery of the bioactive molecule was deeply investigated (Porter et al., 1996). The systems used were developed to predict the physicochemical features indicative of the end phase of lipid digestion and included lipid solution (oleic acid mixture and glycerol monooleate), stabilized emulsion polysorbate 80, and mixed micellar system polysorbate 80. After intraduodenal administration, an increased lymphatic transport was observed, which buoyed the theory that such nutraceutical formulations (dispersed systems) may intensify lymphatic transport. Such improved lymphatic transport has been advocated as a prospective mechanism of increased bioavailability (Kommuru et al., 2001).

1.2.4 Stability Perspectives of Nanoemulsions

The stability of emulsions is governed by selection of surfactants. The stability of nano-emulsion is affected by various phenomena such as creaming or sedimentation (in which two liquid phases separate due difference in their densities); Ostwald ripening (net mass transfer from small to big droplets), and droplet coalescence (assimilation of droplets into a single larger one) (Schmitt et al., 2004). To overcome these destabilizations, specifically droplet coalescence, several interfacial aspects including repulsive contact among adsorbed layers, the capturing of interfacial area, the steric hindrance effects, and the high dilational viscoelasticity of the interfacial layers need to be addressed (Boos et al., 2013; Llamas et al., 2018). A deep understanding of the interfacial phenomenon is crucial to deal with fundamental stability issues of nanoemulsions. As surfactants play a pivotal role in addressing these stability issues. A profound characterization and understanding of the interfacial properties and surfactants used for stabilizing a nanoemulsion at the liquid-liquid interface are important to get insights on stabilizing mechanisms.

1.3 CHARACTERIZATION OF NANOEMULSIONS

From application perspective in food industry, characteristic properties of a nanoemulsion are a very crucial aspect. Optically clear nanoemulsions are preferred in the food industry. Optical clarity is governed by the smaller mean droplet size and narrow droplet-size distribution of the nanoemulsion. Then the rheological properties influence the texture of foodstuffs which are again dependent on the mean droplet size. The viscosity of a nanoemulsion needs to be improved, as it impacts appearance, taste, and stability of beverages.

The physicochemical properties of a nanoemulsion include physical properties, interfacial stability, rheological properties, and microstructure. These characteristic features greatly influence the final texture, taste, flavor, and stability of foodstuffs. A brief overview of characterization phenomena is provided in Figure 1.3.

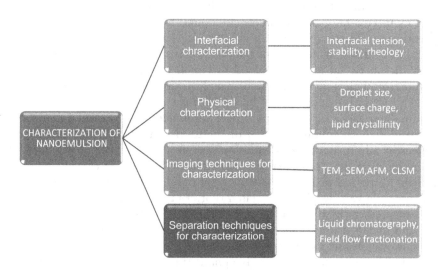

FIGURE 1.3   A brief overview of characterization of nanoemulsion.

TEM: Transmission Electron Microscopy; SEM: Scanning Electron Microscopy; AFM: Atomic Force Microscopy; CLSM: Confocal Laser Scanning Microscopy.

1.3.1 Interfacial Characterization

The interfacial characteristics of the nanoemulsions exhibit a pivotal role in the formulation process as well as the stability of the nanoemulsion. The characterization of the interfacial properties, such as thermodynamic stability, interfacial tension, rheology, and viscosity becomes necessary for the nanoemulsion stability because early estimation could avoid the problem of the instability because of coalescence and phase separation. The nanoemulsions are exposed to distinct thermodynamic stability tests to confirm stability of the prepared nanoemulsion under stress conditions. The investigations comprise of successive heating and cooling of nanoemulsions where they are subjected to several cycles of cooling (4°C) and heating (45°C). In thermodynamic stability study, nanoemulsions are exposed to freeze-thaw cycles and variations in the size, shape, or phase are observed. Kinetic instability, e.g. creaming, settling or phase separation is examined by centrifugation of the nanoemulsion at 3000–4000 rpm (Singh et al., 2017). The nanoemulsions that do not show any sign of creaming and phase separation pass these tests and this thermodynamic stability of nanoemulsion is attributed to the interfacial stability.

Interfacial tension is the surface tension that occurs between two phases. The changes in interfacial tension of nanoemulsions are observed with different surfactant concentrations and at variable temperatures. The surfactant is used as a stabilizing agent for the emulsion and it exerts its stabilizing effect by decreasing the interfacial tension at the oil-water interface. Similarly, film formation avoids the coalescence of oil droplets (Kumar and Mandal, 2018). For the determination of interfacial tension, the measurement of the existence of distinct concentrations of surfactants is done through the inverted drop process with a tracker automatic drop tensiometer. For the determination of the interfacial tension of the nanoemulsion, Du Nouy ring method can be used through monitoring the torsion required to pull the ring from the liquid interfaces continuously. Interfacial rheology establishes a critical relationship between the interfacial stress and interfacial deformation (Murray, 2002). Different types of deformation involved in the interface include stress, dilatants/compression. Dilatational rheology investigates the rheological properties that occur due to alteration in the surface area. In the past few decades, various measuring techniques for the estimation of the interfacial rheological factor have been reported. In the case of shear viscosity, the indirect viscometer, i.e. deep-channel surface viscometer containing two stationary concentric cylinders and a rotating dish, is utilized for the measurement. The channel surface viscometer demonstrated the flow rate through a constricted channel under external pressure (Pelipenko et al., 2012). Viscosity is the measurement of the resistance of a fluid to deform under applied shear stress. It is a critical parameter for the stability of nanoemulsions that depends on the emulsion component (water or oil) concentration and types of surfactant. Brookfield viscometer is the most widely utilized one to measure the viscosity of the nanoemulsion under applied shear stress. The viscosity of the system offers critical insights to observe the phase inversion phenomenon (Marzuki et al., 2019).

1.3.2 Physical Characterization

Characterization of a nanoemulsion comprises evaluation of physicochemical parameters, such as compatibility of the different components of nanoemulsion components, isotropicity, assay, content uniformity, morphological appearance, pH, viscosity, density, conductivity, surface tension, size and size distribution, and zeta potential (ZP) and surface charge with respect to the effect of the composition on physical attributes (Chime et al., 2014; Bourbon et al., 2018; Chung and McClements 2018). The different techniques used in physical characterization of a nanoemulsion include dynamic light scattering (DLS), differential scanning calorimetry (DSC), Fourier transform infrared (FT-IR), X-ray diffraction (X-RD), nuclear magnetic resonance (NMR), small-angle X-ray scattering (SAXS) and sensory analysis.

DLS is one of the most popular methods used to determine the particle size and polydispersity index (PDI) of a nanoemulsion, also known as photon correlation spectroscopy technique using Malvern Zetasizer. This technique is based on the Brownian motion of the particles. In this technique, samples are illuminated with a laser beam and intensity fluctuations in the scattered light are analyzed (Silva et al., 2012) and size measurement is typically carried out at a single scattering angle of 90 degrees (Bourbon et al., 2018). ZP is a measure of the magnitude of the electrostatic or charge repulsion/attraction between particles and is one of the fundamental parameters known to affect stability (Honary and Zahir, 2013; Bourbon et al., 2018). This is one of the significant characterization techniques of nanoemulsion to determine the surface charge, that can be used to understand the physical stability of colloids (Honary and Zahir, 2013; Bourbon et al., 2018; Joseph and Singhvi, 2019). The magnitude of ZP is related to the short-term and long-term physical stability of a nanoemulsion (Joseph and Singhvi, 2019). Nanoemulsion with high ZP (~ ±30 mV) is generally considered to have better physical stability, this is because of a high repulsive force that exceeds the attractive force (Rabinovich-Guilatt et al., 2004; Joseph and Singhvi, 2019). DSC is a fundamental tool in thermal analysis. In this technique differential heat requirement to increase the temperature of reference and sample is quantified as a function of temperature and materials heat capacity is evaluated (Koshy et al., 2017; Shah et al., 2017). FT-IR analysis can be carried out for the assessment of drug excipient interaction, polymerization, crosslinking as well as drug loading in the formulation. This technique is based on the absorption of infrared (IR) radiation when it passes through a sample i.e. nanoemulsion. The spectrum resulting from FT-IR represents the molecular absorption and transmission, creating a molecular fingerprint of the sample. The fingerprint from each sample is the characteristics absorption peak that represents the corresponding vibration between bonds of the atoms from each constituent. X-RD is a non-destructive analytical technique that discloses the crystallographic details and physicochemical properties and chemical composition of the sample under observation. This technique measures the scattered intensity of beam of X-ray striking the materials as a function of incident and scattered angles, polarization, and wavelength (Azároff et al., 1974; Silva et al., 2012). X-RD is mainly used for the analysis of crystalline compounds because of their diffraction pattern.

NMR is a potential analytical tool that analyzes the compound in either solid or liquid state (McClements, 2007). More recently, the use of NMR techniques has gained attention in the characterization of a colloidal system (Hathout and Woodman, 2012). This is one of the most widely used techniques for measuring the solid contents of a food emulsion (McClements, 2007). This instrument can rapidly analyze the concentrated and optically opaque emulsion, without any further sample preparation (Dickinson and McClements, 1995). The NMR technique utilizes interactions between radio waves and the nuclei of hydrogen atoms to obtain information about the structural formula and stereochemistry of molecules (Dickinson and McClements, 1995; Rouessac and Rouessac, 2013). SAXS is a technique used for the characterization of the colloidal dispersion for its shape, size, and structure (nanostructure) (Li et al., 2016). SAXS is a non-destructive method of analysis with minimum sample preparation. In this method, elastic scattering of X-rays by a sample of nanometric range is recorded at very low angles (typically 0.1–10°) (Gradzielski, 2008; Borthakur et al., 2016). This low angular range gives the information about size and shape, characteristic distance, pore sizes, etc. (Aswathanarayan and Vittal, 2019).

1.3.3 Imaging Techniques for Characterization

Shape and surface properties (morphology) of nanoemulsions play a significant role in delivery of bioactives/nutraceuticals. The shape of the particles and its local geometry and orientation plays an important role in determining whether or not the particle is phagocytosed. Microscopy can be used to study the details of the nanoemulsions' size, shape, and aggregation status. Nowadays, transmission electron microscopy (TEM), scanning electron microscopy (SEM) and atomic force microscopy (AFM) are widely employed in studying the size and morphological characteristics of nanoemulsions (Turner et al., 2007; Lal et al., 2010; Sitterberg et al., 2010). Methods of electron microscopy are extensively utilized for the characterization of nanostructures using an accelerated beam focused via magnetic lenses as a lighting source (Ahmad et al., 2019). Characterization of surface properties and structure via electron microscopy is immensely helpful in determining physicochemical characteristics of nanoemulsions which will in turn help in developing stable nanoformulations. TEM is an imaging technique with a resolution of 0.2 nm (Luykx et al., 2008; Wang, 2000; Ahmad et al., 2019). It has wide application in analyzing samples in material science/metallurgy and biological sciences research studies. SEM is a visualization technique of the surface involving the detection of scattered electrons produced from the surface of the nanobiomaterials like nanoemulsions. In SEM, electrons are scanned all over the surface of the particles and produce high-resolution images of the particle. The nanoemulsion droplets less than 100 nm in size can also be revealed (Reimer, 2000; Luykx et al., 2008). Although images produced by TEM are of higher resolution than that of SEM, images produced by SEM provide a good view of the surface of sample with inclusive depth of field. It can be applied to bulky samples which are impossible to analyze in TEM. This enables SEM to focus on a large number of samples at one time with a significant large depth of field available. SEM can be utilized to produce high-resolution images with higher magnification. All the above-mentioned properties of SEM, such as ease of sample preparation, high magnification with better resolution, and large depth of field make SEM one of the favored techniques in scientific studies (Luykx et al., 2008). The AFM technique has been widely applied to obtain the morphological information of nanoemulsions (Ruozi et al., 2005; Edwards and Baeumner, 2006; Luykx et al., 2008). It is capable of resolving surface details down to 0.01 nm and producing a contrasted three-dimensional image of the sample based on the force that acts between the surface of the particles and probing tip. Confocal laser scanning microscopy (CLSM) is an advanced imaging technique in which nanoemulsion is tagged with a fluorescent dye to capture the image. This technique can be utilized to study the pathways of absorption for nanoemulsion across biological membranes. Hence, the fundamentals of imaging techniques like TEM, SEM, AFM, and CLSM are discussed in Chapter 8. In recent years, many research studies utilized these imaging techniques for characterization of nanoemulsions containing bioactive/nutraceuticals.

1.3.4 Separation Techniques for Characterization

The separation technology is usually employed to converts a mixture of substances into two or more distinct product mixtures, generally referred to as *fractions*. Rarely will the mixture be separated completely into pure components through separation techniques. In the colloidal food science separation technology, the active ingredients of the various compounds are primarily classified and quantified. However, *in situ* detection of nanoemulsions in food matrices is often not possible; the key problem is their characterization in foodstuffs. The separation techniques are needed to isolate the nanoemulsions from food products earlier to their characterization. The specific analytical techniques can be used for *in situ* identification of nanoemulsions. While some research methods can be used to detect a new delivery system (NDS) *in situ*, this is not feasible in most situations because the nutritional content interferes. The presence and recognition of protein-based NDSs can interact, for example, with proteins in protein-containing foods. Separation methods can also be used before classification to separate the NDS from fruits. While the nanoemulsions can be somewhat categorized by different detectors coupled with specific separation techniques (Robertson et al., 2016), the key separation techniques include field flow fractionation and liquid chromatography, and these are discussed in detail in Chapter 9.

1.4 APPLICATIONS OF NANOEMULSION IN FOOD TECHNOLOGY

This section is mainly devoted to the discussion of the applicability of the nanoemulsion technique in food science. The technological limitations of developing functional foods are low solubility, stability, and bioavailability of the bioactive compounds. However, nanotechnology has been explored to address these issues in food industries (Weiss et al., 2006). Development of nanoemulsion-based food products has improved the stability, solubility, and bioavailability of nutraceuticals (Huang et al., 2010). Recently, the bottom-up technology has immensely impressed food technology and nutraceutical industry by improving different aspects of food processing techniques to ensure safety and molecular blending of products (Chen et al., 2006). It has improved texture, taste, durability, and appearance of food products. Nutraceuticals or functional foods act as antioxidants, antimicrobials, and health promoting agents. They are also used as preservative, coloring, and flavoring agents. Incorporation of functional ingredients in foods depends upon their physiochemical compatibility with the food matrix. However, most of the foodstuffs show poor solubility, volatility, low bioavailability, and chemical and biological degradation, making it difficult to directly incorporate them into the food matrix. Development of nanoemulsions overcomes all these shortcomings. Nanoemulsions are capable of stabilizing functional foods. Apart from transporting, bioactive lipids nanoemulsions efficiently protect the nutraceuticals from biological and chemical degradation. Overall, nanoemulsions significantly enhance the health factor of nutraceuticals and functional foods. Nanoemulsions have been extensively applied as delivery agents for phytosterols, vitamins of lipophilic nature, carotenoids, polyunsaturated fatty acids, etc. Nanoemulsions provide enhanced nutraceutical loading capacity, improved solubility, bioavailability, controlled release, and protection against degradation by enzymes.

1.4.1 Nanoemulsions as Delivery Vehicles for Nutraceuticals and Improving Food Nutritional Properties

Resveratrol is a naturally obtained polyphenol present in raspberries, blueberries, and the skin of grapes. It is found to have anticancer, antioxidant, and anti-obesity effects. Resveratrol is encapsulated into a nanoemulsion system. It is achieved by exploiting the spontaneous emulsification technique. It is developed using 10% grape seed and orange oil as the oil phase, 10% Tween 80 as the surfactant, and remaining 80% as the aqueous phase. The nanoemulsion thus formed was found to have a droplet size of 100 nm and encapsulation efficiency of 120 ± 10 µg/ml. The developed formulation was reported to be stable against UV-light degradation (Davidov-Pardo and McClements, 2015). Cholecalciferol or Vit-D-based o/w edible nanoemulsion has been used for fortification of dairy emulsions. Polysorbate 20 was used as an emulsifying agent, soybean lecithin was mixed with cocoa butter to develop nanoemulsions having a mean droplet diameter of less than 200 nm using the high-pressure homogenization technique. Nanoemulsions loaded with Vit-D3 in its oily core were developed. Fortification of whole fat milk was done with nanoemulsions of Vit-D3. This was found to be stable for 10 days against gravitational separation and particle growth (Golfomitsou et al., 2018). A kenaf seed oil based o/w nanoemulsion was developed using sodium caseinate as the emulsifying agent, Tween 20 and β-cyclodextrin, and it has shown to improve physicochemical and in vitro stability of the antioxidant and bioactive ingredients. The researchers reported that the developed product was found to be stable for 8 weeks at 4°C with enhanced retention of phytosterols and Vit-E (Cheong et al., 2018). Photostabilization of the astaxanthin nanoemulsion was achieved to make it able to be used in foods. The carrageenan coating and chitosan coating were found to protect astaxanthin from UV-light and prevent photodegradation of the same (Alarcon-Alarcon et al., 2018). The bioaccessibility of hydrophobic bioactive compounds has been reported to be increased by the use of nanoemulsions. Nanoemulsions of curcumin prepared using alginate and chitosan were found to improve the antioxidant property of curcumin and were also found to control the lipid digestibility of the curcumin. Thus, these bioactive nanoemulsion systems can be used to target obesity (Silva et al., 2018). Curcumin nanoemulsions stabilized with lecithin have been formulated and were reported with 75% encapsulation efficiency and were found to be stable for 86 days (Bhosale et al., 2014).

1.4.2 Nanoemulsions Utilized to Improve Food System Color, Flavor, and Texture

Ketones, aldehydes, and esters are present as coloring and flavoring agents in food but are susceptible to photolytic and oxidative degradation. Encapsulating these materials into nanoemulsion systems may prevent the degradative effect and enhance the shelf life (Goindi et al., 2016). Citral is used as a flavoring agent in cosmetics and food industry which easily gets degraded forming bad flavor compounds. Citral developed as an o/w nanoemulsion system with additionally added β-carotene or black tea extract as antioxidant has shown enhanced stability (Yang et al., 2011). Ubiquinol-10 and citral nanoemulsion system has shown enhanced chemical stability. Ubiquinol-10 was found to protect citral from chemical and oxidative degradation. It reduces the production of off-flavor compounds (Zhao et al., 2013). In a similar study, Tween 20 and gelatin were successfully used as emulsifying agents and were found to stabilize citrate from acidic degradation in the nutraceutical and food industry (Tian et al., 2017). β-carotene is a naturally occurring coloring agent and widely used antioxidant in the food and nutraceutical industry. However, it is susceptible to degradation by light, oxygen, and heat. A nanodispersion system of β-carotene was prepared by exploiting the emulsification-evaporation method which was reported with enhanced stability. Emulsification of organic β-carotene solution was done in aqueous phase having an emulsifying agent. The mean diameter of droplets for the developed nanodispersion system of β-carotene was found to be 60–140 nm (Tan and Nakajima, 2005a). An o/w β-carotene nanodispersion system was developed using polyglycerol esters of fatty acids as non-ionic emulsifying agents. Researchers concluded that the mean diameter of the droplets of the developed nanodispersion system was found to be 85–132 nm with physical and physicochemical stability (Tan and Nakajima, 2005b). The β-carotene nanodispersion system stabilized with protein, and developed by the emulsification-evaporation method was found to have a mean diameter of nanodispersion droplets of 17 nm. Sodium caseinate used as an emulsifying agent was found to decrease the particle size at higher concentrations. It also helps in improving the PDI of the nanodispersion system (Chu et al., 2007). Tween 20 as an emulsifying agent was used to develop an o/w β-carotene nanodispersion system exploiting the high-pressure homogenization technique. Researchers concluded that the developed nanodispersion system was having droplets of mean diameter in the range of 130–185 nm. The formulation was found to be stable and showed only 25% degradation after 4 weeks at 4 and 25°C (Yuan et al., 2008). Tween 20, sucrose fatty acid ester, decaglycerol monolaurate, and sodium caseinate are used as emulsifying agents to prepare a β-carotene nanodispersion system. The developed formulation showed 30–210 nm of mean diameter. Nanodispersion systems stabilized by starch casein were reported to be more stable against oxidation. It works by creating a physical barrier and protects by antioxidant property of caseins (Yin et al., 2009). Spray-dried powders and modified starch are added to the β-carotene nanoemulsion. These were reported to have enhanced storage stability. Starch shows low oxygen permeability and thus assists in β-carotene retention during storage (Liang et al., 2013).

1.4.3 Essential Oil Nanoemulsions in Food Preservation

The essential oils derived from plants exhibit substantial antimicrobial potential against different foodborne pathogens. The essential oils including thyme, oregano, clove, and orange and their components, such as thymol, carvacrol, eugenol, limonene, and cinnamon possess potent antimicrobial activity and can be used as natural preservatives. However, their application is restricted in foodstuffs because of their lipophilic nature. Here, nanoemulsions can provide a solution by overcoming their poor solubility issue. Encapsulated essential oils in nanoemulsions can be efficiently used in foods as natural preservatives (Alexandre et al., 2016). There are a lot of studies which have demonstrated that the essential oil encapsulated in a nanoemulsion exhibits potential antimicrobial activity. The oregano oil nanoemulsions impeded the growth of foodborne bacteria *Listeria monocytogenes*, *Salmonella typhimurium*, and *Escherichia coli* O157:H7 on fresh lettuce by disrupting bacterial membranes (Bhargava et al., 2015). Likewise, orange oil encapsulated nanoemulsions averted the decay of apple juice by *Saccharomyces cerevisiae* (Sugumar et al., 2015). Moreover, cinnamaldehyde nanoemulsions contained in pectin edible films hampered the growth of *E. coli*, *Salmonella enterica*, *Listeria monocytogenes*, and *Staphylococcus aureus* (Otoni et al., 2014a). Nanoemulsions of clove bud and oregano essential oil having 180–250 nm droplet size incorporated into edible methyl cellulose films revealed enhanced antimicrobial activity by inhibiting the growth of yeasts and molds and increasing the shelf life of sliced bread (Otoni et al., 2014b). Nanoemulsions of ginger essential oil integrated into gelatin-based films have demonstrated enhanced physical attributes of the active food packaging films (Alexandre et al., 2016). Curcumin nanoemulsions having a mean globule size of 40 nm with pectin edible coatings boosted the shelf life of chilled chicken for long time span. The nanoemulsion mitigated microbial spoilage by impeding the growth of psychrophiles, yeast, and mold (Abdou et al., 2018). Nanoemulsions of ginger essential oil with sodium caseinate coating increased the shelf life of chicken breast filets by decreasing growth of psychrophilic bacteria in refrigerated chicken filets for 12 days (Noori et al., 2018).

1.4.4 Nanoemulsion-Based Food Packaging Materials

Nanoemulsion systems have been extensively used in coatings and films for possible application in nutraceutical and food packaging. The continuous phase of films and coatings is made up of a biopolymer matrix which provides them with monodispersity index and stability. The coalescence of droplets tends to decrease with the increase in the viscosity of the continuous phase (Artiga-Artigas et al., 2017). Bioedible nanoemulsion films include the dispersion of bioactive agents into the continuous phase which tends to form the matrix of the packaging film. High- or low-energy methods are used to formulate food-grade emulsion. These methods are exploited to It homogenize the formulation which is then casted into films having controlled thickness and is further dried. The developed films are then characterized for their morphological, structural, mechanical, thermal, and functional properties (Otoni et al., 2014a). Edible nanoemulsion films of clove and cinnamaldehyde using biopolymer pectin were produced with enhanced antimicrobial activity. Pectin films have low hydrophilic and hydrophobic ratio which allows decreased vapor and water permeability (Otoni et al., 2014a). Oregano and clove essential oil based nanoemulsions have been developed using cellulose and its derivatives (Otoni et al., 2014b). Essential oils of mandarin, lemon, bergamot, and carvacrol have been developed into nanoemulsion coatings and films using chitosan and were shown to have enhanced antimicrobial activity (Severino et al., 2015). Essential oils of lemongrass, corn sage, and thyme were developed in nanoemulsion films using sodium alginate (Acevedo Fani et al., 2015; Artiga-Artigas et al., 2017). In a similar study, canola oil-based nanoemulsion films were developed using porcine gelatin (Alexandre et al., 2016). Food coatings with nanoemulsions have been carried out and tested for their effective ability to protect food material from degradation. Chitosan coating containing nanoemulsions of essential oils of mandarin on green bean was found to inhibit the growth of *Listeria innocua* (Severino et al., 2014). Thymol nanoemulsion coated with chitosan and quinoa protein edible coating containing thymol nanoemulsion were found to significantly inhibit the growth of fungi on cherry tomatoes after 7 days at a storage temperature of 5°C (Robledo et al., 2018).

1.4.5 Nanoemulsions Utilized to Preserve/Process Bioactive and Nutritional Food Compounds

Functional foods with bioactive compounds are a bit difficult to be designed due to the low solubility and bioavailability and stability issues. These components are easily susceptible to degradation and deterioration due to oxidation (Shahidi and Zhong, 2010). Some of the bioactive compounds show poor solubility and get metabolized frequently, eventually reducing its bioavailability (Jin et al., 2016). A nanoemulsion system provides a novel platform for encapsulating bioactive compounds into an oily phase using an emulsifying agent that ensures the solubility, stability, release ability, and bioavailability of the bioactive compound (McClements et al., 2007). For successful development of nanoemulsions for functional foods, there should be compatibility between the components and the organoleptic properties must not be hampered. Encapsulation protects the bioactive materials from different processing conditions and inhibits the degradation of the material from oxidation, temperature variations, light, pH, and other manufacturing and storage conditions. However, economic feasibility for industrial scale-up production must also be considered (Pathak, 2017).

Numerous functional foods have been developed as nanoemulsion systems and various *in vitro* and *in vivo* studies have also been reported. But still their application in commercial functional food products is extremely limited. Sustainable food processing can be done by developing a nanoemulsion system. Different nutraceutical industries like Nestle and Unilever are extensively exploiting nanoemulsion systems in their nutraceutical products (Salvia-Trujillo et al., 2017). W/o nanoemulsions are being developed by Nestle. They have also patented micelle forming and polysorbate emulsions for uniform and rapid thawing of frozen foods in the microwave (Möller et al., 2009). Unilever used nanoemulsion system to reduce fat content in ice creams (Aswathanarayan et al., 2019). NutraLease Ltd. has developed a nanoencapsulation technique which is known as nano-sized self-assembled structured liquids (NSSL). Minute micelles known as nanodrops are developed which serve as carriers for lipophilic bioactive agents. They protect the bioactive agents from degradation in the digestive tract and assist in absorption. These are used in developing beverages having functional compounds. With the NSSLtechnology, omega-3 fatty acids, phytosterols, β-carotene, vitamins, and isoflavones have been incorporated in the food products and were found to have enhanced bioavailability and improved shelf-life (Garti et al., 2007; NutraLease, 2007; Aswathanarayan et al., 2019). NovaSol beverages having nanoemulsions of curcumin, sweet pepper, apocarotenal, chlorophyll, lutein have been developed by Aquanova with enhanced stability (AquaNova, 2011; Aklakur et al., 2016; Aswathanarayan et al., 2019). Similarly, bottled water enriched with electrolytes and flavors has also been developed using the nanoemulsion technology (Piorkowski and McClements, 2014). Natural antioxidants extracted from fruits, flowers, cereals, and vegetables have been developed into a nanoemulsion system to be added as food preservatives. This technique has been patented. These systems were freeze dried and were later used for the preservation of foods. Nanoemulsions are applied as a thin layer over the food to inhibit fluid and gas exchange with the outer environment of the food product and thus keeps it stable for a longer duration of time. It was also suggested that upon thawing such frozen foods show better organoleptic properties (Malnati et al., 2019).

1.5 CONCLUSIONS AND FUTURE DIRECTIONS

In the last decade, numerous studies have been conducted to understand the advantages of nanoemulsions for encapsulating bioactive compounds for functional foods. Nanoemulsions were found to increase the bioavailability of nutraceutical compounds and were confirmed through different in vitro studies. However, still very limited results are available for reporting the actual health benefits of nanoemulsions in food. A number of studies have been conducted for understanding and evaluating the application of low- and high-energy approaches to successfully formulate nanoemulsions. Few reports are available on reducing the cost of nanoemulsion production. Similarly, the risk associated with the application of nanoemulsion techniques in the food and nutraceutical industry must be evaluated. No elucidation is provided for potential biological and toxicological fate of nanoemulsions after digestion.

Nanoemulsions have shown great potential for encapsulating foods and bioactive compounds and thus opened numerous applications in the nutraceutical industry. Coating of foods and bioactive compounds with an edible nanoemulsion coating has been found to enhance the stability, quality, and shelf life. However, their potential application in the food industry is dependent on the cost effectiveness in its development and their safety profile, which needs to be considered before commercialization of the product. Thus, it becomes of utmost importance to optimize the encapsulated bioactive component for scaling up the industrial production. The future of this innovative technique depends upon the risk assessment, safety profile, and cost effectiveness in producing and commercializing nanoemulsion-based food and nutraceutical products.

CONFLICT OF INTEREST

The authors declare no conflict of interest.

REFERENCES

Abdou, E. S., Galhoum, G. F., and Mohamed, E. N. (2018). Curcumin loaded nanoemulsions/pectin coatings for refrigerated chicken fillets. Food Hydrocoll. 83, 445–453.

Acevedo-Fani, A., Salvia-Trujillo, L., Rojas-Graü, M. A., and Martín-Belloso, O. (2015). Edible films from essential oil loaded nanoemulsions: Physicochemical characterization and antimicrobial properties. Food Hydrocoll. 47, 168–177.

Ahmad, M., Mudgil, P., Gani, A., Hamed, F., Masoodi, F. A., and Maqsood, S. (2019). Nano-encapsulation of catechin in starch nanoparticles: Characterization, release behavior and bioactivity retention during simulated in-vitro digestion. Food Chem. 270, 95–104.

Aklakur, M., Asharf Rather, M., Kumar, N. (2016). Nanodelivery: an emerging avenue for nutraceuticals and drug delivery. Crit. Rev. Food Sci. Nutr. 56(14), 2352–2361.

Alarcon-Alarcon, C., Inostroza-Riquelme, M., Torres-Gallegos, C., Araya, C., Miranda, M., Sánchez-Caamaño, J.C., Moreno-Villoslada, I., Oyarzun-Ampuero F.A. (2018). Protection of astaxanthin from photodegradation by its inclusion in hierarchically assembled nano and microstructures with potential as food. Food Hydrocoll. 83, 36–44.

Alexandre, E. M. C., Lourenço, R. V., Bittante, A. M. Q. B., Moraes, I. C. F., and Sobral, P. J. A. (2016). Gelatine based films reinforced with montmorillonite and activated with nanoemulsion of ginger essential oil for food packaging applications. Food Pack. Shelf Life. 10, 87–96.

Amado, F. M. L., Vitorino, R. M. P., Domingues, P. M. D. N., Lobo, M. J. C., and Duarte, J. A. R., (2005). Analysis of the human saliva proteome. Expert Rev. Proteomics. 2, 521–539.

Anton, N., and Vandamme, T. F. (2009). The universality of low-energy nanoemulsification. Int. J. Pharm. 377, 142–147. doi: 10.1016/j.ijpharm.2009.05.014.

AquaNova (2011). Available online at: http://www.aquanova.de/media/public/pdf_produkte~unkosher/NovaSOL_beverage. (accessed June 15, 2021).

Artiga-Artigas, M., Acevedo-Fani, A., and Martín-Belloso, O. (2017). Effect of sodium alginate incorporation procedure on the physicochemical properties of nanoemulsions. Food Hydrocoll. 70, 191–200.

Aswathanarayan J. B., and Vittal R. R. (2019). Nanoemulsions and their potential applications in food industry. Front Sustain Food Syst. 3, 95.. doi.org/10.3389/fsufs.2019.00095

Azároff L. V., Kaplow R., Kato N., Weiss R. J., Wilson A. J. C., and Young R. A. (1974). X-ray Diffraction: McGraw-Hill, New York.

Basit, A.W. (2005). Advances in colonic drug delivery. Drugs 65, 1991–2007.

Bhargava, K., Conti, D. S., da Rocha, S. R. P., and Zhang, Y. (2015). Application of an oregano oil nanoemulsion to the control of foodborne bacteria on fresh lettuce. Food Microbiol. 47, 69–73.

Bhosale, R. R., Osmani, R. A., Ghodake, P. P., Shaikh, S. M., and Chavan, S. R. (2014). Nanoemulsion: A review on novel profusion in advanced drug delivery. Indian J. Pharm. Biol. Res. 2, 122–127.

Boos, J., Preisig, N., and Stubenrauch, C. (2013). Dilational surface rheology studies of n-dodecyl-β-d-maltoside, hexaoxyethylene dodecyl ether, and their 1:1 mixture. Adv. Colloid Interface Sci. 197–198, 108–117.

Borthakur P., Boruah P. K., Sharma B., and Das M. R. (2016). Nanoemulsion: Preparation and its application in food industry. In: Grumezescu A. M., editor. Emulsions. Academic Press. pp. 153–191.

Bourbon A. I., Gonçalves R. F., Vicente A. A., and Pinheiro A. C. (2018). Characterization of particle properties in nanoemulsions. In: Seid Mahdi J., and David Julian M., editors. Nanoemulsions: Elsevier. pp. 519–546.

Charman, W. N. A., and Stella, V. J. (1986). Estimating the maximal potential for intestinal lymphatic transport of lipophilic drug molecules. Int. J. Pharm. 34(1–2), 175–178.

Chen, H., Weiss, J., and Shahidi, F. (2006). Nanotechnology in nutraceuticals and functional foods. Food Tech. 60, 30e36.

Chen, Q.-H., Zheng, J., Xu, Y.-T., Yin, S.-W., Liu, F., and Tang, C.-H. (2018). Surface modification improves fabrication of Pickering high internal phase emulsions stabilized by cellulose nanocrystals. Food Hydrocoll. 75, 125–130. doi: 10.1016/j.foodhyd.2017.09.005.

Cheong, A. M., Tan, C. P., and Nyam, K. L. (2018). Stability of bioactive compounds and antioxidant activities of kenaf seed oil-in-water nanoemulsions under different storage temperatures. J. Food Sci. 83, 2457–2465.

Chime S., Kenechukwu F., Attama A. (2014). Nanoemulsions—advances in formulation, characterization and applications in drug delivery. In: Sezer A., editor. Application of Nanotechnology in Drug Delivery. Croatia, InTech. pp. 77–126.

Chu, B. S., Ichikawa, S., Kanafusa, S., and Nakajima, M. (2007). Preparation of protein-stabilized β-carotene nanodispersions by emulsification–evaporation method. J. Am. Oil Chem. Soc. 84, 1053–1062.

Chung C., McClements D. J. (2018). Characterization of physicochemical properties of nanoemulsions: Appearance, stability, and rheology. In: Seid Mahdi J., and David Julian M., editors. Nanoemulsions. Elsevier. pp. 547–576.

Davidov-Pardo, G., and McClements, D. J. (2015). Nutraceutical delivery systems: Resveratrol encapsulation in grape seed oil nanoemulsions formed by spontaneous emulsification. Food Chem. 167, 205–212.

Dickinson E., and McClements D. J. (1995). Advances in Food Colloids: Springer Science & Business Media.

Duffus, L. J., Norton, J. E., Smith, P., Norton, I. T., and Spyropoulos, F. (2016). A comparative study on the capacity of a range of food-grade particles to form stable O/W and W/O Pickering emulsions. J. Colloid Interface Sci. 473, 9–21. doi: 10.1016/j.jcis.2016.03.060.

Edwards, K. A., and Baeumner, A. J. (2006). Liposomes in analyses. Talanta 68(5), 1421–1431.

Garti N, Aserin A, Spernath A, Amar I, inventors; Nutralease Ltd, assignee. Nano-sized self-assembled liquid dilutable vehicles. United States Patent US 7,182,950. 2007 Feb 27.

Glatter, O., and Glatter, I. (2013). Water-in-oil emulsions and methods for their preparation. EP2604253-A1. Available online at: https://patentimages.storage.googleapis.com/44/dc/df/b6657c09437977/EP2604253A1.pdf

Goindi, S., Kaur, A., Kaur, R., Kalra, A., and Chauhan, P. (2016). Nanoemulsions: An emerging technology in the food industry. Emulsions 3, 651–688.

Golfomitsou, I., Mitsou, E., Xenakis, A., and Papadimitriou, V. (2018). Development of food grade O/W nanoemulsions as carriers of vitamin D for the fortification of emulsion based food matrices: A structural and activity study. J. Mol. Liq. 268, 734–742.

Gradzielski, M. (2008). Recent developments in the characterisation of microemulsions. Current Opin. Colloid Interface Sci. 13(4), 263–269.

Hathout R. M., and Woodman T. J. (2012). Applications of NMR in the characterization of pharmaceutical microemulsions. J. Control. Release. 161(1), 62–72.

Honary S., and Zahir F. (2013). Effect of zeta potential on the properties of nano-drug delivery systems-a review (Part 2). Trop. J. Pharm. Res. 12(2), 265–273.

Huang, Q., Yu, H., and Ru, Q., (2010). Bioavailability and delivery of nutraceuticals using nanotechnology. J. Food Sci. 75, R50eR57.

Imeson, A. (2011). Food Stabilisers, Thickeners and Gelling Agents: John Wiley & Sons.

Jin, W., Xu, W., Liang, H., Li, Y., Liu, S., and Li, B. (2016). Nanoemulsions for food: Properties, production, characterization, and applications. Emulsions 3, 1–36.

Joseph E., and Singhvi G. (2019). Multifunctional nanocrystals for cancer therapy: A potential nanocarrier. Nanomaterials for Drug Delivery and Therapy. Elsevier. pp. 91–116.

Kabalnov, A. S., Makarov, K. N., Pertzov, A. V., and Shchukin, E. D. (1990). Ostwald ripening in emulsions: 2. Ostwald ripening in hydrocarbon emulsions: Experimental verification of equation for absolute rates. J. Colloid Interface Sci. 138, 98–104.

Kalantzi, L., Goumas, K., Kalioras, V., Abrahamsson, B., Dressman, J. B., and Reppas, C. (2006). Characterization of the human upper gastrointestinal contents under conditions simulating bioavailability/bioequivalence studies. Pharm. Res. 23, 165–176.

Kommuru, T. R., Gurley, B., Khan, M. A., and Reddy, I. K. (2001). Self-emulsifying drug delivery systems (SEDDS) of coenzyme Q10: Formulation development and bioavailability assessment. Int. J. Pharm. 212, 133–246.

Koshy, O., Subramanian, L., and Thomas, S. (2017). Chapter 5—Differential scanning calorimetry in nanoscience and nanotechnology. In: Thomas S., Thomas R., Zachariah A. K., and Mishra R. K., editors. Thermal and Rheological Measurement Techniques for Nanomaterials Characterization. Elsevier. pp. 109–122.

Kumar, N., and Mandal, A. (2018). Surfactant stabilized oil-in-water nanoemulsion: Stability, interfacial tension, and rheology study for enhanced oil recovery application. Energy Fuels. 32(6), 6452–6466. https://doi.org/10.1021/acs.energyfuels.8b00043.

Lal, R., Ramachandran, S., and Arnsdorf, M. F. (2010). Multidimensional atomic force microscopy: A versatile novel technology for nanopharmacology research. AAPS J. 12(4), 716–728.

Li, T., Senesi, A. J., and Lee, B. (2016). Small angle X-ray scattering for nanoparticle research. Chem. Rev. 116(18), 11128–11180.

Liang, R., Huang, Q., Ma, J., Shoemaker, C. F., and Zhong, F. (2013). Effect of relative humidity on the store stability of spray-dried beta-carotene nanoemulsions. Food Hydrocoll. 33, 225–233.

Liu, H.-X., Adachi, I., Horikoshi, I., and Ueno, M. (1995). Mechanism of promotion of lymphatic drug absorption by milk fat globule membrane. Int. J. Pharm. 118, 55–64.

Llamas, S., Santini, E., Liggieri, L., Salerni, F., Orsi, D., Cristofolini, L., and Ravera, F. (2018). Adsorption of sodium dodecyl sulfate at water-dodecane interface in relation to the oil in water emulsion properties. Langmuir. 34, 5978–5989.

Luykx, D. M., Peters, R. J. B., van Ruth, S. M., and s Bouwmeester, H. (2008). A review of analytical methods for the identification and characterization of nano delivery systems in food. J. Agric. Food Chem. 56(18), 8231–8247.

Malnati, R. M. E. J., Du-Pont, M. X. A., and Morales, D. A. O. (2019). Method for Producing a Nanoemulsion with encapsulated natural Antioxidants for Preserving Fresh and Minimally Processed Foods, and the Nanoemulsion thus Produced. WIPO WO2019039947A1. Available online at: https://patentimages.storage.googleapis. com/ca/2d/9e/67dbe574133d47/WO2019039947A1.pdf.

Mandalari, G., Adel-Patient, K., Barkholt, V., Baro, C., Bennett, L., Bublin, M., Gaier, S., Graser, G., Ladics, G. S., and Mierzejewska, D. (2009). In vitro digestibility of B-casein and B-lactoglobulin under simulated human gastric and duodenal conditions: A multi-laboratory evaluation. Regul. Toxicol. Pharmacol. 55, 372–381.

Marzuki, N. H. C., Wahab, R. A., and Hamid, M. A. (2019). An overview of nanoemulsion: Concepts of development and cosmeceutical applications. Biotechnol. Biotechnol. Equip. 33(1), 779–797. https://doi.org/10.1080/13102818.2019.1620124.

McClements, D. J. (2015). Food Emulsions: Principles, Practices, and Techniques. CRC Press, Boca Raton, FL.

McClements D. J., Decker, E. A., and Park, Y. (2008). Controlling lipid bioavailability through physicochemical and structural approaches. Crit. Rev. Food Sci. Nutr. 49, 48–67.

McClements, D. J., Decker, E. A., Park, Y., and Weiss, J. (2009). Structural design principles for delivery of bioactive components in nutraceuticals and functional foods. Crit. Rev. Food Sci. Nutr. 49, 577–606.

McClements, D. J., Decker, E. A., and Weiss, J. (2007). Emulsion-based delivery systems for lipophilic bioactive components. J. Food Sci. 72, 109–124.

McClements, D. J., and Rao, J. (2011). Food-grade nanoemulsions: Formulation, fabrication, properties, performance, biological fate, and potential toxicity. Crit. Rev. Food Sci. Nutr. 51, 285–330. doi: 10.1080/10408398.2011.559558.

McClements D. J., and Xiao H. (2012). Potential biological fate of ingested nanoemulsions: Influence of particle characteristics. Food Funct. 3, 202–220.

Möller, M., Eberle, U., Hermann, A., Moch, K., and Stratmann, B. (2009). Nanotechnology in the Food Sector. TA-SWISS, Zürich.

Murray, B. S. (2002). Interfacial rheology of food emulsifiers and proteins. Curr. Opin. Colloid Interface Sci. 7(5–6), 426–431. https://doi.org/10.1016/S1359-0294 (02)00077-8.

Nickerson, K. P., Chanin, R., and Mcdonald, C. (2015). Deregulation of intestinal antimicrobial defense by the dietary additive, maltodextrin. Gut Microbes 6, 78–83.

Nickerson, K. P., Homer, C. R., Kessler, S. P., Dixon, L. J., Kabi, A., Gordon, I. O., Johnson, E. E., Carol, A., and Mcdonald, C. (2014). The dietary polysaccharide maltodextrin promotes Salmonella survival and mucosal colonization in mice. PLoS One 9(7), e101789. doi: 10.1371/journal.pone.0101789.

Noori, S., Zeynali, F., and Almasi, H. (2018). Antimicrobial and antioxidant efficiency of nanoemulsion-based edible coating containing ginger (Zingiber officinale) essential oil and its effect on safety and quality attributes of chicken breast fillets. Food Control 84, 312–320.

NutraLease (2007). Available online at: https://www.0nanotechproject.tech/cpi/products/ nano-sized-self-assembled-liquid-structures-nssl-supplements/ (accessed June 15, 2021).

Otoni, C. G., de Moura, M. R., Aouada, F. A., Camilloto, G. P., Cruz, R. S., Lorevice, M. V., de FF Soares, N., Mattoso, L.H. (2014a). Antimicrobial and physical-mechanical properties of pectin/papaya puree/cinnamaldehyde nanoemulsion edible composite films. Food Hydrocoll. 41, 188–194.

Otoni, C. G., Pontes, S. F., Medeiros, E. A., and Soares, N. de. F. (2014b). Edible films from methylcellulose and nanoemulsions of clove bud (Syzygium aromaticum) and oregano (Origanum vulgare) essential oils as shelf life extenders for sliced bread. J. Agric. Food Chem. 62, 5214–5219.

Pathak, M. (2017). Nanoemulsions and their stability for enhancing functional properties of food ingredients. Nanotechnology Applications in Food: Flavor, Stability, Nutrition, and Safety, Academic Press, Burlington, MA. pp. 87–106.

Pelipenko, J., Kristl, J., Rošic, R., Baumgartner, S., and Kocbek, P. (2012). Interfacial rheology: An overview of measuring techniques and its role in dispersions and electrospinning. Acta Pharm. 62(2), 123–140. https://doi.org/10.2478/v10007-012-0018-x.

Piorkowski, D. T., and McClements, D. J. (2014). Beverage emulsions: Recent developments in formulation, production, and applications. Food Hydrocoll. 42, 5–41.

Porter, C. J. H., Charman, S. A., and Charman, W. N. (1996). Lymphatic transport of halofantrine in the triple-cannulated anesthetized rat model: Effect of lipid vehicle dispersion. J. Pharm. Sci. 85, 351–356.

Rabinovich-Guilatt, L., Couvreur, P., Lambert, G., Goldstein, D., Benita, S., and Dubernet, C. (2004). Extensive surface studies help to analyse zeta potential data: The case of cationic emulsions. Chem. Phys. Lipids 131(1), 1–13.

Reimer, L. (2000). Scanning electron microscopy: physics of image formation and microanalysis. Meas. Sci. Technol. 11(12), 1826.

Ridel, L., Bolzinger, M. A., Fessi, H., and Chevalier, Y. (2015). Nanopickering: Pickering nanoemulsions stabilized by bare silica nanoparticles. J. Colloid Sci. Biotechnol. 4, 110–116. doi: 10.1166/jcsb.2015.1122.

Robertson, J. D., Rizzello, L., Avila-Olias, M., Gaitzsch, J., Contini, C., and Magon, M. S. (2016). Purification of nanoparticles by size and shape. Sci. Rep. 6, 27494, doi: 10.1038/srep27494.

Robledo, N., Vera, P., López, L., Yazdani-Pedram, M., Tapia, C., and Abugoch, L. (2018). Thymol nanoemulsions incorporated in quinoa protein/chitosan edible films; antifungal effect in cherry tomatoes. Food Chem. 246, 211–219.

Rouessac F., and Rouessac A. (2013). Chemical Analysis: Modern Instrumentation Methods and Techniques. John Wiley & Sons.

Ruozi, B., Tosi, G., Forni, F., Fresta, M., and Vandelli, M. A. (2005). Atomic force microscopy and photon correlation spectroscopy: Two techniques for rapid characterization of liposomes. Eur. J. Pharm. Sci. 25(1), 81–89.

Salvia-Trujillo, L., Soliva-Fortuny, R., Rojas-Graü, M. A., McClements, D. J., and Martín-Belloso, O. (2017). Edible nanoemulsions as carriers of active ingredients: A review. Annu. Rev. Food Sci. Technol. 8, 439–466.

Sams, L., Paume, J., Giallo, J., Carriere, F. (2016). Relevant pH and lipase for in vitro models of gastric digestion. Food Funct. 7, 30–45.

Schmitt, V., Cattelet, C., and Leal-Calderon, F. (2004). Coarsening of alkane-in-water emulsions stabilized by nonionic poly(oxyethylene) surfactants: The role of molecular permeation and coalescence. Langmuir, 20, 46–52.

Schuch, A., Wrenger, J., and Schuchmann, H. P. (2014). Production of W/O/W double emulsions. Part II: Influence of emulsification device on release of water by coalescence. Colloids Surf. A 461, 344–351. doi: 10.1016/j.colsurfa.2013.11.044.

Severino, R., Ferrari, G., Vu, K. D., Donsi, F., Salmieri, S., and Lacroix, M. (2015). Antimicrobial effects of modified chitosan based coating containing nanoemulsion of essential oils, modified atmosphere packaging and gamma irradiation against *Escherichia coli* O157:H7 and *Salmonella Typhimurium* on green beans. Food Control. 50, 215–222.

Severino, R., Vu, K. D., Donsì, F., Salmieri, S., Ferrari, G., and Lacroix, M. (2014). Antibacterial and physical effects of modified chitosan based-coating containing nanoemulsion of mandarin essential oil and three non-thermal treatments against Listeria innocua in green beans. Int. J. Food Microbiol. 191, 82–88.

Shah MR, Imran M, Ullah S., editors. (2017). Chapter 2 – Nanostructured lipid carriers. Lipid-Based Nanocarriers for Drug Delivery and Diagnosis. William Andrew Publishing, Oxford, UK. pp. 37–61.

Shahidi, F., and Zhong, Y. (2010). Lipid oxidation and improving the oxidative stability. Chem. Soc. Rev. 39, 4067–4079.

Silva H. D., Cerqueira M. Â., and Vicente A. A. (2012). Nanoemulsions for food applications: Development and characterization. Food Bioproc. Tech. 5(3), 854–867.

Silva, H. D., Cerqueira, M. A., and Vicente, A. A. (2015). Influence of surfactant and processing conditions in the stability of oil-in-water nanoemulsions. J. Food Eng. 167, 89–98. doi: 10.1016/j.jfoodeng.2015.07.037.

Silva, H. D., Poejo, J., Pinheiro, A. C., Donsi, F., and Vicente, A. A. (2018). Evaluating the behaviour of curcumin nanoemulsions and multilayer nanoemulsions during dynamic in vitro digestion. J. Funct. Foods 48, 605–613.

Singh, Y., Meher, J. G., Raval, K., Khan, F. A., Chaurasia, M., Jain, N. K., and Chourasia, M. K. (2017). Nanoemulsion: Concepts, development and applications in drug delivery. J. Control. Release 252, 28–49. https://doi.org/10.1016/j.jconrel.2017.03.008.

Singh, H., Ye, A., and Horne, D. (2009). Structuring food emulsions in the gastrointestinal tract to modify lipid digestion. Prog. Lipid Res. 48, 92–100.

Sitterberg, J., Özcetin, A., Ehrhardt, C., and Bakowsky, U. (2010). Utilising atomic force microscopy for the characterisation of nanoscale drug delivery systems. Eur. J. Pharm. Biopharm. 74(1), 2–13.

Sugumar, S., Singh, S., Mukherjee, A., and Chandrasekaran, N. (2015). Nanoemulsion of orange oil with non ionic surfactant produced emulsion using ultrasonication technique: evaluating against food spoilage yeast. Appl. Nanosci. 6, 113–120.

Tadros, T., Izquierdo, P., Esquena, J., and Solans, C. (2004). Formation and stability of nano-emulsions. Adv. Colloid Interface Sci. 108, 303–318. doi: 10.1016/j. cis.2003.10.023.

Taisne, L., Walstra, P., and Cabane, B. (1996). Transfer of oil between emulsion droplets. J. Colloid Interface Sci. 184, 378–390.

Tan, C. P., and Nakajima, M. (2005a). β-Carotene nanodispersions: preparation, characterization and stability evaluation. Food Chem. 92, 661–671.

Tan, C. P., and Nakajima, M. (2005b). Effect of polyglycerol esters of fatty acids on physicochemical properties and stability of β-carotene nanodispersions prepared by emulsification/evaporation method. J. Sci. Food Agric. 85, 121–126.

Tian, H., Li, D., Xu, T., Hu, J., Rong, Y., and Zhao, B. (2017). Citral stabilization and characterization of nanoemulsions stabilized by a mixture of gelatin and Tween 20 in an acidic system. J. Sci. Food Agric. 97, 2991–2998.

Tso, P., and Crissinger, K. (2000). Overview of digestion and absorption. In: Stipanuk M.H., Caudill M.A., editors. Biochemical and Physiological Aspects of Human Nutrition. W.B. Saunders, Philadelphia, PA. pp. 75–90.

Turner, Y. T. A., Roberts, C. J., and Davies, M. C. (2007). Scanning probe microscopy in the field of drug delivery. Adv. Drug Deliv. Rev. 59(14), 1453–1473.

Vitorino, R., Lobo, M. J. C., Ferrer-Correira, A. J., Dubin, J. R., Tomer, K. B., Domingues, P. M., and Amado, F. M. L. (2004). Identification of human whole saliva protein components using proteomics. Proteomics 4, 1109–1115.

Wang, Z. L. (2000). Transmission electron microscopy of shape-controlled nanocrystals and their assemblies. J. Phys. Chem. B. 104(6), 1153–1175.

Weiss, J., Takhistov, P., and McClements, D. J. (2006). Functional materials in food nanotechnology. J. Food Sci. 71, 107–116.

Witthayapanyanon, A., Acosta, E. J., Harwell, J. H., and Sabatini, D. A. (2006). Formulation of ultralow interfacial tension systems using extended surfactants. J. Surfactants Deterg. 9, 331–339. doi: 10.1007/s11743-0065011-2.

Wooster, T. J., Golding, M., and Sanguansri, P. (2008). Impact of oil type on nanoemulsion formation and Ostwald ripening stability. Langmuir ACS J. Surf. Colloids 24, 12758–12765. doi: 10.1021/la801685v.

Yang, X., Tian, H., Ho, C. T., and Huang, Q. (2011). Inhibition of citral degradation by oil-in-water nanoemulsions combined with antioxidants. J. Agric. Food Chem. 59, 6113–6119.

Yin, L. J., Chu, B. S., Kobayashi, I., and Nakajima, M. (2009). Performance of selected emulsifiers and their combinations in the preparation of β-carotene nanodispersions. Food Hydrocoll. 23, 1617–1622.

Yuan, Y., Gao, Y., Zhao, J., and Mao, L. (2008). Characterization and stability evaluation of β-carotene nanoemulsions prepared by high pressure homogenization under various emulsifying conditions. Food Res. Int. 41, 61–68.

Zhang R., and McClements D. J. (2018). Characterization of gastrointestinal fate of nanoemulsions. Nanoemulsions. Academic Press, Burlington, MA. 577–602, https://doi.org/10.1016/B978-0-12-811838-2.00018-7.

Zhang, R., Zhang, Z., Zhang, H., Decker, E. A., and McClements, D. J. (2015). Influence of emulsifier type on gastrointestinal fate of oil-in-water emulsions containing anionic dietary fiber (pectin). Food Hydrocoll. 45, 175–185.

Zhao, Q., Ho, C.T., Huang, Q. (2013). Effect of ubiquinol-10 on citral stability and off-flavor formation in oil-in-water (O/W) nanoemulsions. J. Agric. Food Chem. 61(31), 7462–7469.

CHAPTER 2

Ingredients and Composition of Food Grade Nanoemulsions

Gulam Mustafa,[1] Faisal Obaid Alotaibi,[1]
Fahad Khalid Aldhafiri,[2] Mohammed Faiz Arshad[1]

[1]College of Pharmacy (Boys), Al-Dawadmi Campus, Shaqra
University, Riyadh, Kingdom of Saudi Arabia
[2]Public Health Department, College of Applied Medical Sciences,
Majmaah University, Riyadh, Kingdom of Saudi Arabia

CONTENTS

2.1 INTRODUCTION

Since food systems are complex arrays of different ingredients that do not always mix well with each other, oil and water interfaces usually exist and can be stabilized by appropriate emulsifiers or surface stabilizing agents. Thus, an understanding of the science and technology of emulsion systems is vital for food researchers, as many natural and processed foods such as milk, salad dressings, ice cream, soft drinks, and cakes are made partially or exclusively from emulsifiers. An emulsifier composed of different components and processing conditions exhibits a wide range of physical and chemical properties that allow it to provide food systems with important functions in which stability, texture, taste, aroma, appearance, and biological response play a crucial role in achieving the goal of each specific product. Due to its widespread use in food systems, emulsifiers have already become a popular way to utilize functional ingredients, including color, flavor, preservatives, vitamins, minerals, and nutrients. The application of nanotechnology in the agriculture and food industry was first addressed by the US Department of Agriculture in September 2003 [1]. Nanotechnology is expected to change the entire food industry, and to completely change the way a food product is processed, packaged, transported, and consumed in near future. Nanotechnology applications will continue to affect the food industry due to their unique and novel properties. Food nanoemulsions are increasingly being used in the food sector owing to their physicochemical properties for effective packaging, biologically active compound closures, thawing, target delivery, and bioavailability. Increasing the surface area by decreasing the nanoemulsion particle size can lead to improved usability of the phytochemicals contained within the nanoemulsions. An emulsified form of many nutrients and food additives is readily available in the market. An emulsion becomes a nanoemulsion when the diameter of the scattered droplets is about 500 nm or even smaller [2–4]. Nanoemulsions can absorb functional components in their drops, making it easier to reduce their chemical degradation [5]. Varied nanomaterials that have been developed and being used not only have a difference in the organoleptic properties of food, but also the safety and health benefits that food has to offer. Different types of nanoemulsions, which have more complex properties such as multiple nanostructures or multilayer nanoemulsions, carry different loading or packaging capabilities. The typical structure of a nanoemulsion and its three major components which include oil, water, and a surfactant as shown in Figure 2.1.

The precise mix of these major ingredients is very important which decides the existence of true or meta-stable nanoemulsions and hence determines the stability and properties of the emulsion. Apart from the major components, there are minor components that are also very essential for making food-grade nanoemulsions. Proteins, fats, and sugars comprise a portion of the minor components to be emulsified and added in food-grade nanoemulsion to increase the absorption or enhancing the bioavailability of nutritional supplements [6]. Different ingredients including vitamins, minerals, flavors, acidic substances, preservatives, colorants, and antioxidants are also minor components to be included in nutrients for achieving different functions

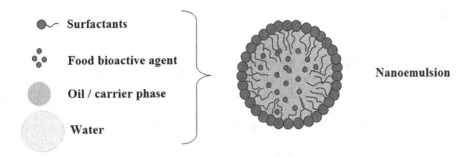

Surfactants

Food bioactive agent

Oil / carrier phase

Water

Nanoemulsion

FIGURE 2.1 A typical structural overview of a food-grade nanoemulsion.

like taste or texture, appearance, food enrichment, and stability of the food products. The edible oils for the food-grade nanoemulsions can be obtained from various natural resources including plants and animal derivatives. But the excessive consumption of some of these oils is also connected with various diseases like obesity, diabetes, and related cardiovascular problems. Therefore, it is a major challenge for food scientists and formulation scientists to focus on the various properties of the oil, like nutritional value, crystallization property, level of unsaturation, oxidative stability, and quality [7–9]. The water and oil ratio decides the texture of nanoemulsions. The organoleptic property of the food is greatly influenced by the unique properties of water and its ratio in the optimized product. The food nanoemulsions also have special consideration regarding the water of crystallization for their stability. On the other hand, smaller water crystals give their distinctive texture and taste, whereas larger crystals give a rough texture to the food products in food-grade emulsions like ice creams. These components must be controlled according to the demand and stability of the food product [10, 11]. The kinetic stability of nanoemulsions can be improved by incorporating stabilizers which include emulsifiers, ripening retarders, weighting agents, or texture modifiers [12]. Nanoemulsion stability is supported by the interfacial film between the oil droplet and the water phase established by the emulsifiers or the surfactants. The formation of interfacial films prevents the flocculation and/or coalescence types of instability issues after subjecting the food products to different mechanical pressure during various processing procedures. Emulsifiers such as small molecule surfactants (Tweens or Spans), phospholipids (soy, lecithin), amphiphilic polysaccharides (modified starch, gum Arabic), and amphiphilic proteins (casein whey protein) are extensively used in the food industry to formulate nanoemulsions. The FDA has approved various types of emulsifiers to be used in the food industry as a single emulsifier cannot be used in all food products because of incompatibility as well as insufficient HLB. Thus, based on the molecular properties, the surfactant is chosen to give the desired effect. Apart from these major components, there are lists of minor components that play an important role to fulfill the various aspects of the food-grade nanoemulsion. The major challenges that food industries face to prepare nanoemulsions include legal acceptability, label-friendly, and economically viable ingredients approved by the native FDA. Toxicological issues are the most important concern in food technology because of the nano-size droplets that could change the normal function of the gastrointestinal tract as well as the encapsulated food ingredients [13]. Therefore, the present chapter is designed to give a detailed account of both the major and minor components which can be utilized to prepare food-grade nanoemulsion with mechanism and criterion affecting their selection.

2.1.1 Applications and Characteristic Features of Nanoemulsion

There are various applications of nanoemulsions in drug delivery, cosmetics, medicines, and food nutrients, making them a global carrier of pharmaceuticals and nutrients. The appeal of nanoemulsions in a variety of applications in personal care products and cosmetics, as well as in healthcare, is due to their following unique properties [14]. A glimpse of the functionality of nanoemulsions is shown in Figure 2.2.

Some of the characteristics of nanoemulsions are as follows:

i. The exceedingly small size of the droplets leads to a significant decrease in the force of gravity, and the Brownian motion is just sufficient to overcome the gravitational pull. This means that no greasing or precipitation occurs during storage.

ii. Also, the small drop size prevents any scarring of the drops. Weak flocculation is prevented, allowing the system to remain dispersed without separation.

iii. It also prevents small drops from sticking together, as these drops are not amenable to deformation and thus prevent surface fluctuations. Also, the large thickness of the surfactant (relative to the radius of the droplets) prevents any thinning or cracking of the liquid film between the drops.

iv. Nanoemulsions are suitable for transdermal drug delivery because of the large surface which helps in rapid and effective penetration of the active substances.

v. The nanometric size range of nanoemulsions helps in penetrating even the "rough" surface of the skin and thus enhances the effectiveness of the active substances.

vi. The transparency, fluidity, and the absence of any thickeners give nanoemulsion an agreeable visual and skin feel.

vii. A high surfactant concentration, usually in the range of 20% and above is required for microemulsions whereas a reasonable ratio of surfactant is required for nanoemulsions. For a nanoemulsion of 20% O/W, a surfactant concentration in the range of 5–10% may be sufficient.

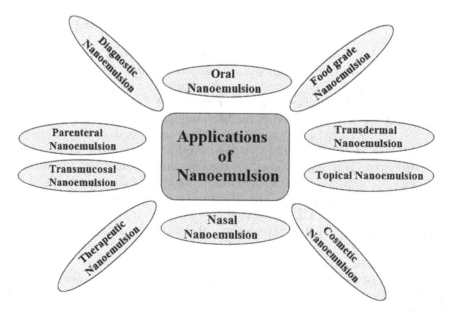

FIGURE 2.2   Illustration showing multiple applications of multifaceted nanoemulsion.

viii. The small size of the droplets allows for even deposition on the substrates. Improved hydration, diffusion, and penetration can also be achieved due to the lower surface voltage of the whole system and the low interface voltage of O/W droplets.

ix. For the incorporation of fragrant oils, nanoemulsions offer a better alternative, in many personal care products. This could also be applied in perfumes, which are desired to be formulated alcohol-free.

x. The nanoemulsion emerged as a better substitute to liposomes and vesicles to resolve the stability issue and also it is possible in some cases to construct laminar liquid crystal phases around the nanoemulsion droplets.

2.2 MAJOR COMPONENTS OF FOOD GRADE NANOEMULSIONS

2.2.1 Oil Phase

There are lots of oil phase ingredients used in the formulation of food-grade nanoemulsions. The oil phase for food-grade nanoemulsions includes a variety of nonpolar molecules, such as free fatty acids (FFA), monoacylglycerols (MAG), diacylglycerols (DAG), triacylglycerols (TAG), waxes, mineral oils, or various lipophilic nutraceuticals [12], and flavor oils (Figure 2.3). The oils extracted from soybeans, flaxseeds, sunflower, olives, saffron, corn, algae, fish are common TAG that are most used in food-grade nanoemulsions

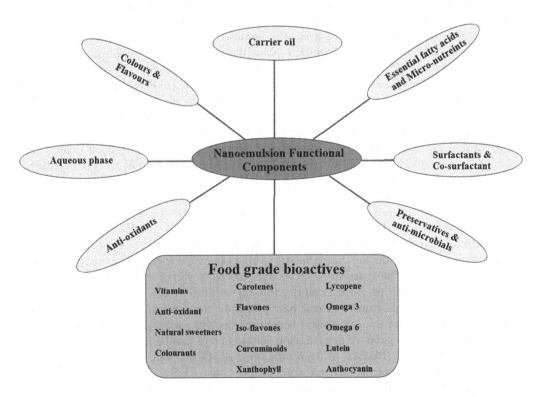

FIGURE 2.3 This diagram shows the functional components used as a nutritional phase required to change into food-grade nanoemulsions to fulfill nutritional demand and various related issues.

due to their cost-effectiveness and nutritional value [12]. However, nutraceutical lipids are susceptible to several factors for their stability, hence their use in food-grade nano-emulsion remains a major challenge. Physicochemical properties of the oil phase like density, viscosity, aqueous solubility, polarity, interfacial tension, refractive index, and chemical stability greatly influence the properties of nanoemulsions [2, 15, 16].

The food market attracted the versatility of nanoemulsions to incorporate dietary fats like fat-soluble antioxidants, fat-soluble vitamins, phytosterols, polyunsaturated fats, carotenoids, and fat-soluble dyes, or it is incorporated by homogenization and/or pack-aging into food. Since dietary fats are susceptible to several factors that lead to their chemical or physical degradation, their use in commercial nano foods remains a major challenge. Specific lipid nutrients have certain confrontations, which ultimately depend on their physical and chemical properties including solubility, chemical stability, melting point, and oil-water separation coefficient [5, 15]. Hence, the nanoemulsions must be manufactured very carefully, and the manufacturing protocol must be chosen wisely to maintain the specific physical and chemical properties of the specific dietary fats used in the nanoemulsions.

2.2.1.1 Essential Oils and Vegetable Oils

Essential oil is a concentrated hydrophobic liquid that contains volatile chemical com-pounds from plants. Essential oils (EO) are also known as volatile oils or simply the plant oils from which they were extracted. Essential oil is "essential" as it contains the "essence" of the scent of the plant; the distinctive smell of the plant from which it is derived. EO derived from edible, medicinal, and herbal plants are well known as a natural food supplement with great therapeutic application potential. Its chemical constituents are among the secondary plant metabolites with lipophilic and highly volatile properties [16]. These herbal products are often used as formulation addi-tives in foods, beverages, perfumes, and cosmetics [16–18]. Vegetable oils are mainly used as an oil phase to prepare food nanoemulsions to meet food demand, not for quarrying. The detailed application of vegetable essential oil is summarized in Table 2.1, which is mainly composed of phytosterol and phytostanol. It is these vegetable fats that demonstrate potential therapeutic activity, especially the lipid profile [19, 20]. EO possess the potential to restrain the cholesterol taken in dietary supplements owing to its ability to reduce low-density cholesterol (which is hazardous to health) as well as total cholesterol [21].

Phytosterol and phytostanol are a major reason for the growing interest in using vegetable oils in foods and beverages, especially in foods based on nanoemulsion, which could work as important nutrients with effective medicinal properties. The dietary phytosterols do not harm health, this is because their intestinal absorption is extremely poor. There are some challenges for phytosterol and phytostanol, which include the tendency to form crystals, low-fat solubility, low water solubility, and high melting point. Because of these barriers, some new technologies, including phy-tosterol and phytostanol, are always recommended for successful use in foods. They are also subject to oxidative degradation. Phytosterols can be esterified with polyun-saturated fatty acids (PUFAs) to overcome these obstacles. The esterified phytosterols are absorbed in the gastrointestinal tract and produce FFA and phytosterols [22–25]. Although phytosterols are prone to oxidative degradation, it is not yet clear whether their oxidized form will exhibit the same biological activity or toxicity. To overcome this, further encapsulation of phytosterol has been proposed. Due to its hydropho-bic nature, it does not dissolve in water and reduces its antimicrobial ability when

TABLE 2.1 Research Updates on Therapeutic Essential Oil-Based Nanoemulsions or Nanoencapsulation

Plant Oil	Active Nutrients	Activity Assessed	References
Sunflower oil	High-oleic and vitamin E	Shelf life and quality of *S. Guttatus* on storage	[33]
Castor oil	Omega 6 essential fatty acids, vitamin E	Non-reactivity with contact lens	[34]
Soybean oil	23% monounsaturated fat, and 58% polyunsaturated fat	Increasing bioavailability of doxorubicin	[35] [36]
Basil oil	Vitamin A, calcium, and antioxidants	Bactericidal potential	[37]
Rosemary oil	1,8-cineole (46.4%), camphor (11.4%), and α-pinene (11.0%)	Larvicidal activity, antioxidant and antimicrobial properties	[38]
Oregano oil	Carvacrol and thymol	Efficiency in controlling foodborne bacteria on fresh lettuce	[20]
Lemon oil	Terpenes, sesquiterpenes, aldehydes, and esters	Stability based on size and turbidity	[39]
Almond oil	Monounsaturated fat, polyunsaturated fat, vitamin E, and phytosterols	Loading efficiency	[40]
Walnut oil	Omega-3 fatty acids and antioxidants	Stability	[41]
Orange oil	Limonene, β-myrcene, α-pinene, citronellol	Sonication and nanoemulsification	[13]
Babchi oil	Psoralen, corylifolin, isopsoralen	Anti-inflammatory and flux	[42]
Palm oil	β-carotene	Plasma cholesterol	[43]
Safflower oil	Omega-6 fatty acids (help to burn fat in the body), copper, tryptophan, vitamin B1, and phosphorus	Bioavailability	[44]
Peanut oil	Monounsaturated fat, polyunsaturated fat, vitamin E, and phytosterols	Bioavailability	[45]

(*Continued*)

TABLE 2.1   (Continued)

Plant Oil	Active Nutrients	Activity Assessed	References
Corn oil	Polyunsaturated fats include omega-6 and omega-3 fats	Stability	[46]
		Loading efficiency	[47]
Coconut oil	Lauric acid, capric acid, caprylic acid, myristic acid, palmitic acid, polyunsaturated fatty acids – linoleic acid and monounsaturated fatty acids – oleic acid, iron, vitamin K, and vitamin E a potent antioxidant	Effect of operating parameters and chemical compositions on stability	[48]
		Fairness index	[49]
Celery essential oil	Limonene, myrcene, beta-pinene	Potential anticancer and antibacterial activity	[50]
Avocado oil	Oleic acid a monounsaturated omega-9 fatty acid with, high omega-6 to omega-3 ratio	Stability	[51]
Rice bran oil	Oryzanol, tocotrienols	Irritation potential and moisturizing	[52]

incorporated into food. The vegetable oils are cheap, easily scalable, environment friendly option for formulating nanoemulsion dosage forms. Second, the nanoemulsion prepared from vegetable oils is of exceedingly small droplets size range, making it kinetically more stable in contrast with large emulsions, which leads to fusion, flocculation, and sedimentation [26].

2.2.1.2 Unsaturated Fats

Dietary fats can be classified into fats and oils, wherein the fats are solid whereas the oils are liquid at room temperature. Fats are generally higher in saturated fatty acids (SFAs), which have a higher melting point than unsaturated fats. Unsaturated fats are good healthy fats because they can improve blood cholesterol levels, reduce inflammation, stabilize heart rhythms, help control diabetes, and play many other beneficial roles. Unsaturated fats are found mainly in plant foods, such as vegetable oils, nuts, and seeds. The health recommendation recently focused on reducing the consumption of solid fats, as this may be an easier means for consumers to identify fats that are high in solid fats. There are two main types of good and healthy unsaturated fats namely monounsaturated and polyunsaturated.

2.2.1.2.1 Monounsaturated Fats In recent years, dietary fatty acids have become a major topic in nutrition research, and the popularity of dietary oils rich in monounsaturated fatty acids (MUFAs), are recommended for culinary applications, as well as in various food products. Recommending the intake of dietary oils containing good fatty acid properties like high in MUFAs and PUFAs is among the key nutritional strategies to safeguard against and cope with chronic disease threats, while reduced consumption is advised for SFAs and trans-fatty acids (TFAs) [27]. In the

United States (USA), vegetable oils are commonly used as edible oils as a major contributor to daily fat and caloric intakes [28]. Besides the amount of fat consumed, dietary fat quality is one of the strongest determinants of health. Newer types of oils rich in oleic acid which encompass more than 90% of all MUFAs consumed in the USA are high oleic canola oil, high oleic blending oil, and high oleic algal oil with the percentage of MUFA 73%, 84%, and 90%, respectively. These have already the potential to considerably modify the dietary fatty acid intake [29]. This type of fat is found in various foods and oils. Studies show that consuming food rich in MUFAs instead of saturated fats improves blood cholesterol levels, which reduces the risk of heart disease and may also help to reduce the risk of developing type 2 diabetes.

Such a transition in community consumption is projected to have a beneficial impact on certain health biomarkers, along with circulating lipid profiles, homeostasis of glucose, and insulin sensitivity [30]. Flavors and aromas are one of the important types of food additives used in food and especially in beverages.

Compounds used in the fragrances have limitations in regard to their application in food products owing to the incompatibilities and solubility issues in most nutrient media, instability, and instability while handling and storage [31]. Thus, emulsifiers are used in beverages to disperse these water-insoluble aromas and odors into an aqueous beverage. Nanoemulsions prepared with extra virgin olive oil had better physical stability and better CoQ10 retention compared to those prepared with limited olive oil [32]. The list of MUFA and its nanoparticles is listed in Table 2.2.

2.2.1.2.2 Polyunsaturated Fats This type of fat is found mostly in plant foods and oils. Evidence suggests that eating foods rich in PUFAs instead of saturated fats improves blood cholesterol levels, which reduce the risk of heart disease and may also help reduce the risk of developing type 2 diabetes. The higher the percentage of unsaturated fats, the faster is the oxidation reaction. The rate of oxidation was reported to be quick with linolenic acid followed by linoleic and oleic acids [53]. Therefore, oxidization happens more quickly in grape seed oil, which has the highest amount of PUFAs (about 68–85%), and the largest part is made up of linoleic acid (about 67%) [54, 55]. The major active components of peanut oil are arachidonic, oleic, and stearic acids [56] besides linoleic acid (21–35%) and α-linolenic acid (0.1–0.4%) [57]. The high content of unsaturated fatty acids in peanut oil might be a health issue and unsaturated fatty acids are prone to oxidation, leading to changes in taste as well as the aroma. The rich concentrations of health-promoting compounds including tocopherol, tocotrienol, phytosterol, and beta-oryzanol in rice bran, attract consumer interest over others [58].

Oxidation stability is one of the issues of edible vegetable oils which has a huge impact in technical processing as well as the shelf life of vegetable oil-based nutraceuticals.

Canola oil contains the lowest level of saturated fats compared to all the other essential oils in the market today and a high protein (28%) content. When the oil is extracted, canola oil is also high in monounsaturated oleic acid. The content of tocopherol, one of the natural antioxidants in rapeseed, is comparable to that of peanuts and palm oil. For oils with a high linolenic acid content, there is an associated advantage that it could decrease expiration date and improve life by mitigating oxidation during storage and processing. Lozenges dosage forms use up to 67% corn oil as oral nutritional supplements. Similarly, because of its greater percentage of unsaturated acids, corn oil is used

TABLE 2.2   MUFAs and PUFAs Oil-Based Nanoemulsions

| Name of Oils | Monounsaturated Fats (MUFAs) | | |
	Application	Purposes for Nanoemulsifications	References
Canola oil	Boost male reproductive system in rats, vitamin E source	Stabilization with food-grade surfactant	[59]
Rice bran oil	Antioxidant and skin cosmetics	Stability enhancement	[60]
Olive oil	Antioxidants, anti-inflammatory, anti-diabetic, Alzheimer's disease	Physical stability and CoQ_{10} retention	[32]
Algae oil	Anticancer and anti-inflammation functions	Cytotoxicity test	[61]
	Polyunsaturated Fats (PUFAs)		
Walnut oil (Omega 3)	Anti-inflammatory, cardioprotective agent	Increased oxidative stability	[41]
Peanut oil	Source of antioxidant vitamin E	Used as a carrier for griseofulvin to increase the bioavailability	[62]
Palm oil	Source of antioxidant vitamin E	Cosmetic	[63]
Corn oil	Cardioprotective (source of vitamin E, linoleic acid, and phytosterols)	Oxidative stability	[64]

as a supplement for fats and oils that have a high proportion of fatty acids in the diets of patients with hypercholesterolemia. The key elements of oleic acid are (Z)-9-octadecenoic acid with varying quantities of saturated and other unsaturated acids [65, 66]. It may contain suitable antioxidants. In the food technology and pharmaceutical industry, it is used extensively as an emulsifying agent and also to improve systemic bioavailability [65–67].

The health benefits of PUFAs especially ω-3 fatty acids are very well known in society, hence it is the major driving force to use in the food and beverage sector [68]. Due to its possible well-known medicinal properties against many diseases such as inadequate growth of the infant, cardiovascular, neurological, and immune system-related disorders, the use of Omega 3 fatty acids is widespread [69]. These properties have attracted food manufacturers to include such PUFAs and to promote nanoemulsion-based foods, as they will provide effective delivery and activities in nano-droplet size ranges. It can be seen that PUFAs are highly susceptible to oxidative degradation and ultimately pose long-term storage problems, in addition to being very sensitive at smaller nano-sizes. It is recommended by valid scientific study to follow the precautionary measures and the guidelines to increase the stability of PUFA constituted nanoemulsions by controlling the quality of the initial ingredient, interfacial engineering, and stabilization of nanoemulsions.

2.2.2 Co-Solubilizer Phase/Carrier Oil

Certain essential oils are highly insoluble to convert into nanoemulsions, therefore additional carrier oils are required to increase miscibility. Various nutritional products utilized the carrier oils to formulate food-grade nanoemulsions (Table 2.3). Metaphase lipid transporters are among the most important excipients in nanoemulsion not only because high dissolving power of therapeutics, but also because of the increase in lipophilic portion helping bioactive molecules getting transported easily through the intestinal lymphatic system [70, 71]. Modified or hydrolyzed vegetable oils have been widely used since these excipients form good emulsification systems with a large number of surfactants and exhibit better drug solubility properties [72, 73]. They provide structural and physiological benefits, and their breakdown products are similar to the natural end products of intestinal digestion. Because of biosimilar metabolites, medium-chain, semi-synthetic derivatives are progressively and effectively replacing regular medium-chain triglycerides in the food and pharmaceutical industries [72, 74]. Solvent capacity for less hydrophobic bioactive molecules can be improved by blending triglycerides with other oily excipients, which include mixed monoglycerides and diglycerides.

The nature of the oil component also affects the bending or curvature which affects the ability to penetrate the tail region of the surfactant monolayer. Short-chain oils penetrate to a greater extent than long-chain alkanes-based oils in the tail group region of surfactant, amplifying this region to a greater extent of flexibility and increased negative bending. Several long and medium-chain triglycerides including Labrafac, Lauroglycol, Labrafil M1944CS, and olive oil have been reported to follow this mechanism [75].

Example of certain synthetic carrier oils that have been used in many works of literature as a co-solubilizer as well as nanoregion enhancers is olive oil, castor oil, corn oil, sunflower oil, safsol-218®, sefsol-228®, labrafac lipo®, miglyol 812®, IPM, IPP, oleic acid, homotex PT®, lauroglycol 90®, captex 355®, triacetin, capryol 90®, and capmul MCM®.

2.2.3 Emulsifier Phase

The molecules and ions that are absorbed on the interfaces are called surfactants or surface-active agents. The alternative term is amphiphile, which indicates that a molecule or ion has a certain affinity for both polar and non-polar solvents. Depending on the number and nature of polar and non-polar groups present, amphiphiles may be predominantly hydrophilic, lipophilic (oily), or relatively well balanced between these two extremes. For example, alcohols, amines, and straight-chain acids are amphiphiles that vary from predominantly hydrophilic to lipophilic with an increase in the number of carbon atoms in the alkyl chain [87]. The surfactant chosen must be able to lower interfacial tension to a very small value to assist the dispersion process during the preparation of the nanoemulsion [23]. It is found that the physical properties of the oil phase and the nature of the surfactant layer have a great influence on the formation and fixation of the nanoemulsion. Long-chain triglycerides (LCTs) are high-viscosity oils, while

TABLE 2.3   Lists of the Formulations Used Co-Solvent or Co-Solubilizers as an Oil Carrier Phase for Various Nutrients to Convert into Nanoemulsions

Bioactive	Functions	Oil Carrier	Nature of Oils	Activity Access	References
Coenzyme Q10	Commonly used for conditions of heart well being	Olive-pomace oil	Medium-chain triacylglycerols or orange oil	Solubility enhancement	[32]
β-carotene	Used in certain cancers, heart disease, cataracts	Corn oil	Medium-chain triacylglycerols or orange oil	Types of oil on bioavailability	[76]
		Corn oil and Miglyol 812		Bioavailability	[77]
Curcumin	Antioxidant, anti-inflammatory, analgesics	Corn oil	Long, medium, and short-chain triacylglycerols	Effect of chain length on bioavailability	[78]
Vitamin E	Immune booster, antioxidant, anti-inflammatory, skin health, and beauty		Medium-chain triglycerides	Bioavailability	[79]
Thyme essential oil	Antimicrobial, antifungal, antioxidant, antiviral, anti-cancer, and anti-inflammatory	Corn oil	Medium-chain triacylglycerols	Antimicrobial efficacy	[80]
D-limonene	Anti-inflammatory, antioxidant, lipid-lowering, anti-diabetic, immunosuppressive	Sunflower and palm oil	16-carbon saturated fatty acids	Antimicrobial efficacy	[81]
Peppermint essential oil	Irritable bowel syndrome, anti-itching, flavoring agent in foods	Carrier oil not mentioned	Medium-chain triglyceride	Antibacterial activity	[82]
Lemongrass essential oil	Anti-inflammatory, anti-fungal	Carrier oil not mentioned	Medium-chain triglyceride (MCT) to long-chain triglyceride (LCT)	Nanoemulsions size and antibacterial activity	[77]
Oil of Nigella sativa	Antioxidant, anti-inflammatory	Capryol 90	Propylene glycol monocaprylate	Increased bioavailability and enhance anti-inflammatory activity	[83]

(*Continued*)

TABLE 2.3 (Continued)

Bioactive	Functions	Oil Carrier	Nature of Oils	Activity Access	References
Oil of resveratrol	Antioxidant, anti-aging, anti-diabetic	Sefsol 218	Propylene glycol mono caprylic ester	The antioxidant activity showed high scavenging efficiency	[84]
Silymarin	Hepatoprotective, antioxidant, anti-inflammatory, anticancer, and cardioprotective activities	Sefsol 218	Propylene glycol mono caprylic ester	Enhanced stability of silymarin	[85]
Eugenol	Anti-inflammatory properties, as well as antioxidant effect	Sesame oil	Blends of linoleic, oleic, palmitic, and stearic fatty acids	Nanoemulsion treatment resulted in alteration of membrane permeability	[86]

short-chain triglycerides (SCTs) are low-viscosity oils. The improvement of the surfactant structure and differential viscosity resulted in the formation of significantly small droplet-sized nanoemulsions [88]. As a consequence, essential oil carrier in nanoemulsions is mostly used to achieve relatively high stability. Emulsifiers are interface-active molecules that are primarily used to facilitate the droplet breakdown from the top-down approach, thus helping to form smaller droplet sizes, and also inhibit the aggregation of these droplets. These combined effects help to maintain long-term stability. During the production of nanoemulsions, the emulsions are absorbed in the O/W interface, which ultimately leads to a decrease in the interface voltage, which leads to the ease of breaking the droplets [89]. After facilitating the destruction of the drops, the emulsions should act as a protective layer around the oil droplets to prevent build-up and clotting. It is important to note that the emulsifier concentration should be sufficient to cover all O/W interfaces, and the coating rate around the oil droplets of the emulsions should be faster than that of the coalescence [90]. The nanoemulsions consisting of various essential oils that are easily emulsified by non-ionic emulsifying agents are healthy, stable with excellent biocompatibility [91]. Many review articles have been published dealing with emulsifiers, their chemical nature, mode of operation, factors affecting the selection of emulsifiers, legal and health issues [92–94].

Natural polymers can replace synthetic surfactants used in the emulsion formulation. Biopolymers are currently used in the food industry to stabilize emulsions and appear as promising candidates in the nutraceutical field. All proteins and some sugars can be absorbed on the surface of the pellets, thus reducing the interface tension and increasing the flexibility of the interphase. Proteins and polysaccharides can also be bonded via covalent bonding or electrostatic interactions. The combination of the properties of these biopolymers under suitable conditions increases the stability of the emulsion. Alternative layers of countercharged biopolymers may also be formed around the pellets to form multilayer "films". These layers can provide electrostatic and static stability, thus improving thermal stability and resistance to external treatment [95].

2.2.4 Co-Emulsifier Phase

The liquid interface film is again achieved by adding a co-surfactant. In the absence of a surfactant, an extremely hard film of the surfactant is formed, and thus a nanoemulsion is obtained in a very limited concentration range. The presence of the interface film co-surfactant allows sufficient flexibility to absorb the various curves required to form a nanoemulsion in a wide range of compositions [96].

In the development of nanoemulsions, it was observed that in most cases, single-chain surfactants alone are incapable to decrease the oil/water (O/W) interfacial tension to enable nanoemulsification [97, 98]. Co-surfactants are often needed to make the primary surfactant sufficiently soluble and help in making the interface sufficiently flexible. Medium-chain alcohols, usually added as modulating agents, have the consequence of reducing the interface voltage while increasing the smoothness of the interface, thus increasing the entropy of the system [96, 97]. Medium-chain alcohol also increases the movement of the hydrocarbon tail and also allows more oil to enter the area. It has also been recommended that some oils, such as ethyl esters of fatty acids, also act as modulating depressants, penetrating the hydrophobic chain region of the monolayer of surfactants [99]. All of the mechanisms described above are assumed to promote the formation of nanoemulsions. The presence of alkanol also decreases repulsive interactions between some of the main charged groups in the case of nanoemulsions stabilized by ionic surfactants.

Sodium caseinate (NaCas) is frequently used in formulating delivery systems as a co-emulsifier. Qu and Zhong reported that when proteins are conjugated by reducing saccharides such as maltodextrins (MD), it can increase the emulsification and stabilization functions of O/W emulsions [100]. Zhang and coworkers reported the oil-in-water emulsions freeze-thaw stability with sodium caseinate–maltodextrin conjugates which brought about a substantial improvement in freeze-thaw stability as compared with the mixture of control protein and protein–sugar [101].

2.3 MINOR COMPONENTS OF FOOD GRADE NANOEMULSION

2.3.1 Ostwald Ripening Inhibitor or Molecular Diffusion Inhibitor

The challenges in establishing stable nanoemulsions are the ability to create an initial emulsion in which the total droplet size distribution is less than 100 nm, followed by subsequent stabilization against Ostwald ripening, flocculation, fusion, and phase separation (Figure 2.4). Ostwald ripening or molecular diffusion arises from multiple dispersion of the emulsion and the difference in solubility between small and large droplets. Ostwald's maturity was introduced in 1900 by the Gibbs-Thomson law and it is based on changes in the solubility of nanoparticles according to their size. This is the main destabilizing mechanism of nanoemulsion [102]. In the meantime, large particles tend to grow inappropriately because of the low level of surface energy [103, 104]. According to this theory, the rate of Ostwald ripening in O/W emulsions is directly proportional to the solubility of the oil in the aqueous phase. It indicates that Ostwald's ripeness can be used as a tool for evaluating the thermodynamics of an oil-in-water solution [105]. The physical decomposition of the emulsions is due to the spontaneous tendency to the minimum intermediate zone between the dispersion phase and the dispersion medium. Reduction of the interstitial zone is mainly accomplished by two mechanisms: coagulation the first, possibly fusion followed, and the second by Ostwald maturation. Assessment of Ostwald ripening

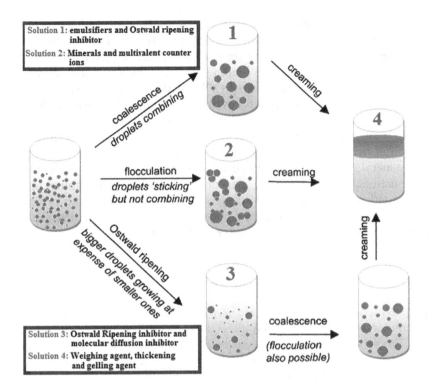

FIGURE 2.4   Possible instability issues in nanoemulsions and their solutions using some vital minor components.

in mixed oil-based nanoemulsions showed that the entropy gain due to oil demixing provided a thermodynamic barrier which inhibits the growth of oil droplets and hence high stability [88]. Theoretically, Ostwald's ripening should result in condensation of all droplets in one drop, resulting in phase separation, as shown in Figure 2.4.

Studies indicate that nanoemulsions containing three oils of LCTs do not undergo Ostwald maturation and are physically stable for more than 3 months. Ostwald maturation is inhibited by the large molar volume of LCTs, rendering them insoluble in water, thus providing a kinetic barrier to Ostwald maturation. Ostwald ripening and coalescence are often considered the most important destabilizing mechanism that leads to dispersion twisting and can be prevented by careful selection of stabilizers. One technique for improving the stability of NE is the use of maturation inhibitors. When separating the phases, the development of nanoemulsions is observed as being particularly unstable which tends to ripen Ostwald due to the high solubility of water in the oil. Accordingly, this can be reduced with corn oil, which contains a ripening inhibitor that has an oil phase. Vegetable oils, when used as an emulsifier maturation inhibitor, increase the droplet size.

2.3.2 Minerals and Chelating Agents

Minerals are found in foodstuffs in various forms, such as ions, compounds, or chelates. The solubility and other functional properties of the mineral differ with each form. Many diseases are associated with a deficiency or overuse of minerals. Hence, consumers prefer to choose a food product that contains a sufficient amount of minerals. It should be noted that changing the mineral composition of food emulsifiers to improve their nutritional aspects can cause unwanted changes in their physical, chemical, and sensory properties.

The electrostatic sorting and ion-binding effects at high metal concentrations can adversely affect the aggregate stability of O/W emulsions droplets (Figure 2.4). In the presence of multilayer ions, these effects have been observed at comparatively small metal concentrations. A few metal ions may also encourage adverse chemical reactions that allow the substance to decompose, such as iron and copper ions, which may promote the oxidation of lipids. In these systems, it is usually necessary to add chelating agents to retain the metal ions and to prevent chemical instability. Some types of minerals also affect the functional properties of other nutrients. For example, the ability of many biopolymers to condense or gel a solution strongly depends on the type and concentration of metal ions present. Careful selection and control of the metal ions present in the edible emulsifier are important in designing a successful product [92].

2.3.3 Thickening Agents and Gelling Agents

These are polymeric compounds that are used in nanoemulsions to increase the viscosity of the aqueous phase, which ultimately alters the mouth-filling while reducing the growth rate of the droplets (Figure 2.4). Substances are also known as tissue modifiers usually participate in the continuous phase of emulsion to produce variable rheological properties, including thickeners and gelling agents. Thickeners typically include soluble polymers that have rectangular structures and can achieve a high solution viscosity because they can alter the liquid flow profile. These are biocompatible polymers that will dissolve in the aqueous phase and cover a volume larger than the original volume of the polymer chain. Crystallization agents can form a chemical or physical bond with their neighbors and transfer the solid properties to a nanoemulsion solution. This can be explained by the improved stability of the nanoparticles utilizing tissue modifiers, which inhibit the movement of the droplets and thus slow down the gravitational separation [23]. The oxidative stability of the polymer and the molecular characteristics are the main criteria that should be mainly considered while selecting the thickening agents. Many natural and semisynthetic polysaccharides, such as alginates, pectin, chitosan, carboxymethylcellulose, and so on, are usually considered as an ideal thickening agent and gelling agent [106].

2.3.4 Weighing Agent

Weighing agents are excipients that are added to nanoemulsions to reduce the release of oil droplets. It should be noted that the density and aroma of vegetable oils are much lower than that of the aqueous phase, as well as aqueous solutions of sugar. The nanoemulsion containing tiny oil droplets tends to move upwards. The upward movement of the oil droplets in aqueous solutions leads to resonance or fusion and phase separation, as shown in Figure 2.4. The best way to reduce resonance can be possible by using hydrophobic materials with a higher density of oil droplets.

This high-density material is the weighing agent, which increases the density of the oil droplets, which coincides with the aqueous phase. A few commercially available weighting agents are sucrose acetate isobutyrate (SAIB), brominated vegetable oil (BVO), base resins, and ester, among others. SAIB was synthesized by esterification of sucrose with acetic hydride and isobutyric. SAIB is a transparent, highly viscous liquid that can be mixed with an oil phase before homogenization. It has good stability for lipolysis and oxidation. The Food and Drug Administration (FDA, USA) has set the SAIB level at 300 ppm. BVO is manufactured after adding bromine to cottonseed, corn, olive, and/or soybean oil by forming a double bond between molecules of bromine oils and triglycerides with USFDA approval for the use of 15 ppm but has been banned in Asia and the European Union. Rubber ester is a hydrophobic polymer that is derived from natural resources. It is obtained by esterification of glycerol and rosin resin which is mainly obtained from pine trees. It should be noted that due to its delivery in crystalline form and esterification stages, it is not considered a natural weight agent or natural food additive. According to the FDA, the permitted level for ester gum is 100 ppm per serving in the USA. Dammar gum is a natural weighting agent obtained from their excretions mainly. However, it has recently been approved in the US and thus lacks GRAS status [107].

2.3.5 Sweetening Agents

Nanoemulsions are sweetened with natural or synthetic compounds that interact with the taste buds of the tongue to impart sensitivity to sweetness and sugar-based compounds are the compounds that give a perception of sweetness to mammals [90]. Among these compounds, sucrose is 100% sweet and can be used to analyze sweetness using carefully controlled sensory tests. The correct direction of binding fructose and its volume is the main reason it is the sweetest compound. Additionally, many other natural and synthetic compounds are available to make it sweeter than sucrose. Thus, it can be concluded that the amount of needed sugars decreases with increasing relative sweetness. The relative sweetness, chemical and physical stability, and aroma of the sugars are the factors that determine their use in nanoemulsions as a sweetener. Once compared to sugars (glucose, fructose, sucrose, etc.) and sugar alcohols (sorbitol, xylitol, mannitol, and erythritol), they are then considered beneficial for developing a low-calorie product. The reason for this saying is that the human body breaks down sugar alcohol slower and less efficiently, resulting in fewer calories per gram in it. In addition to sugar alcohols, there are many high-intensity sweeteners with preferred flavors, low/medium-calorie nanoemulsions, such as natural sweeteners like stevia, and artificial sweeteners like saccharin, aspartame, cyclamate, sucrose, neotame, and acesulfame K [108].

2.3.6 Vitamins

Nowadays, nanoemulsion-based food products incorporating vitamins mainly Vit-A, D, E, and K are being added to increase the nutritive value [109]. Vitamin E is used in various forms in which α-tocopherol is the most bio-active form extensively used in pharma, cosmetics, and food industries [110]. The therapeutic potential of vitamin E leading to antioxidant activity for the management of diabetes, cancer, cardiovascular diseases is very well known [111]. Because of this, there has been increased demand and recommendation by various food scientists in fortifying many foods and beverages with emulsified vitamin E to increase their bioavailability.

However, most of the nano-vitamin emulsion products are found in the cosmetic industry, not in the food industry. Quotations for almost all vitamins exist, but vitamin E is most frequently used in cosmetic nanoemulsion. The nano-stabilized emulsions loaded with vitamin E are prepared using common water-soluble glycerin. It was formed by titration of a mixture of vitamin E acetate, carrier oil, and a non-ionic surfactant (Tween 80) in an aqueous solution of glycerol with continuous mixing [112]. The other method is the washing method, in which nanoemulsion vitamin E is developed using rapeseed oil as an emulsifying agent. This method includes pre-dissolving of the surfactants and vitamin E into the oil phase. The aqueous phase is constantly mixed with the oil phase that contains the surfactant and active ingredient. The required emulsification temperature is generally set to 74°C with continuous stirring [112–114].

Khalid et al. formulated a nanoemulsion of ascorbic acid by homogenization method by using food-grade emulsifier and soybean oil. The obtained droplet diameter of the developed nanoemulsion was 2000–3000 nm with stability of more than 30 days [115, 116]. Similarly, Ziani et al. have developed vitamin D, E, and lemon oil-based nanoemulsions, as an approach to incorporate vitamins with flavor [91]. A US patent (US20130189316 A1) has been granted for nanoemulsion of vitamin K. Therapeutic applicability of vitamin K nanoemulsion was found significantly higher than the Phytonadione Injectable Emulsion. The drug sometimes causes hypersensitive reactions if injected intravenously or intramuscularly [117].

Vitamin E is diminished during the processing, use, and storage of commercial foods and drinks for which oxidation is the main cause of vitamin E instability [118]. To achieve high oxidative stability, it is recommended to use vitamin E acetate in commercial foods and beverages instead of vitamin E. It is observed that the activity of pancreatic esterase breaks down vitamin E acetate into vitamin E in the digestive system [79, 119]. Vitamin E's higher lipophilic behavior limits its immediate dispersion into aqueous solutions. It should be transported using a colloidal method as a powerful emulsion-based delivery system. Vitamin E acetate is recommended to be used in commercial foods and beverages in place of vitamin E to achieve high oxidative stability. It is noted that the digestive system's pancreatic esterification effect breaks down vitamin E acetate into vitamin E microemulsions and nanoemulsions [59, 120, 121]. It has also been suggested that the increased bioavailability of vitamin E can be achieved by using a colloidal delivery system rather than its bulk form. Recently, Nandita and colleagues developed a grade nanoemulsion with increased encapsulation efficiency [112]. To increase the encapsulation efficiency and delivery of vitamin E by altering colloidal solutions, further research is required.

2.3.7 Antioxidants

The free radical scavenging ability of the extracts was tested by DPPH (1,1-diphenyl-2-picrylhydrazyl) radical scavenging assay as described by Blois [122] and Desmarchelier et al. [123]. The effect of antioxidants on DPPH is thought to be due to their hydrogen donating ability [124]. Free radical scavenging of DPPH is an acceptable mechanism for screening the antioxidant activity of plant extracts. In the DPPH assay, the DPPH violet solution was reduced to a yellow product, diphenylpicryl hydrazine after adding the extract. Free radical scavenger is known as a lipophilic antioxidant in food technology and is mainly used in nanoemulsions oil in water to slow down the oxidative decomposition of sensitive nutrients, ultimately helping to extend the shelf life of food-based on nanoemulsion. The chemical degradation of the aforementioned sensitive foods – mainly PUFAs and carotenoids – can be avoided by the use of lipophilic antioxidants [125]. Also, fat-soluble antioxidants are used in food for their health benefits to make foods with antioxidant properties. Some of the active lipophilic antioxidants used at industrial scale are Ascorbyl palmitate, rosemary extracts, gallic acid, tert-butylhydroquinone (TBHQ), butyl hydroxytoluene (BHT), and butylated hydroxyanisole (BHA) [126, 127]. It has been experimentally proven in the laboratory that, these are the most active fat-soluble antioxidants and that one molecule of them can clean out free radicals before they become inactive. Similarly, the late phase of the oxidative reaction can be increased with the help of this antioxidant. It can be seen that the use of antioxidants in nanoemulsions has not been studied much and is still in the native phase.

Ferulic acid is a strong antioxidant with verified skincare efficacies. The nanoencapsulation approach has been utilized to improve its stability. The developed nano-ferulic acid indicates that together with nanoencapsulation and low pH (less than pKa of ferulic acid) of the hydrogel was crucial for both product appearance and chemical stability of ferulic acid [128].

The addition of pomegranate peel phenolics at 0.5 and 1.0% (w/w) showed significant improvement of the antioxidant and α-glucosidase inhibitory activities of the enriched ice creams compared with the control sample. Grape wine lees significantly increase the DPPH radical scavenging activity and reduce the power of ice cream, and also its inhibitory effect on the oxidation of the human erythrocyte membrane [36]. Naturally, several lipophilic dyes are available which include paprika, lycopene, β-carotene, and various other carotenoids. Due to their presence from natural sources, they can be used as dyes or colors in food [129, 130].

Also, many of these dyes have medicinal properties, such as carotenoids, as described above, and thus also function as nutraceuticals. Most of these lipophilic colorants from natural sources are highly unstable and may chemically degrade, resulting in rapid color fading during storage [76]. To use these natural lipophilic colorants in nanoemulsions, the mechanism of their chemical degradation must be understood for each type of colorant. Also, it has been proposed that the key factors responsible for chemical degradation (e.g. oxygen, pH, pro-oxidants, light, etc.) should be recognized to prepare an effective distribution scheme and ensure the exact shelf life of the nutraceuticals.

Nowadays, the trend of using functional foods is increasing due to the health benefits of these foods. In this regard, the use of carotenoids as a functional ingredient in these foods is attracting a lot of attention. Carotenoids are a group of pigmented compounds derived from plants and microorganisms and are responsible for the red, orange, and yellow colors in fruits and vegetables. The red color and antioxidant activity of lycopene are responsible for this extended framework of conjugated double bonds. Among carotenoids, this carotenoid is the most potent natural antioxidant and is also found in tomatoes [131]. However, lycopene is increasingly involved in the use of this carotene as a functional component in food fortification due to its antioxidant activity and functional effects [132].

One of the innovative strategies to overcome the instability issues is nanoencapsulation. This technology allows protecting the bioactive material from adverse conditions and improves the solubility and bioavailability of this compound. Also, nanoscale encapsulation allows rate controlled and targeted release of biologically active compounds. Nanoencapsulation of bioactive compounds and drugs is performed using lipid pulp systems and produces nanocarriers with various applications in the food industry field.

Products with high antioxidant activity tend to have a bitter taste, which reduces their acceptance by consumers. Techniques emerging in the food industry include removal of hydrophobic peptides by chromatography, activated carbon adsorption of bitter peptidome, or selective extraction with alcohol. These techniques can lead to a loss of vital activity since most of the bioactive amino acids and peptides are of a hydrophobic nature.

Micro and nanoencapsulation of polyphenols such as chloroquine phosphate can help reduce odor, astringency, and bitterness, and increase loading efficiency and bioavailability [133].

2.3.8 Colors

Even though there are many types of natural pigments from various microbial sources, the commercial development of natural pigments as food colorants is challenging. There are high regulatory barriers to developing any new food use compounds, including colors. Large quantities of raw materials are needed to produce equal amounts of natural colors from artificial colors. Usually, higher doses are required for the natural color of the desired shade, thus increasing the cost. Carotenoids are organic pigments that are naturally synthesized by microorganisms and plants. Replacing artificial colors with natural colors in the food industry is a challenge, especially given the relatively low range of natural colors approved for use in food. Also, natural dyes are usually more sensitive to light, pH, UV rays, temperature, oxygen, and heat, resulting in loss of color due to fading and reduced life span. The color of carotenoids can be divided – from yellow to red, with carotene and xanthophylls [134]. A list of natural coloring agents converted successfully into the nanoemulsion dosage forms (Table 2.4).

Several epidemiological studies have shown that diets rich in carotenoid-containing foods are associated with a reduced risk of certain types of chronic diseases, such as cancer, cardiovascular disease, macular degeneration, and age-related cataracts [135, 136]. Carotenoids are insoluble in water and only slightly soluble in oil at room temperature, which greatly limits their use. However, due to the presence of long-chain conjugated double bonds, the carotenoids are liable to degrade when exposed to oxygen, heat, light, and acid, which in turn leads to reduced bioavailability in vivo [137]. Therefore, to improve its bioavailability, emulsification is a promising solution. Many forms of carotenoids have been emulsified and one carotenoid, lycopene, has also been available commercially. Lycopene carotenoid nanoparticles (trademark LycoVit®) are manufactured by BASF. A European company claims that the product is used as a food additive or food enhancer. This product has been categorized as GRAS (generally regarded as safe) by the US FDA (BASF US Patent US5968251) [138].

β-carotene has been emulsified using different formulation strategies such as using lipid carrier/liposomes casein micelles or by β-lactoglobulin complexes [139–141]. Other carotenoids being emulsified include lycopene, lutein, and astaxanthin. Using a high-pressure homogenizer, lycopene emulsions were prepared and the thermal stability of lycopene in emulsion systems was evaluated [142, 143]. Astaxanthin was also prepared by high-pressure homogenization with a scattered mean diameter varying between 160 and 190 nm [54, 144]. Chitosan is used to produce nanoencapsulated lutein to enhance its bioavailability. The absorption of lutein from nanocapsules is greater than that of mixed micelles [145]. The study was designed and evaluated to include curcumin-loaded nanoemulsions obtained by inversion of a point emulsion in pineapple ice cream to replace synthetic yellow dyes. The results showed that the inclusion of curcumin in the nanoemulsion is a potential alternative to reduce the use of artificial colors, as ice creams exhibit the same physical, chemical, and rheological properties [146].

TABLE 2.4 Nanoemulsion Formulations Encapsulating Natural Coloring Agent

Pigment	Color	Bioactivity	Purpose of Nanoemulsifications	References
Astaxanthin	Pink-red	Antioxidant photo protectant, anticancer, anti-inflammatory	Increased BA by sublingual absorption	[149]
β-carotene	Red	Anticancer, antioxidant suppression of cholesterol synthesis, source of provitamin-A	Enhance functionality, preventing degradation, and increasing bioaccessibility	[150, 151]
Lutein	Yellow	Antioxidant, supplements for aged eyes, cardioprotective, anticancer	Bioavailability enhancement	[152]
			Improved solubility issue in water and instability	[151]
Phycoerythrin	Red	Antioxidant, antitumor activity, immunoregulatory	Fluorescent nanoemulsions as a theranostic agent	[153]
Curcuminoid	Yellow color	Antioxidant, anti-inflammatory, and immune-regulatory activities, anti-cancers, diabetes, as well as the liver, cardiovascular CNS disorders	Enhanced phototoxic effect of the curcumin	[154]
			Inhibition of lung cancer	[155]
Anthocyanin	Red to blue (pH dependent)	Antioxidant, anti-inflammatory, anti-viral, and anti-cancer	Improving anthocyanin stability and bioavailability	[155]
Lycopene	Red	Antioxidant, anti-inflammatory, and anti-cancer	Enhance suppression of colon cancer cell growth	[156]
			UVA radiation protective agents	[157]
			Enhance stability	[158]

2.3.9 Flavors

For aromatic plants, such as oregano, tea tree oil, lemon, and thyme, essential oils (EOs) have antimicrobial activity, which can be used as a natural substitute. Secondary EO constituents, such as monoterpene, γ-terpinene, and p-cinnamic hydrocarbons, may contribute to the antibacterial activity of these oils [147]. Some studies show that a few EOs have anti-acne activity; tea tree oil is being commercially used [148]. Several other anti-acne potentials of seven EOs used in Mediterranean folk medicine: mentha (*Mentha piperita*), lemongrass (*Cymbopogon citratus*), lavender (*Lavandula angustifolia*), and chamomile (*Matricaria recutita*), oregano (*Origanum vulgare*), thyme (*Thymus vulgaris*), tea tree (*Melaleuca alternifolia*).

The Mediterranean region is one of the largest producers of these aromatic plants. The favorable climatic and agricultural conditions are improving the quality of EOs and they usually act as a natural antimicrobial agent through their severe lethality against a wide range of microbes and also have potential health benefits [159]. EO is a volatile molecule that is derived as secondary metabolites from various plant parts. EO has been used to flavor foods because of its volatile and aromatic nature [16, 17].

2.4 OVERVIEW OF THE DEVELOPED FOOD GRADE NANOEMULSIONS

2.4.1 Nano-Curcumin

Curcumin is a polyphenol extracted from the rhizome of the turmeric plant, *Curcuma longa* [160]. In its composition, this compound has three chemical components, including one moiety of diketone and two phenolic groups. Electron transfer and hydrogen abstraction processes can oxidize the active functional groups of curcumin [161]. Along with antioxidant properties, this polyphenol compound exhibits anti-inflammatory, anti-neurodegenerative, and anticancer activities [162]. The potential of curcumin in preventing and treating various aging-associated pathological conditions has been repeatedly reported [163]. Oxidative stress, inflammation, atherosclerosis, cardiovascular and neurodegenerative disorders, type 2 diabetes, osteoporosis, rheumatoid arthritis, kidney, and ocular age-related diseases, as well as cancer, are among these conditions. The health-promoting potential of this compound has been comprehensively explored in clinical trials over the last decade [164]. Its bioavailability is, however, limited through its low water solubility and gastrointestinal stability [165]. Innovative nano delivery strategies are currently developed to overcome these limitations [166]. Evidence of combined *in vitro* and *in vivo* shows that nano-curcumin-based nutraceuticals have stronger health-promoting properties compared to crude ones. Nanoencapsulation with lipid or polymeric nanoparticles, nanogels, and dendrimers and even conjugation with metal oxide nanoparticles have been shown to greatly improve the water solubility and bioavailability of such bioactive compound [167]. The oral bioavailability and brain distribution of curcumin were shown to be significantly enhanced in N-trimethyl chitosan surface-modified fat nanoemulsion compared to those of native curcumin [168]. Evidence was also obtained by *in vitro* experiments with a Caco-2 cell line that bovine serum albumin dextran nanoparticles (up to 200 nm in size) loaded with curcumin can have substantial cellular antioxidant activity [169].

2.4.2 Nano-Quercetin

Quercetin is a bioactive flavonoid with strong antioxidant properties, including its effects on the levels of reactive oxygen species (ROS) and also on various pathways of cell signal transduction and antioxidant enzyme activity. The ability to avoid the oxidation of low-density lipoproteins by scavenging free radicals and chelating transition metal ions has been shown in particular. This polyphenolic flavonoid compound's antioxidant activity is thought to be primarily due to its metal ion complexes and complex ions [170].

In addition to antioxidant properties, quercetin is known to demonstrate anti-inflammatory, anti-obesity, anti-diabetic, anti-atherosclerotic, anti-hypercholesterolemic, and antihypertensive activities [171]. However, its health benefits are limited due to low bioavailability [111, 172]. Over the last years, innovative nanotechnology-based approaches have been developed to enhance quercetin bioavailability. Among them, quercetin-loaded solid fat nanoemulsions have been recently developed that exhibit a significantly improved bioavailability compared to pure quercetin powders [173]. It has been found that the same doses of this nanoformulation used in the fifth-day dosing regimen are adequate to produce effects close to those of daily doses of oral quercetin suspension. Related effects on CAT and SOD activities were also observed in the pancreas and kidneys. In rat models, quercetin self-emulsifying nanoformulation also exhibited significantly higher antioxidant capacity compared to free quercetin when assessed as a function of the ability to combat cardiotoxicity and nephrotoxicity induced by doxorubicin and cyclosporin A, respectively [174].

2.4.3 Nano-*Nigella Sativa*

For natural foods that are rich in nutrients may have biological functions, because of which modern consumers demand are increasing over the globe. This emerging market stimulates food producers and researchers to launch new ice cream formulations that are enriched with various ingredients. Commercial ice-cream products, however, are low sources of these nutrients [175]. Several previous studies recommended adding oils that are rich in phytochemicals to enhance the nutritional value of ice-cream products [176], hazelnut oil and olive oil [177], and flaxseed oil [178]. *Nigella sativa* oil has been suggested in food applications as a functional ingredient [83]. Nigella sativa oil and its biological activities are attributed to the presence of thymoquinone [179]. Because of the high hydrophobicity and oil separation in the finished product, this causes poor acceptance by the food industry. A variety of methods for adding oil to the food system have also been proposed which include solid dispersions, liposomes, amorphous solid shape, melt extrusion, and nanocarriers.

2.4.4 Nano-Genistein

Genistein (4′,5,7-trihydroxyisoflavone or 5,7-dihydroxy-3-(4-hydroxyphenyl) chromen-4-one) is a soy phytoestrogenic isoflavone having a potent antioxidant activity [180]. It shows effectiveness against many age-related disorders, including neurodegenerative diseases, osteoporosis, obesity, type 2 diabetes, and cancer [181, 182]. However, because of its low bioavailability, the therapeutic use of this compound is often limited. Also, cause endocrine-disrupting and various toxic effects at high doses [183]. To overcome these potential side effects, innovative nanotechnological solutions have been recently proposed for decreasing the dose size [184]. A significantly improved oral bioavailability of genistein was also reported by solid fat nanoemulsion compared to that for its suspensions or bulk powders [185].

2.4.5 Nano-Purple Rice Bran Oil

A frozen yogurt (FY) fortified with a nano-purple rice bran oil (NPRBO) was developed and evaluated. To produce a FY containing NPRBO (FYNRO), NPRBO with a fat droplet size range of 150–300 nm was mixed with the FY ingredients. Pure frozen yogurt

(PFY), sodium caseinate frozen yogurt (FYSC), and FYNRO were close in hardness. The apparent viscosity of the FYNRO mix, with values of 0.19 and 0.17 Pa•s, was similar to the PFY mix, respectively. Compared to FYSC and PFY, the FYNRO micrograph demonstrated a more compact and dense structure. This research showed that FY could be fortified with NPRBO. With nanoemulsion containing purple rice bran oil, FY can be reinforced. NPRBO is capable of affecting the FY structure. NPRBO cannot affect *Streptococcus thermophilus* and *Lactobacillus bulgaricus* survival in the FY [52, 133].

2.4.6 Nano-Resveratrol

Resveratrol (3,5,4′-trans-trihydroxystilbene) is a polyphenol compound found in grapes skin and seeds, and, in lesser amounts, in several other plant sources [186]. It functions as a phytoalexin in plants, protecting them from such pathogens as fungi and bacteria. Potent antioxidant properties of resveratrol have been repeatedly documented in a variety of animal models. Resveratrol can also reduce the development of mitochondrial superoxide by stimulating mitochondrial biogenesis, preventing the generation of superoxide from uncoupled endothelial nitric oxide synthase by upregulating the GTP cyclohydrolase I tetrahydrobiopterin synthesizing enzyme, and also increasing the expression of various antioxidant enzymes [187]. Also, there have been repeated studies of its anti-inflammatory, cardioprotective, neuroprotective, and anticancer properties. It is also known as one of the most promising natural anti-aging compounds at present [188]. For various aging-related pathological conditions, such as metabolic syndrome, obesity, type 2 diabetes, cardiovascular disorders, hypertension, stroke, chronic kidney, and inflammatory diseases, dementia, and even breast and colorectal cancers, the therapeutic potential of resveratrol has been identified [189]. The therapeutic applicability of resveratrol is, however, substantially restricted through its extensive hepatic and presystemic metabolism [84]. Also, the water solubility of this phytochemical is very poor, allowing low bioavailability on oral administration. Because of this, several preclinical and clinical studies are now underway to establish structurally modified resveratrol derivatives with higher bioavailability upon ingestion [190]. Any such resveratrol-loaded nanosized formulations have recently been investigated for their possible clinical utility. The bioavailability of orally administered trans-resveratrol loaded into solid fat-core nanocapsules was found to be twice as high in the brain, kidney, and liver of male Wistar rats as compared to free trans-resveratrol [191]. Resveratrol loaded in nanoliposome carriers (size from 103 to 134 nm) also exhibited a more pronounced radical scavenging effect when compared to pure resveratrol [192]. High ROS scavenging efficiency was also demonstrated for the vitamin E-loaded resveratrol nanoemulsion (an average globule diameter of about 100 nm) in patients with Parkinson's disease [84]. Endogenous antioxidant enzyme activity, including SOD and GSH levels, was shown to be significantly higher, and malondialdehyde levels were significantly lower in the nanoemulsion-administered resveratrol community.

2.4.7 Nano-Ferulic Acid

This unbelievably common oil has all sorts of advantages. This subtle floral fragrance will allow people to relax and sleep. Also, breathing it in has been found to help alleviate headaches, while topically using the oil will help decrease the itching and swelling from bug bites. Ferulic acid belongs to the category of phenolic acids typically found in plant tissues (4-hydroxy-3-methoxycinnamic acid, FA) [193]. Sangeeta and co-workers have reported the ability of FA to promote tissue regeneration in the skin of diabetic rats, which could be related to its ability to inhibit lipid peroxidation and increase catalase, superoxide dismutase, and glutathione [194]. The use of EOs has grown as medical indications in aromatherapy, indicating the potential of EOs in topical applications because of their many activities (such as antibacterial, antifungal, or antioxidant) and because they can help increase the effectiveness of drugs. Multiple physiological roles include anti-inflammatory, antimicrobial, anticancer, anti-arrhythmic, and anti-thrombotic activities, as well as antidiabetic and immuno-stimulant properties, hence FA is considered a superior antioxidant with low toxicity [195, 196]. This work aimed at developing nano-carriers for the combined delivery of ferulic acid and *Lavandula* EO, for increasing the wound-healing processes. The co-presence of ferulic acid and *Lavandula* EO as compared to synthetic isopropyl myristate-based nano-carriers increased nanoparticles' stability, due to higher ordering chains. A confirmatory in vitro experiment on fibroblasts showed, enhanced cytocompatibility of combined mixture ferulic acid and *Lavandula* EO. Also, the combined delivery of ferulic acid and *Lavandula* EO substantially facilitated higher cell migration efficiency compared to free drug solutions. Ferulic acid is an effective antioxidant with clinically proven efficacy for skincare and cosmetics. To enhance its stability, nanoencapsulation methods have been introduced. The formed nano-ferulic acid indicates that along with the nanoencapsulation and the low pH (less than ferulic acid pKa) of the hydrogel, both the physical and the chemical stability of the ferulic acid are important. However, this aggressive volatility in skincare products has limited its wide application in the beauty and skincare industries [128].

2.5 CONCLUSIONS

In conclusion, nanotechnology has become progressively important in the food industry. Food innovation is observed as one of the major sectors in which nanotechnology including nanoemulsions will play a major part in the forthcoming session. The nanoemulsion is highly stable and easy to formulate the food ingredients and nutraceuticals using ingredients of the GRAS category. To fulfill the public health demand along with leadership in the food and food-processing industry, one must work with nanotechnology science which includes polymeric nanoparticles, solid lipid nanoparticles, nanoemulsions, nanolipidic carriers, and nano-bioinformation in the future. The future belongs to new products, innovative ideas, and new processes intending to improve the food product's nutritive value. Improving the safety and quality of food is one of the major demands of food technology. Finally, nanotechnology enables to change of the existing food systems and processing to ensure product safety, creating a healthy food culture, and enhancing the nutritional quality of food.

CONFLICT OF INTEREST

The authors declare no conflict of interest.

LIST OF ABBREVIATIONS

BHA	Butylated hydroxyanisole
BHT	Hydroxytoluene butyl
BVO	Brominated vegetable oil
CAT	Catalase
DAG	Diacylglycerols
DPE	D-phase emulsification
DPPH	2,2-diphenyl-1-picrylhydrazyl
EC	Essential oils
EIP	Emulsion-inversion point
Er	Enhancement ratio
FA	Ferulic acid
FDA	US Food and Drug Administration
FFA	Free fatty acids
GRAS	Ingredients considered safe by FDA
GSH	Glutathione peroxidase
HLB	Hydrophilic balance system
IPM	Isopropyl myristate
IPP	Isopropyl palmitate
Jss	Steady-state flux
Kp	Permeability coefficient
LCT	Long-chain triglyceride
MAG	Monoacylglycerols
MCT	Medium-chain triglycerides
MD	Maltodextrins
ME	Membrane emulsification
mN/m	milli-Newton/meter
MUFAs	Monounsaturated fatty acids
NaCas	Sodium caseinate
NE	Nanoemulsions
O/W	Oil-in-water emulsions
PDI	Polydispersity index
PIC	Phase-inversion composition
PIT	Phase-inversion temperature
PUFAs	Polyunsaturated fatty acids
ROS	Reactive oxygen species
SAIB	Sucrose acetate isobutyrate
SDS	Sodium dodecyl sulphate
SEDDS	Self-emulsifying drug-delivery systems
SFAs	Saturated fatty acids
SNEDDS	Self-nanoemulsifying drug delivery systems
SOD	Superoxide dismutase
TAG	Triacylglycerols
TBHQ	Tert-butylhydroquinone
TFAs	Unsaturated fatty acids
W/O	Water-in-oil emulsions

REFERENCES

1. Rashidi L, Khosravi-Darani K (2011) The applications of nanotechnology in food industry. Crit Rev Food Sci Nutr 51:723–730.
2. Gupta A, Eral HB, Hatton TA, Doyle PS (2016) Nanoemulsions: formation, properties and applications. Soft Matter 12(11):2826–2841.
3. Mason TG, Wilking JN, Meleson K, Chang CB, Graves SM (2006) Nanoemulsions: formation, structure, and physical properties. J Phy: Cond Matt 18(41):R635–R666.
4. Gutiérrez JM, González C, Maestro A, Solè I, Pey CM, Nolla J (2008) Nano-emulsions: new applications and optimization of their preparation. Curr Opin Coll Interf Sci 13(4):245–251.
5. McClements DJ, Decker EA, Park Y, Weiss J (2009) Structural design principles for delivery of bioactive components in nutraceuticals and functional foods. Crit Rev Food Sci Nutr 49:577–606. https://doi.org/10.1080/10408390902841529
6. Ostertag F, Weiss J, McClements DJ (2012) Low-energy formation of edible nano-emulsions: factors influencing droplet size produced by emulsion phase inversion. J Colloid Interface Sci 388:95–102. https://doi.org/10.1016/j.jcis.2012.07.089
7. Walstra P (1999) Food emulsions: principles, practice, and techniques. Trends Food Sci Technol 10:241.
8. Mizrahi M, Friedman-Levi Y, Larush L, et al (2014) Pomegranate seed oil nano-emulsions for the prevention and treatment of neurodegenerative diseases: the case of genetic CJD. Nanomed Nanotechnol Biol Med 10:1353–1363. https://doi.org/10.1016/j.nano.2014.03.015
9. Xue J, Zhong Q (2014) Thyme oil nanoemulsions coemulsified by sodium caseinate and lecithin. J Agric Food Chem 62:9900–9907.
10. Chee CP, Djordjevic D, Faraji H, et al (2007) Sensory properties of vanilla and strawberry flavored ice cream supplemented with omega-3 fatty acids. Milchwissenschaft 62:66–69.
11. Soukoulis C, Fisk ID, Bohn T (2014) Ice cream as a vehicle for incorporating health-promoting ingredients: conceptualization and overview of quality and storage stability. Compr Rev Food 13(4): 627–655.
12. McClements DJ, Rao J (2011) Food-grade nanoemulsions: formulation, fabrication, properties, performance, biological fate, and potential toxicity. Crit Rev Food Sci Nutr 51(4):285–330.
13. Sugumar S, Singh S (2016) Nanoemulsion of orange oil with nonionic surfactant produced emulsion using ultrasonication technique: evaluating against food spoilage yeast. Appl Nanosci 6(1):113–120.
14. Tadrosa T, Izquierdob P, Esquenab J, Solansb C (2004) Formation and stability of nano-emulsions. Adv Coll Interf Sci 108–109: 303–318.
15. Silva HD, Cerqueira MA, Vicente AA (2012) Nanoemulsions for food applications: development and characterization. Food Bioprocess Tech 5(3):854–867.
16. Aziz ZAA, Ahmad A, Setapar SHM et al (2018) Essential oils: extraction techniques, pharmaceutical and therapeutic potential – a review. Curr Drug Metab 19:1100–1110.
17. Dhakad AK, Pandey VV, Beg S et al (2018) Biological, medicinal and toxicological significance of Eucalyptus leaf essential oil: a review. J Sci Food Agric 98:833–848.
18. Teixeira B, Morques A, Ramos C et al (2013) Chemical composition and antibacterial and antioxidant properties of commercial essential oils. Ind Crop Prod 43:587–595.

19. Otoni CG, Pontes SFO, Medeiros EAA, Soares N de FF (2014) Edible films from methylcellulose and nanoemulsions of clove bud (Syzygium Aromaticum) and oregano (Origanum Vulgare) essential oils as shelf life extenders for sliced bread. J Agric Food Chem 62:5214–5219.

20. Bhargava K, Conti DS, da Rocha SRP, Zhang Y (2015) Application of an oregano oil nanoemulsion to the control of foodborne bacteria on fresh lettuce. Food Microbiol 47:69–73.

21. Ofokansi KC, Chukwu KI, Ugwuanyi SI (2009) The use of liquid self-microemulsifying drug delivery systems based on peanut oil/tween 80 in the delivery of griseofulvin. Drug Dev Ind Pharm 35:185–191. https://doi.org/10.1080/03639040802244292

22. Smoliga JM, Blanchard O (2014) Enhancing the delivery of resveratrol in humans: if low bioavailability is the problem, what is the solution? Molecules 19:17154–17172.

23. McClements DJ (2015) Enhancing nutraceutical bioavailability through food matrix design. Curr Opin Food Sci 4:1–6.

24. Yao M, McClements DJ, Xiao H (2015) Improving oral bioavailability of nutraceuticals by engineered nanoparticle-based delivery systems. Curr Opin Food Sci 2:14–19.

25. Zou L, Liu W, Liu C, et al (2015) Utilizing food matrix effects to enhance nutraceutical bioavailability: increase of curcumin bioaccessibility using excipient emulsions. J Agric Food Chem 63:2052–2062. https://doi.org/10.1021/jf506149f

26. Weiss J, Gaysinsky S, Davidson M, McClements J (2009) Nanostructured encapsulation systems: food antimicrobials. Global issues in food science and technology (First edition, Ch 24). Academic Press. pp. 425–479.

27. Vannice G, Rasmussen H (2014) Position of the academy of nutrition and dietetics: dietary fatty acids for healthy adults. J Acad Nutr Diet 114(1):136–153.

28. US Department of Agriculture, Agricultural Research Service. Nutrient Intakes from Food and Beverages: Mean Amounts Consumed per Individual, by Gender and Age, What We Eat in America, NHANES. 2011–2012. https://www.ars.usda.gov/ARSUserFiles/80400530/pdf/1112/Table_1_NIN_GEN_11.pdf. Published 2014.

29. USDA Food Composition Databases (2015). Washington, DC: US Department of Agriculture, Agricultural Research Service. https://ndb.nal.usda.gov/ndb/

30. Abdullah MMH, Jew S, Jones PJH (2017) Health benefits and evaluation of healthcare cost savings if oils rich in monounsaturated fatty acids were substituted for conventional dietary oils in the United States. Nutr Rev 75(3):163–174.

31. Hashtjin AM, Abbasi S (2015) Optimization of ultrasonic emulsification conditions for the production of orange peel essential oil nanoemulsions. J Food Sci Technol 52(5):2679–2689.

32. Katsouli M, Tzia C (2019) Development and stability assessment of coenzyme Q10-loaded oil-in-water nanoemulsions using as carrier oil: extra virgin olive and olive-pomace oil. Food Bioprocess Technol 12:54–76.

33. Joe MM, Chauhan PS, Bradeeba K, et al (2012) Influence of sunflower oil based nanoemulsion (AUSN-4) on the shelf life and quality of Indo-Pacific king mackerel (Scomberomorus guttatus) steaks stored at 20° C. Food Control 23:564–570.

34. Katzer T, Chaves P, Bernardi A, et al (2014) Castor oil and mineral oil nanoemulsion: development and compatibility with a soft contact lens. Pharm Dev Technol 19:232–237.

35. Jiang SP, He SN, Li YL, et al (2013) Preparation and characteristics of lipid nanoemulsion formulations loaded with doxorubicin. Int J Nanomedicine 8:3141–3150.

36. Hwang TL, Fang CL, Chen CH, Fang JY (2009) Permeation enhancer-containing water-in-oil nanoemulsions as carriers for intravesical cisplatin delivery. Pharm Res 26:2314–2323.

37. Ghosh V, Mukherjee A, Chandrasekaran N (2013) Formulation and characterization of plant essential oil based nanoemulsion: evaluation of its larvicidal activity against Aedes aegypti. Asian J Chem 25:18–20.

38. Duarte JL, Amado JRR, Oliveira AEMFM, et al (2015) Evaluation of larvicidal activity of a nanoemulsion of Rosmarinus officinalis essential oil. Rev Bras Farmacogn 25, 189–192.doi:https://doi.org/10.1016/j.bjp.2015.02.010

39. Rao J, McClements DJ (2011) Food-grade microemulsions, nanoemulsions and emulsions: fabrication from sucrose mono palmitate & lemon oil. Food Hydrocoll 25:1413–1423.

40. Ahmadi Lakalayeh G, Faridi-Majidi R, Saber R, et al (2012) Investigating the parameters affecting the stability of superparamagnetic iron oxide-loaded nanoemulsion using artificial neural networks. AAPS Pharm Sci Tech 13:1386–1395.

41. Homayoonfal M, Khodaiyan F, Mousavi SM (2014) Optimization of walnut oil nanoemulsions prepared using ultrasonic emulsification: a response surface method. J Dispers Sci Technol 35:685–694.

42. Ali J, Akhtar N, Sultana Y, Baboota S, Ahuja A (2008). Antipsoriatic microemulsion gel formulations for topical drug delivery of babchi oil (Psoralea corylifolia). Methods Find Exp Clin Pharmacol 30(4):277–285.

43. Gadkari PV, Balaraman M (2015) Catechins: sources, extraction and encapsulation: a review. Food Bioprod Process 93:122–138.

44. Vyas TK, Shahiwala A, Amiji MM (2008) Improved oral bioavailability and brain transport of Saquinavir upon administration in novel nanoemulsion formulations. Int J Pharm 347:93–101.

45. Ofokansi KC, Chukwu KI, Ugwuanyi SI (2009) The use of liquid self-microemulsifying drug delivery systems based on peanut oil/tween 80 in the delivery of griseofulvin. Drug Dev Ind Pharm 35:185–191.

46. Tabibiazar M, Davaran S, Hashemi M, et al (2015) Design and fabrication of a food-grade albumin-stabilized nanoemulsion. Food Hydrocoll 44:220–228.

47. Nguyen MH, Hwang IC, Park HJ (2013) Enhanced photoprotection for photolabile compounds using double-layer coated corn oil-nanoemulsions with chitosan and lignosulfonate. J Photochem Photobiol B Biol 125:194–201.

48. Ramisetty KA, Pandit AB, Gogate PR (2015) Ultrasound assisted preparation of emulsion of coconut oil in water: understanding the effect of operating parameters and comparison of reactor designs. Chem Eng Process Process Intensif 88:70–77.

49. Al-Edresi S, Baie S (2009) Formulation and stability of whitening VCO-in-water nano-cream. Int J Pharm 373:174–178.

50. Nirmala M J, Durai L, Gopakumar V, Nagarajan R (2020) Preparation of celery essential oil-based nanoemulsion by ultrasonication and evaluation of its potential anticancer and antibacterial activity. Int J Nanomedicine 15:7651–7666.

51. Salama M, Mustafa MEA (2013) Formulation and evaluation of avocado oil nanoemulsion hydrogels using sucrose ester laureate. Adv Mater Res 812:246–249, 5.

52. Alfaro L, Hayes D, Boeneke C, et al (2015) Physical properties of a frozen yogurt fortified with a nano-emulsion containing purple rice bran oil. LWT Food Sci Technol 62:1184–1191.

53. Liu H-R, White PJ (1992) Oxidative stability of soybean oils with altered fatty acid. J Am Oil Chem Soc 69:528–532.

54. Kim DM, Hyun SS, Yun P, et al (2012) Identification of an emulsifier and conditions for preparing stable nanoemulsions containing the antioxidant astaxanthin. Int J Cosmet Sci 34:64–73. https://doi.org/10.1111/j.1468-2494.2011.00682.x

55. Fasina OO, Craig-Schmidt M, Colley Z, Hallman H (2008) Predicting melting characteristics of vegetable oils from fatty acid composition. LWT 41:1501–1505.

56. Guillén MD, Ruiz A (2005) Study by proton nuclear magnetic resonance of the thermal oxidation of oils rich in oleic acyl groups. J Am Oil Chem Soc 82:349–355.

57. St-Onge M-P, Travers A (2015) Fatty acids in corn oil: role in heart disease prevention. In: Handbook of lipids in human function, by Ronald Watson and Fabien Demeester. Academic Press and AOCS Press (Elsevier), NY, USA, First edition, pp. 131–140.

58. Dauqan E, Sani HA, Abdullah A, Muhamad H, Top ABGM (2011) Vitamin E and beta carotene composition in four different vegetable oils. Am J Appl Sci 8:407–412. doi: 10.3844/ajassp.2011.407.412

59. Mehmood T (2015) Optimization of the canola oil based vitamin E nanoemulsions stabilized by food grade mixed surfactants using response surface methodology. Food Chem 183:1–7.

60. Bernardi DS, Pereira TA, Maciel NR, et al (2011) Formation and stability of oil-in-water nanoemulsions containing rice bran oil: in vitro and in vivo assessments. J Nanobiotechnol 9:44.

61. Yang C-C, Hung C-F, Chen B-H (2017) Preparation of coffee oil-algae oil-based nanoemulsions and the study of their inhibition effect on UVA-induced skin damage in mice and melanoma cell growth. Int J Nanomedicine 12:6559–6580.

62. Ofokansi KC, Chukwu KI, Ugwuanyi SI (2009) The use of liquid self-microemulsifying drug delivery systems based on peanut oil/tween 80 in the delivery of griseofulvin. Drug Dev Ind Pharm 35:185–191.

63. Chong W-T, Tan C-P, Cheah Y-K, Lajis AFB, Dian NLHM, Kanagaratnam S, Lai O-M (2018) Optimization of process parameters in preparation of tocotrienol-rich red palm oil-based nanoemulsion stabilized by Tween80-Span 80 using response surface methodology. PLoS One 13(8).

64. Tabibiazar M, Davaran S, Hashemi M, et al (2015) Design and fabrication of a food-grade albumin-stabilized nanoemulsion. Food Hydrocoll 44:220–228.

65. Francoeur ML, Golden GM, Potts RO (1990) Oleic acid: its effects on stratum corneum in relation to (trans)dermal drug delivery. Pharm Res 7:621–627.

66. Niazy EM (1991) Influence of oleic acid and other permeation promoters on transdermal delivery of dihydroergotamine through rabbit skin. Int J Pharm 67:97–100.

67. Mustafa G, Khan ZI, Bansal T, Talegaonkar S (2009) Preparation and characterization of oil in water nano-reservoir systems for improved oral delivery of atorvastatin. Curr Nano Sci 5:428–440.

68. Sánchez-Salcedo EM, Sendra E, Carbonell-Barrachina ÁA, et al (2016) Fatty acids composition of Spanish black (Morus nigra L.) and white (Morus alba L.) mulberries. Food Chem 190:566–571.

69. Vauzour D, Martinsen A, Layé S (2015) Neuroinflammatory processes in cognitive disorders: is there a role for flavonoids and ω-3 polyunsaturated fatty acids in counteracting their detrimental effects. Neurochem Int. 89:63–74.

70. Gershanik T, Benita S (1996) Positively-charged self-emulsifying oil formulation for improving oral bioavailability of progesterone. Pharm Dev Technol 1:147–157.

71. Charman WN, Stella VJ (1991) Transport of lipophilic molecules by the intestinal lymphatic system. Adv Drug Del Rev 7:1–14.

72. Constantinides PP (1995) Lipid microemulsions for improving drug dissolution and oral absorption and biopharmaceutical aspects. Pharm Res 12(11):1561–1572.

73. Kimura M, Shizuki M, Miyoshi K, Sakai T, Hidaka H, Takamura H, Matoba T (1994). Relationship between the molecular structures and emulsification properties of edible oils. Biosci Biotech Biochem 58:1258–1261.

74. Pouton CW (2000) Lipid formulations for oral administration of drugs: non-emulsifying, self-emulsifying and 'self-microemulsifying' drug delivery systems. Eur J Pharm Sci 11(2):S93–S98.

75. Cortesi R, Nastruzzi C (1999) Liposomes, micelles and microemulsions as new drug delivery systems for cytotoxic alkaloids. Pharm Sci Tech Today 2(7):288–298.

76. Qian C, Decker EA, Xiao H, McClements DJ (2012) Physical and chemical stability of b-carotene-enriched nanoemulsions: influence of pH, ionic strength, temperature, and emulsifier type. Food Chem 132:1221–1229. https://doi.org/10.1016/j.foodchem.2011.11.091

77. Salvia-Trujillo L, Rojas-Graü MA, Soliva-Fortuny R, Martín-Belloso O (2013) Effect of processing parameters on physicochemical characteristics of microfluidized lemongrass essential oil-alginate nanoemulsions. Food Hydrocoll 30:401–407.

78. Yousef SA, Mohammed YH, Namjoshi S, Grice JE, Benson HAE, Sakran W, Roberts MS (2019) Mechanistic evaluation of enhanced curcumin delivery through human skin in vitro from optimised nanoemulsion formulations fabricated with different penetration enhancers. Pharmaceutics 11(12):639.

79. Mayer S, Weiss J, McClements DJ (2013) Behavior of vitamin E acetate delivery systems under simulated gastrointestinal conditions: lipid digestion and bioaccessibility of low-energy nanoemulsions. J Colloid Interface Sci 404:215–222. https://doi.org/10.1016/j.jcis.2013.04.048

80. Ziani K, Chang Y, McLandsborough L, McClements DJ (2011) Influence of surfactant charge on antimicrobial efficacy of surfactant stabilized thyme oil nanoemulsions. J Agric Food Chem 59(11): 6247–6255.

81. Donsì F, Annunziata M, Sessa M, Ferrari G (2011) Nanoencapsulation of essential oils to enhance their antimicrobial activity in foods. LWT Food Sci Technol 44:1908–1914.

82. Liang R, Xu S, Shoemaker CF, et al (2012) Physical and antimicrobial properties of peppermint oil nanoemulsions. J Agric Food Chem 60:7548–7555.

83. Alotaibi FO, Mustafa G, Ahuja A (2018) Study of enhanced anti-inflammatory potential of Nigella sativa in topical nanoformulation. Intern J Pharm Pharma Sci 10(7):41–51.

84. Pangeni R, Sharma S, Mustafa G, Ali J, Baboota, S (2014) Vitamin E loaded resveratrol nanoemulsion for brain targeting for the treatment of Parkinson's disease by reducing oxidative stress. Nanotechnology 25:485102. doi:10.1088/0957-4484/25/48/485102

85. Parveen R, Baboota S, Ali J, Ahuja A, Ahmad S (2015) Stability studies of silymarin nanoemulsion containing Tween 80 as a surfactant. J Pharm Bioallied Sci 7(4):321–324.

86. Vijayalakshmi G, Mukherjee A, Chandrasekaran N (2014) Eugenol-loaded antimicrobial nanoemulsion preserves fruit juice against, microbial spoilage. Colloids Surf B Biointerfaces 114:392–397.

87. Weiss J, Decker EA, McClements DJ, Kristbergsson K, Helgason T, Awad T (2008) Solid lipid nanoparticles as delivery systems for bioactive food components. Food Biophysics 3:146–154.

88. Wooster TJ, Golding M, Sanguansri P (2008) Impact of oil type on nanoemulsion formation and Ostwald ripening stability. Langmuir 24(22):12758–12765. doi:10.1021/la801685v

89. Dasgupta N, Ranjan S, Chakraborty AR, et al (2016) Nanoagriculture and water quality management. In: Ranjan S, Nandita D, Lichtfouse E (eds) Nanoscience in food and agriculture 1. Springer, Berlin/Heidelberg.

90. Komaiko J, Sastrosubroto A, McClements DJ (2016) Encapsulation of w-3 fatty acids in nanoemulsion-based delivery systems fabricated from natural emulsifiers: sunflower phospholipids. Food Chem 203:331–339. https://doi.org/10.1016/j.foodchem.2016.02.080

91. Ziani K, Fang Y, McClements DJ (2012) Fabrication and stability of colloidal delivery systems for flavor oils: effect of composition and storage conditions. Food Res Int 46:209–216. https://doi.org/10.1016/j.foodres.2011.12.017

92. McClements DJ, Li Y (2010) Structured emulsion-based delivery systems: controlling the digestion and release of lipophilic food components. Adv Colloid Interf Sci 159:213–228. https://doi.org/10.1016/j.cis.2010.06.010

93. Amenta V, Aschberger K, Arena M, et al (2015) Regulatory aspects of nanotechnology in the agri/feed/food sector in EU and non-EU countries. Regul Toxicol Pharmacol 73:463–476. https://doi.org/10.1016/j.yrtph.2015.06.016

94. Attwood D, Mallon C, Ktistis G, Taylor CJ (1992) A study on factors influencing the droplet size in nonionic oil-in-water microemulsions. Int J Pharm 88:417–422.

95. Bouyer E, Mekhloufi G, Rosilio V, Grossiord J (2012) Proteins, polysaccharides, and their complexes used as stabilizers for emulsions: alternatives to synthetic surfactants in the pharmaceutical field? Int J Pharm 436:359–378.

96. Lawrence MJ, Rees GD (2000) Microemulsion-based media as novel drug delivery systems. Adv Drug Deliv Rev 45:89–121.

97. Attwood D (1994) Microemulsions. In: Kreuter J (ed.) Colloidal drug delivery systems. Dekker, New York, pp. 31–71.

98. Eccleston J (1994) Microemulsions. In: Swarbrick J, Boylan JC (eds) Encyclopedia of pharmaceutical technology Vol. 9. Marcel Dekker, New York, pp. 375–421.

99. Warisnoicharoen W, Lansley AB, Lawrence MJ (2000) Light-scattering investigations on dilute nonionic oil-in-water microemulsions. AAPS Pharm Sci 2:16.

100. Qu B, Zhong Q (2017) Casein-maltodextrin conjugate as an emulsifier for fabrication of structured calcium carbonate particles as dispersible fat globule mimetics. Food Hydrocoll 66:61–70.

101. Zhang Z, Wang X, Yu J, Chen S, Ge H, Jiang L (2017) Freeze-thaw stability of oil-in-water emulsions stabilized by soy protein isolate dextran conjugates. LWT – Food Sci Tech 78:241–249.

102. Verma S, Kumar S, Gokhale R, Burgess DJ (2011) Physical stability of nanosuspensions: investigation of the role of stabilizers on Ostwald ripening. Int J Pharm 406(1–2):145–152.

103. Wagner C (1961) Theorie der alterung von niederschlägen durch umlösen (Ostwaldreifung). Zeitschrift für Elektrochemie, Berichte der Bunsengesellschaft für physikalische Chemie 65(7–8):581–591.

104. Lifshitz IM, Slyozov VV (1961) The kinetics of precipitation from supersaturated solid solutions. J Phys Chem Solids 19(1–2): 35–50.

105. Taylor P (2003) Ostwald ripening in emulsions: estimation of solution thermodynamics of the disperse phase. Adv Colloid Interface Sci 106:261–285.

106. Saha D, Bhattacharya S (2010) Hydrocolloids as thickening and gelling agents in food: a critical review. J Food Sci Technol 47:587–597.

107. Jain A, Shivendu R, Nandita D, Ramalingam C (2016) Nanomaterials in food and agriculture: an overview on their safety concerns and regulatory issues. Crit Rev Food Sci Nutr 1. https://doi.org/10.1080/10408398.2016.1160363

108. Baker JR Jr, Hemmila MR, Wang SC, et al (2015) Nanoemulsion therapeutic compositions and methods of using the same. Patent no. US8962026B2.

109. Hormann K, Zimmer A (2016) Drug delivery and drug targeting with parenteral lipid nanoemulsions – a review. J Control Release 223:85–98.

110. Campardelli R, Reverchon E (2015) α-tocopherol nanosuspensions produced using a supercritical assisted process. J Food Eng 149:131–136. https://doi.org/10.1016/j.jfoodeng.2014.10.015

111. Li YJ, Li LY, Li JL, et al (2015) Effects of dietary supplementation with ferulic acid or vitamin E individually or in combination on meat quality and antioxidant capacity of finishing pigs. Asian Australas J Anim Sci 28:374–381. https://doi.org/10.5713/ajas.14.0432

112. Nandita D, Ranjan S, Mundra S, et al (2016) Fabrication of food grade vitamin E nanoemulsion by low energy approach, characterization and its application. Int J Food Prop 19:700–708. https://doi.org/10.1080/10942912.2015.1042587

113. Morais DJM, Burgess J (2014) Vitamin E nanoemulsions characterization and analysis. Int J Pharm 465:455–463. https://doi.org/10.1016/j.ijpharm.2014.02.034

114. Zheng N, Gao Y, Ji H, et al (2016) Vitamin E derivative-based multifunctional nanoemulsions for overcoming multidrug resistance in cancer. J Drug Target 24:663–669.

115. Khalid N, Kobayashi I, Neves MA, et al (2013a) Preparation and characterization of water-in-oil emulsions loaded with high concentration of l-ascorbic acid. LWT Food Sci Technol 51:448– 454. https://doi.org/10.1016/j.lwt.2012.11.020

116. Khalid N, Kobayashi I, Neves MA, et al (2013b) Preparation and characterization of water-in-oil-in-water emulsions containing a high concentration of L-ascorbic acid. Biosci Biotechnol Biochem 77:1171–1178. https://doi.org/10.1271/bbb.120870

117. Argenta DF, Bidone J, Misturini FD, et al (2016). In vitro evaluation of mucosa permeation/retention and antiherpes activity of genistein from cationic nanoemulsions. J Nanosci Nanotechnol 16:1282–1290.

118. Cheng K, Niu Y, Zheng XC, et al (2016) A comparison of natural (D-α-tocopherol) and synthetic (DL-α-tocopherol acetate) vitamin E supplementation on the growth performance, meat quality and oxidative status of broilers. Asian Australas J Anim Sci 29:681.

119. Yang Y, McClements DJ (2013) Vitamin E bioaccessibility: influence of carrier oil type on digestion and release of emulsified α-tocopherol acetate. Food Chem 141:473–481. https://doi.org/10.1016/j.foodchem.2013.03.033

120. Gonnet M, Lethuaut L, Boury F (2010) New trends in encapsulation of liposoluble vitamins. J Control Release 146:276–290.

121. Zhou Q, Xu J, Yang S, et al (2015) The effect of various antioxidants on the degradation of O/W microemulsions containing esterified astaxanthins from Haematococcus pluvialis. J Oleo Sci 64(5):515–525.

122. Blois MS (1958) Antioxidant determinations by the use of a stable free radical. Nature 181:1199–1200.

123. Desmarchelier C, Bermudez MJN, Coussio J, Ciccia G, Boveris A (1997) Antioxidant and prooxidant activities in aqueous extract of Argentine plants. Int J Pharmacog 35:116–120.

124. Baumann J, Wurn G, Bruchlausen FV (1979) Prostaglandin synthetase inhibiting O2 radical scavenging properties of some flavonoids and related phenolic compounds. Deutsche Pharmakologische Gesellschaft abstracts of the 20th spring meeting, Naunyn-Schmiedebergs abstract no: R27 cited. Arc Pharmacol 307:R1–77.

125. Sotomayor-Gerding D, Oomah BD, Acevedo F, et al (2016) High carotenoid bioaccessibility through linseed oil nanoemulsions with enhanced physical and oxidative stability. Food Chem 199:463–470. https://doi.org/10.1016/j.foodchem.2015.12.004

126. Budilarto ES, Kamal-Eldin A (2015) The supramolecular chemistry of lipid oxidation and antioxidation in bulk oils. Eur J Lipid Sci Technol 117:1095–1137. https://doi.org/10.1002/ejlt.201400200

127. Mohammadi A, Jafari SM, Esfanjani AF, Akhavan S (2016) Application of nano-encapsulated olive leaf extract in controlling the oxidative stability of soybean oil. Food Chem 190:513–519. https://doi.org/10.1016/j.foodchem.2015.05.115

128. Das S, Wong ABH (2020) Stabilization of ferulic acid in topical gel formulation via nanoencapsulation and pH optimization. Sci Rep 10:12288.

129. Meléndez-Martínez AJ, Mapelli-Brahm P, Benítez-González A, Stinco CM (2015) A comprehensive review on the colorless carotenoids phytoene and phytofluene. Arch Biochem Biophys 572:188–200.

130. Kiokias S, Proestos C, Varzakas T (2016) A review of the structure, biosynthesis, absorption of carotenoids-analysis and properties of their common natural extracts. Curr Res Nutr Food Sci J 4:25–37.

131. Shi J, Maguer ML (2000) Lycopene in tomatoes: chemical and physical properties affected by food processing. Crit Rev Food Sci Nutr 40(1):1–42.

132. Rao AV, Rao LG (2007) Carotenoids and human health. Pharmacol Res 55(3):207–216.

133. Adjonu R, Doran G, Torley P, Agboola S (2014) Whey protein peptides as components of nanoemulsions: a review of emulsifying and biological functionalities. J Food Eng 122:15–27.

134. Britton G, Khachik F (2009) Carotenoids in food. In: Britton G, Pfander H, Liaaen-Jensen S (eds) Carotenoids. Birkhauser Publishers, Basel, pp. 45–66.

135. Tang G (2010) Bioconversion of dietary provitamin A carotenoids to vitamin A in humans. Am. J Clin Nutr 91:1468S–1473S.

136. Ciccone MM, Cortese F, Gesualdo M et al (2013) Dietary intake of carotenoids and their antioxidant and anti-inflammatory effects in cardiovascular care. Mediators Inflamm 2013:782137 doi: 10.1155/2013/782137.

137. Lee MT, Chen BH (2002) Stability of lycopene during heating and illumination in a model system. Food Chem 78:425–432.

138. Gutiérrez FJ, Albillos SM, Casas-Sanz E, et al (2013) Methods for the nanoencapsulation of β-carotene in the food sector. Trends Food Sci Technol 32:73–83.

139. Pardeike J, Hommoss A, Müller RH (2009) Lipid nanoparticles (SLN, NLC) in cosmetic and pharmaceutical dermal products. Int J Pharm 366:170–184.

140. Dalgleish DG (2011) On the structural models of bovine casein micelles – review and possible improvements. Soft Matter 7:2265.

141. Ron N, Zimet P, Bargarum J, Livney YD (2010) Beta-lactoglobulin – polysaccharide complexes as nanovehicles for hydrophobic nutraceuticals in non-fat foods and clear beverages. Int Dairy J 20:686–693.

142. Ax K, Mayer-Miebach E, Link B, et al (2003) Stability of lycopene in oil-in-water emulsions. Eng Life Sci 3:199–201. https://doi.org/10.1002/elsc.200390028

143. Boon CS, Xu Z, Yue X, et al (2008) Factors affecting lycopene oxidation in oil-in-water emulsions. J Agric Food Chem 56:1408–1414. https://doi.org/10.1021/jf072929

144. Kim S, Ng WK, Shen S, Dong Y, Tan RBH (2009) Phase behavior, microstructure transition, and antiradical activity of sucrose laurate/propylene glycol/the essential oil of Melaleuca alternifolia/water microemulsions. Colloids Surf A Physicochem Eng Asp 348:289–297.

145. Arunkumar R, Prashanth KVH, Baskaran V (2013) Promising interaction between nanoencapsulated lutein with low molecular weight chitosan: characterization and bioavailability of lutein in vitro and in vivo. Food Chem 141:327–337.

146. Borrin TR, Georges EL, Moraes ICF, Pinho SC (2016) Curcumin-loaded nanoemulsions produced by the emulsion inversion point (EIP) method: an evaluation of process parameters and physico-chemical stability. J Food Eng 169:1–9. doi:10.1016/j.jfoodeng.2015.08.012

147. Burt S (2004) Essential oils: their antibacterial properties and potential applications in foods – A review. Int J Food Microbiol 94:223–253.

148. Sinha P, Srivastava S, Mishra N, Yadav NP (2014) New perspectives on antiacne plant drugs: contribution to modern therapeutics. BioMed Res Int 2014:301304.

149. Fratter A, Biagi D, Cicero AFG (2019) Sublingual delivery of astaxanthin through a novel ascorbyl palmitate-based nanoemulsion: preliminary data. Mar Drugs 17(9):508.

150. Teixé-Roig J, Oms-Oliu G, Ballesté-Muñoz S, Odriozola-Serrano I, Martín-Belloso O (2020) Improving the in vitro bioaccessibility of β-carotene using pectin added nanoemulsions. Foods 9(4):447.

151. Zhou X, Wang H, Wang C, Zhao C, Peng Q, Zhang T, Zhao C (2018) Stability and in vitro digestibility of beta-carotene in nanoemulsions fabricated with different carrier oils. Food Sci Nutr 6(8):2537–2544.

152. Yoo J, Baskaran R, Yoo BK (2013) Self-nanoemulsifying drug delivery system of lutein: physicochemical properties and effect on bioavailability of warfarin. Biomol Ther (Seoul) 21(2): 173–179.

153. Patel SK, Beaino W, Anderson CJ, Janjica JM (2015) Theranostic nanoemulsions for macrophage COX-2 inhibition in a murine inflammation model. Clin Immunol 160(1): 59–70.

154. de Matos RPA, Calmon MF, Amantino CF, Villa LL, Primo FL, Tedesco AC, Rahal P (2018) Effect of curcumin-nanoemulsion associated with photodynamic therapy in cervical carcinoma cell lines. Biomed Res Int 2018:4057959.155.

155. Chen B-H, Inbaraj BS (2019) Nanoemulsion and nanoliposome based strategies for improving anthocyanin stability and bioavailability. Nutrients 11(5):1052.

156. Huang Rwei-Fen S, Yi-Jun W, Stephen IB, Bing-Huei C (2015) Inhibition of colon cancer cell growth by nanoemulsion carrying gold nanoparticles and lycopene. Int J Nanomedicine 10: 2823–2846.

157. Butnariu MV, Giuchici CV (2011) The use of some nanoemulsions based on aqueous propolis and lycopene extract in the skin's protective mechanisms against UVA radiation. J Nanobiotech 9:3.

158. Choi SJ, McClements DJ (2020) Nanoemulsions as delivery systems for lipophilic nutraceuticals: strategies for improving their formulation, stability, functionality and bioavailability. Food Sci Biotechnol 29(2):149–168.

159. Bezerra Filho CM, da Silva LCN, da Silva MV, et al (2020) Antimicrobial and Antivirulence Action of Eugenia brejoensis Essential Oil in vitro and in vivo Invertebrate Models. Front Microbiol 11:424. doi: 10.3389/fmicb.2020.00424.

160. Hewlings SJ, Kalman DS (2017) Curcumin: a review of its' effects on human health. Foods 6:E92. doi:10.3390/foods6100092

161. Priyadarsini KI (2014) The chemistry of curcumin: from extraction to therapeutic agent. Molecules 19:20091–22112. doi:10.3390/molecules191220091

162. Sarker MR, Franks SF (2018) Efficacy of curcumin for age-associated cognitive decline: a narrative review of preclinical and clinical studies. Geoscience 40:73–95. doi:10.1007/s11357-018-0017-z

163. Sundar DKS, Houreld NN, Abrahamse H (2018) Therapeutic potential and recent advances of curcumin in the treatment of aging-associated diseases. Molecules 23:E835. doi:10.3390/molecules23040835

164. Salehi B, Stojanovic-Radic Z, Matejic JN, et al (2019) The therapeutic potential of curcumin: a review of clinical trials. Eur J Med Chem 163:527–545. doi:10.1016/j.ejmech.2018.12.016

165. Kumar A, Ahuja A, Ali J, Baboota S (2010) Conundrum and therapeutic potential of curcumin in drug delivery. Crit Rev Ther Drug Carrier Syst 27:279–312. doi:10.1615/CritRevTherDrugCarrierSyst.v27.i4.10

166. Flora G, Gupta D, Tiwari A (2013) Nanocurcumin: a promising therapeutic advancement over native curcumin. Crit Rev Ther Drug Carrier Sys 30:331–368. doi:10.1615/CritRevTherDrugCarrierSyst.2013007236

167. Shome S, Talukdar AD, Choudhury MD, Bhattacharya MK, Upadhyaya H (2016) Curcumin as potential therapeutic natural product: a nanobiotechnological perspective. J Pharm Pharmacol 68:1481–1500. doi:10.1111/jphp.12611

168. Ramalingam P, Ko YT (2015) Enhanced oral delivery of curcumin from N-trimethyl chitosan surface-modified solid lipid nanoparticles: pharmacokinetic and brain distribution evaluations. Pharm Res 32:89–402. doi:10.1007/s11095-014-1469-1

169. Fan Y, Yi J, Zhang Y, Yokoyama W (2018) Fabrication of curcumin-loaded bovine serum albumin (BSA)-dextran nanoparticles and the cellular antioxidant activity. Food Chem 239:1210–1218. doi:10.1016/j.foodchem.2017.07.075

170. Xu D, Hu MJ, Wang YQ, Cui YL (2019) Antioxidant activities of quercetin and its complexes for medicinal application. Molecules 24:E1123. doi:10.3390/molecules24061123

171. Anand David AV, Arulmoli R, Parasuraman S (2016) Overviews of biological importance of quercetin: a bioactive flavonoid. Pharmacogn Rev 10:84–89. doi:10.4103/0973-7847.194044

172. Kawabata K, Mukai R, Ishisaka A (2015) Quercetin and related polyphenols: new insights and implications for their bioactivity and bioavailability. Food Funct 6:1399–1417. doi:10.1039/C4FO01178C

173. Vijayakumar A, Baskaran R, Jang YS, Oh SH, Yoo BK (2016) Quercetin-loaded solid lipid nanoparticle dispersion with improved physicochemical properties and cellular uptake. AAPS Pharm Sci Tech 18:875–883. doi:10.1208/s12249-016-0573-4

174. Jain S, Jain AK, Pohekar M, Thanki K (2013) Novel self-emulsifying formulation of quercetin for improved in vivo antioxidant potential: implications for drug-induced cardiotoxicity and nephrotoxicity. Free Radic Biol Med 65:117–130. doi:10.1016/j.freeradbiomed.2013.05.041

175. Sun-Waterhouse D, Edmonds L, Wadhwa SS, Wibisono R (2013) Producing ice cream using a substantial amount of juice from kiwifruit with green, gold or red flesh. Food Res Inter 50:647–656.

176. Ullah R, Nadeem M, Imran M (2017) Omega-3 fatty acids and oxidative stability of ice cream supplemented with olein fraction of chia (Salvia hispanica L.) oil. Lip Health Dis 16:34.

177. Güven M, Kalender M, Taşpinar T (2018) Effect of using different kinds and ratios of vegetable oils on ice cream quality characteristics. Foods 7:104.

178. Gowda A, Sharma V, Goyal A, Singh AK, Arora S (2018) Process optimization and oxidative stability of omega-3 ice cream fortified with flaxseed oil microcapsules. J Food Sci Tech 55:1705–1715.

179. Hassanien MFR, Assiri AMA, Alzohairy AM, Oraby HF (2015) Health-promoting value and food applications of black cumin essential oil: an overview. J Food Sci Tech 52:6136–6142. doi:10.1007/s13197-015-1785-4

180. Spagnuolo C, Russo GL, Orhan IE, et al (2015) Genistein and cancer: current status, challenges, and future directions. Adv Nutr 6:408–419. doi:10.3945/an.114.008052

181. Saha S, Sadhukhan P, Sil PC (2014) Genistein: a phytoestrogen with multifaceted therapeutic properties. Mini Rev Med Chem 14:920–940. doi:10.2174/1389557514666141029233442

182. Ganai AA, Farooqi H (2015) Bioactivity of genistein: a review of in vitro and in vivo studies. Biomed Pharmacother 76:30–38. doi:10.1016/j.biopha.2015.10.026

183. Patisaul HB (2017) Endocrine disruption by dietary phyto-oestrogens: impact on dimorphic sexual systems and behaviours. Proc Nutr Soc 76:130–144. doi:10.1017/S0029665116000677

184. Rassu G, Porcu EP, Fancello S, et al (2018) Intranasal delivery of genistein-loaded nanoparticles as a potential preventive system against neurodegenerative disorders. Pharmaceutics 11:E8. doi:10.3390/pharmaceutics11010008

185. Kim JT, Barua S, Kim H, et al (2017) Absorption study of genistein using solid lipid microparticles and nanoparticles: control of oral bioavailability by particle sizes. Biomol Ther 25:452–459. doi:10.4062/biomolther.2017.095

186. Salehi B, Mishra AP, Nigam M, et al (2018) Resveratrol: a double-edged sword in health benefits. Biomedicines 6:E91. doi:10.3390/biomedicines6030091

187. Xia N, Daiber A, Förstermann U, Li H (2017) Antioxidant effects of resveratrol in the cardiovascular system. Br J Pharmacol 174:1633–1646. doi:10.1111/bph.13492

188. Wahl D, Bernier M, Simpson SJ, de Cabo R, Le Couteur DG (2018) Future directions of resveratrol research. Nutr Healthy Aging 4:287–290. doi:10.3233/NHA-170035

189. Singh AP, Singh R, Verma SS, et al (2019) Health benefits of resveratrol: evidence from clinical studies. Med Res Rev 39:1851–1891. doi:10.1002/med.21565

190. Popat R, Plesner T, Davies F, et al (2013) A phase 2 study of SRT501 (resveratrol) with bortezomib for patients with relapsed and or refractory multiple myeloma. Br J Haematol 160:714–717. doi:10.1111/bjh.12154

191. Frozza RL, Bernardi A, Paese K, et al (2010) Characterization of trans-resveratrol-loaded lipid-core nanocapsules and tissue distribution studies in rats. J Biomed Nanotechnol 6:694–703. doi:10.1166/jbn.2010.1161

192. Vanaja K, Wahl MA, Bukarica L, Heinle, H (2013) Liposomes as carriers of the lipid soluble antioxidant resveratrol: evaluation of amelioration of oxidative stress by additional antioxidant vitamin. Life Sci 93:917–923. doi:10.1016/j.lfs.2013.10.019

193. Zduńska K, Dana A, Kolodziejczak A, Rotsztejn H (2018) Antioxidant properties of ferulic acid and its possible application. Skin Pharmacol Physiol 31(6):332–336.

194. Dwivedi S, Singh D, Deshmukh PT, Soni R, Trivedi R (2015) Healing potential of ferulic acid on dermal wound in diabetic animals. Asian J Mol Model 1:1–16.

195. Bakkali F, Averbeck S, Averbeck D, Idaomar M (2008) Biological effects of essential oils – a review. Food Chem Toxicol 46(2):446–475.

196. Bona E, Cantamessa S, Pavan M, Novello G, Massa N, Rocchetti A, Berta G, Gamalero E (2016) Sensitivity of Candida albicans to essential oils: are they an alternative to antifungal agents? J Appl Microbiol 121(6):1530–1545.

CHAPTER 3

Recent Advances in the Preparation Method of Nanoemulsions

Gulam Mustafa,[1] Javed Ahmad,[2] Leo M.L. Nollet[3]

[1]College of Pharmacy (Boys), Al-Dawadmi Campus, Shaqra University, Riyadh, Kingdom of Saudi Arabia
[2]Department of Pharmaceutics, College of Pharmacy, Najran University, Najran, Kingdom of Saudi Arabia
[3]University College, Ghent, Belgium

CONTENTS

3.1 INTRODUCTION

Various diseases and health-related complications give rise to safer, healthier, cost-effective, and improved nutritive valued foodstuff. Food products therefore typically have health promotion or disease prevention values in addition to their nutritional value. On the other side, their advantages could be overshadowed by the poor bioavailability or inefficient long-term durability of these health-promoting items.

Nanoemulsion (NE) drug delivery systems are an effective means for administering and enhancing the bioavailability of water-insoluble drugs and bioactive food components

DOI: 10.1201/9781003121121-3

in the blood. A few decades earlier, nanotechnology has attracted the attention of various global industries, mainly food-based to deliver hydrophobic nutritive including flavors and colors. To enhance the bioavailability, secure encapsulation, and targeted delivery, nanoemulsion-based food products are being increasingly used nowadays [1–3].

NEs are emulsions that have a particle size at the nanometer range of 20–500 nm [4–6]. NEs are very distinct in size, shape, and kinetic stability from the conventional macroemulsions and microemulsions. Microemulsions are thermodynamically stable, whereas there is thermodynamic instability with nanoemulsions NEs [7, 8]. A typical NE component and its organization are depicted in the diagram for quick understanding (Figure 3.1).

The nanometric size ranges along with some unique characteristics, such as large unit size, high physical and chemical stability, optically transparent, and desired rheology making NE an ideal approach in the food industry [3, 9, 10]. By adding stabilizers, such as emulsifiers, ripening retarders, weighting agents, or texture modifiers, the kinetic stability of NEs can be enhanced [3]. The selection of suitable emulsifiers, which influence not only the lowest possible size but also the stability and surface characteristics of emulsion droplets, is an essential feature of the emulsion system [11–14]. An emulsion exhibits various physical and chemical properties that meet the requirements for target applications in food or many other fields [15]. A temporary emulsion is created when the oil and water are combined. However, when left to stand, the mixture would be separated into two separate phases due to the coalescence of the scattered pellets [16]. High concentrations of emulsifiers and surfactants from ingredients generally recognized as safe (GRAS) by FDA are included in the NE. Destabilizing processes such as Oswald ripening can be avoided if surfactants are combined with oil and other ingredients in appropriate proportions [17, 18].

NEs are explored in various fields such as drug delivery, foodstuffs, cosmetics, pharmaceuticals, and material synthesis. High- and low-energy methods are used to prepare NEs, including high-pressure homogenization, ultrasound, phase inversion temperature (PIT), and emulsion inversion point (EIP), in addition to recently

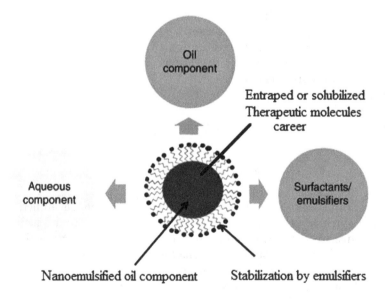

FIGURE 3.1   Structure organization of nanoemulsion and their basic structural comments.

developed methods such as the bubble burst method. In this review article, we summarize the main methods for preparing NEs, theories of droplet size prediction, physical conditions and chemical additives affecting droplet stability, as well as recent applications.

3.2 PREPARATION OF NANOEMULSIONS

Spontaneous formation of microemulsions from systems of three components that includes oil phase, emulsifier phase, and aqueous phase can occur at very low interconnection values of 0.10–0.01 mN/m (Figure 3.2). The composition and function of the interfacial layer play an essential role in reaching certain boundary values. The microemulsions exhibit a strong affinity for both the aqueous and hydrocarbon phases under ideal conditions. Droplets with a stable NE size (10–200 nm) may develop as long as there is an excess of surfactant in the system and as the amount of water in the system increases, the number of droplets will only increase. If the surfactant is not enough to shield the droplets' wide surface area, it will coalesce, and the emulsion will become more dispersed.

Through altering the phase and modifying the temperature, a fine emulsion can be formulated. When modifying the temperature or composition, it is very convenient to switch from one form of microemulsion to another (phase inversion). These temperatures are called PIT or phase inversion formation. The PIT is sometimes called the hydrophilic equilibrium temperature. The HLB refers to a hydrophilic–lipophilic balance responsible for initiating the process of phase inversion. Phase inversion describes the process of converting the oil-in-water emulsion when agitated oil reverts to water-in-oil and vice versa. Phase reversal emulsification is commonly used in the preparation of cosmetic products, pharmaceutical products, foodstuffs, and detergents.

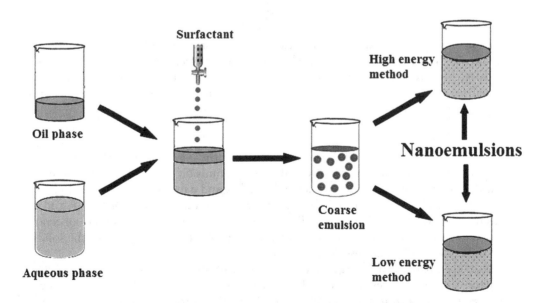

FIGURE 3.2 Typical diagram showing the steps for the preparation of NEs.

3.2.1 Theories of Nanoemulsion

Many factors must be regulated to achieve a stable, reproducible emulsion. This requires the selection of the right composition, the regulation of the ingredients' adding order, and the application of shear in a way that tears the droplets effectively. NEs have some additional criteria. In the continuous phase, the molecules of the dispersed phase must be significantly insoluble so that, despite the very high Laplace pressure, Ostwald's ripening and development do not happen rapidly. It is possible to avoid Ostwald's ripening in another way but choosing a very insoluble liquid for the dispersed phase is the easiest method [19].

The second condition is the choice of components, especially surfactants, which do not lead to the formation of "microemulsion" phases of the crystalline liquid. Systems containing short-chain alkanes, alcohols, water, and surfactants are known to form these phases [20]. The third condition is that the continuous phase has a significant increase in surfactant. This excess makes it possible to rapidly cover the new surface of the nanoparticles during emulsification, consequently reducing the shear-induced cohesion. This excess is typically in a continuous process in the form of surfactant micelles. These micelles dissociate into rapidly absorbing monomers on the surfaces of the newly formed droplets. The fourth condition is to apply some form of extreme shear to tear the fine droplets into nanoscale droplets. Usually, the pressure level should reach the Laplace pressure of droplets with the required volume, usually in the range of 10–100 atm.

Energy is required to produce an emulsion, apart from the basic ingredients oil, water, surfactant, and oil. This can be explained by considering the energy needed for the interface to expand, $\Delta A\gamma$ (where ΔA is the increase in the interfacial area when the bulk oil with area A1 produces a large number of droplets with area A2; A2≫A1, γ is the interfacial tension). Since γ is positive, the energy to expand the interface is large and positive [21]. This energy term cannot be compensated by the small entropy of dispersion TΔS and the total free energy of formation of an emulsion, ΔG is positive,

$$\Delta G = \Delta A\gamma - T\Delta S$$

From the equation given above, it is clear that the formation of an emulsion is non-spontaneous, and energy is required. The formation of micro-size droplet emulsions is quite easy, so high-speed mixing instruments such as Ultraturrax or Silverson Mixer are required for emulsion production. On the other hand, nanosized droplets emulsion is complex and this needs a large amount of surfactant and/or energy [21].

3.2.1.1 Solubilization Theory

This theory determines the stability and solubility of the NE and measures it to be both kinetically stable and or dynamically stable microemulsion in a single-phase solution. The difference in solubility between normal micelles and nano micelles can be explained by this theory. This defines the relationship between the oil-in-water (O/W) NE and the micellar isotropic region. The oil solubility is lower in normal micelles and the concentration of all aqueous concentrations is critical. In the context of hydrocarbons, the solubility of the mycelium in a nanoscale droplet increases.

Also, the creation of a large number of low-curvature intermediate nano-micellar and nano-sized structures assists in the nanoemulsification process [22]. This concept can indeed be justified in understanding the rheological variation in NEs dependent on droplet size [23–25].

3.2.1.2 Mixed Film Theory

The formulation of nanoparticles, interfacial films, and extremely low surface tension are defined in this hypothesis. This spontaneous droplet formation is due to the complex membrane formation at the oil-water interface of the surfactant and/or co-surfactant, which subsequently reduces the voltage of the oil-water interface to exceptionally low values (from zero to negative values). In mixed film theory, it must be assumed that the NE is liquid in nature, has different properties in water and oily surfaces. The two-dimensional spreading pressure (π) generated to affect the overall biphasic tension (γi) which can be denoted by the following equation:

$$\gamma_i = \gamma_{o/w} - \pi_i$$

where $\gamma o/w$ is oil-water interfacial tension in the absence of the film. It can be seen that the use of a large amount of surfactants and/or surfactants increases the diffusion pressure, which is greater than $\gamma o/w$, which leads to negative surface tension increasing the interfacial area. Increasing the intermediate layer eventually reduces the droplet size. Using a recently mixed theory, a stable nutrient-rich mustard oil NE enriched with vitamin E acetate was developed using Tween as a surfactant [26]. Additionally, numerous stable NEs of food-grade have been produced, such as d-limonene organogel NE [27], vitamin E-enriched NE [28], polyunsaturated NE (omega 3) NE [29], D-limonene o/w NE [30].

3.2.1.3 Thermodynamic Theory

This theory states that the dependency of oil droplet formation forms a bulk oil phase on the interfacial surface area (A), interfacial energy ($\gamma\Delta A$), and system entropy ($T\Delta S$). The interfacial area and interfacial energy subsequently increase during the top-down approach of decreasing oil droplet diameter. The expression is given by the equation:

$$\Delta G = \Delta A\gamma - T\Delta S,$$

where ΔG is the free energy of NE formulation, γ is the surface tension of the oil-water interface and T is the temperature. It should be noted that during the development of NEs, very large changes in ΔA can occur; however, γ remains positive all the time. Interestingly, the entropic components would be the key parameter due to the substantial change in entropy found during the creation of several smaller-diameter droplets. Thus, the negative free energy help in nanoemulsification and hence it is stated that the process of NE fabrication is spontaneous, and the resulting end product NE is thermodynamically stable.

　　　Pascual-Pineda and co-workers [31] evaluated the thermodynamic stability of paprika oil NE. Similarly, Fotticchia and co-workers [32] evaluated the thermodynamic stability for nanoemulsion formation using isothermal titration calorimetry. The effect of alcohol on the thermodynamic stability of protein-stabilized NEs was evaluated for successful calorimetry of titration [33].

3.2.2 Methods of Nanoemulsification

NE is a nonequilibrium system that needs an external or internal source of energy to be a successful formulation [34]. NEs can be fabricated using many approaches that are high-energy or low-energy approaches. The given pictorial view presented the different aspects of emulsification approaches (Figure 3.3). The NE processing techniques used

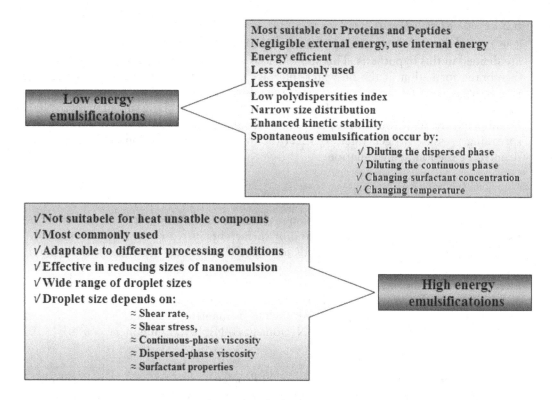

FIGURE 3.3 Two major types of nanoemulsification techniques used in food and pharmaceutical industries and their specifications.

have a significant influence on the size of the droplet and consequently affect the stability mechanisms of the emulsion system through the operating conditions and composition of the emulsion system. In general, NE formulation uses lower surfactant concentrations than microemulsion concentrations [21].

In high-energy approaches to mixing and disrupting oil/water phases, mechanical equipment that can generate powerful disruptive forces are used to create small oil droplets [35, 36]. On the other hand, low-energy methods rely on the automatic formation of small drops of emulsion in water-mixed emulsion systems when the solution or environmental conditions, such as temperature and composition, change [37, 38]. The method used in the formation of NEs, along with the operating conditions and device structure, affects the size of the droplets formed. Throughout this paragraph, we have a summary of the high-energy and low-energy approaches most widely used for the formation of NEs.

3.2.2.1 Low-Energy Emulsification

The low-energy methods are dependent on the internal chemical energy of the system [21]. Due to the change in environmental parameters such as temperature or composition, the NEs here are formed as a result of phase transitions that occur during the emulsification process [39], implementing constant temperature and modifying the composition or using constant composition and changing temperature [40, 41]. In addition to the temperature, natural frequency, and stirring velocities of the environmental factors, the composition of the emulsion, such as surfactant-oil-water ratio, surfactant form, and ionic strength significantly affects the droplet size [40].

Low-energy techniques can create smaller droplet sizes than high-energy approaches, but for limited categories of oils and emulsifiers, low-energy methodologies can be applied. For example, proteins or polysaccharides cannot be used as emulsifiers; alternatively, high concentrations of synthetic surfactants should be used to form NEs by low-energy approaches. This factor limits the use of such approaches in many food applications [3]. Various low-energy approaches are listed in the coming section and represented in Figure 3.4.

3.2.2.1.1 Spontaneous Emulsification Method This technique involves the impulsive formation of a NE as a result of the movement of water-miscible from the organic phase to the aqueous phase after mixing together [42]. The aqueous phase consists of water and hydrophilic surfactant whereas the organic phase is usually a homogeneous solution of oil, a lipophilic surfactant, and water-miscible solvent [38]. Spontaneous emulsification is therefore triggered by various mechanisms, including two phasic diffusions of solutes, interstitial disturbance, surface tension gradient, dispersion mechanism, or condensation mechanism. These mechanisms are strongly subjective to the composition of the system and their physical and chemical properties (oil phase and surfactants) [38]. The droplets of the developed NEs produced depend on varying the compositions of dispersed phases and dispersion medium, as well as the mixing parameters [3].

Several spontaneous emulsification techniques can be used in the fabrication of NE [43]. If two immiscible phases are in contact with each other, such as water and oil, and one of the phases is partly soluble with two phases, such as an amphiphilic surfactant. In this scenario, both the phases with partly miscible components will shift from their initial phase to the other, causing the interface of oil-water, interstitial disruptions, and increase the spontaneous formation of droplets. In this method, variation in the formulations of the first two steps and mixing conditions can control the volume of droplets produced. McClements and Rao showed in their experiment that when two immiscible

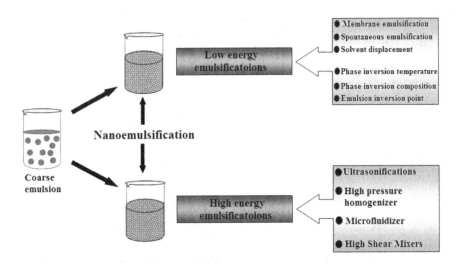

FIGURE 3.4 Summarizes the low-energy emulsification and high-energy emulsification, the two major types of nanoemulsification techniques used in food and pharmaceutical industries, and their specifications.

phases such as water and oil come into contact and one of the phases contains a component that is partially miscible with the two phases such as amphiphilic or surfactant [16]. In this scenario, those elements that are partly blended with the two phases can transfer from their main phase to the other, causing an increase in the area between the surface of the oil and water, periodic turbulence, and spontaneous droplet formation. In this method, differences in the composition of the first two phases and the mixing conditions can control the volume of droplets formed. Droplets with a diameter of around 110 nm were formed by the microfluidization process, while the spontaneous emulsification method could produce droplets with a diameter of about 140 nm [16]. Various realistic interventions using varying oil and surfactant showed that the spontaneous emulsification method could be used to develop NEs, given that the system composition surfactant, oil, and water content, was optimized (Table 3.1).

This process itself increases entropy and thus decreases the Gibbs free energy of the system [42]. The spontaneous emulsification process systems are referred to either self-emulsifying drug delivery systems (SEDDS) or as self-nano-emulsifying drug delivery systems (SNEDDS).

3.2.2.1.2 Self-Nanoemulsification Method (SNEDDS) By using the self-emulsification process, NE formation is accomplished by altering the spontaneous curvature of the surfactant. Compounds of emulsifiers quickly diffuse from the dispersed phase to the continuous phase, bringing movement used to develop nano-sized droplets emulsion [5, 46]. SNEDDS can be characterized as an isotropic aggregate of various components like oils, surfactants, co-surfactants, and drugs devoid of an aqueous dispersion medium [47]. After being diluted by aqueous fluids either in vitro or in vivo, the preconcentrated

TABLE 3.1 Nanoemulsion Prepared Using Spontaneous Emulsification, a Low-Energy Emulsification Technique

Active Nutrients	Emulsifiers	Formulation Outcomes	References
Lipophilic food components as oil phase	Span 80, PGPR	Droplet sizes depend on surfactant adsorption density	[44]
Alpha-tocopherol	Span 80 & 85 Tween 20 & 80	Smallest droplets size: 171 ± 2 nm, the solvent-acetone proportion 15/85% (v/v) leading to a fine nanoemulsion	[38]
Oil of *Nigella sativa*	Tween 80 PEG 400	Nanoemulsification increased the therapeutic potential	[12]
Babchi oil	Tween 80 transcutol-P	Nanoemulsifications increased flux (Jss) and permeability coefficient	[10]
Palm oil	Pluronic F-68	Nanoemulsion obtained with small particle size (94.21 nm) and least Ostwald ripening	[45]

mixture forms fine and optical transparent O/W NE [48]. Diffusion of the hydrophilic co-solvent or co-surfactant from the organic phase into the aqueous phase and transient negative or ultra-low interfacial tensions are the two most widely recorded mechanisms of nanoemulsion formation from SNEDDS [46-49]. SNEDDS is perhaps one of the most promising methods of enhancing the bioavailability of innovative hydrophobic bioactive molecules [50, 51].

3.2.2.1.3 Membrane Emulsification Method Membrane emulsification (ME) was proposed by Nakashima and coworkers in 1988 [52]. The microporous membranes were used in the ME method to make droplets by squeezing the dispersed phase immediately into the continuous phase through the pores. The transformation of a dispersed phase (droplets) into a continuous phase via a membrane involves this technique. The benefits of ME are the low-energy input to produce smaller droplets and size distribution compared to traditional emulsification, hence the narrow polydispersity index (PDI) (Table 3.2). Unfortunately, during the scale-up of this process, the low flux of the dispersed phase through the membrane is a strong limitation [53]. The deciding factor of droplet diameter depended on the pore size of membranes, and the narrow coefficient of variation on the size of the droplets produced (10%). In contrast to high-energy techniques, it needs less surfactant and creates emulsions with a small size distribution range.

3.2.2.1.4 Solvent Displacement This approach is based on the rapid diffusion of a water-miscible organic solvent containing a lipophilic functional compound that encourages NE formation during the aqueous phase. This rapid diffusion enables one-step preparation of NE at a low-energy input with a high encapsulation yield. In the end, from the nanodispersion, the organic solvent is vacuum-evaporated. The use of this technique, however, is restricted to water-miscible solvents [60]. The phase inversion methods that use the chemical energy released as a result of the phase transitions that occur during

TABLE 3.2 Different Types of Membrane Emulsification Systems and Applications

Active Nutrients	Emulsifiers	Membrane Type and Material	Process Design	References
Sunflower oil	Tween 20	Tube membrane (stainless steel)	Vibrating and rotating	[54]
Corn oil	Tween 20, tween 80	Flat membrane (SPG, polyethersulfone)	Premix	[55]
Pumpkinseed oil/ sunflower oil	Tween 20, pluronic® F-68	Flat membrane (micro-porous glass)	Stirred cell	[56]
Paraffin	Tween 20	Flat membrane (stainless steel)	Rotating flat	[57]
Sunflower oil	Tween 20, 80, Lecithin, WPI	Tube membrane (SPG, ceramic)	Crossflow	[58]
Sunflower oil	Tween 20	Tube membrane (micro-porous glass)	Stirred cell	[59]

emulsification are another low-energy strategy. By inducing phase inversion in emulsion from a W/O to O/W shape or vice versa, NEs were produced either by adjusting the temperature in the PIT, the composition in the PIC, or the EIP [5].

3.2.2.1.5 PIT Method

This method is based on the assumption that nonionic surfactants change their affinities with a fixed composition of water and oil by changing the temperature, by adjusting the optimum circumference (molecular geometry), or the comparative solubility of nonionic surfactants [61, 62]. By varying the temperature-time profile of certain oil, water, and non-ionic surfactant mixtures, NEs are spontaneously formed. By using the PIT process, NEs are formed by suddenly breaking up the microemulsions retained at the phase inversion point by rapid cooling [63] or by dilution in water or oil [42]. The regulated conversion of W/O emulsion to O/W emulsion or vice versa via an intermediate crystalline or bicontinuous liquid microemulsion process is also involved in PIT [3].

In water, the surfactant particles tend to bond together to form a monolayer due to the hydrophobic effects, and these monolayers have perfect bending, resulting in more efficient packing of particles. If $p < 1$, the optimum curvature is convex and the surfactant favors the formation of O/W emulsions, the optimum curvature is concave for W/O emulsions for $p > 1$, while monolayers have zero curvature for $p = 1$, where either O/W or W/O systems are not favored by surfactants and instead contribute to the formation of surfactants, the packing parameter (p) determines the optimum curvature of the monolayer surfactant.

The temperature-induced changes in the surfactant's physicochemical properties are the key explanation behind this phase inversion. Here the molecular geometry of a surfactant is dependent on the packing parameter, $p = aT/aH$, where aT is the cross-sectional area of the lipophilic tail-group and aH is the cross-sectional area of the hydrophilic head-group.

Due to the physicochemical properties and packing parameters (p) of nonionic surfactants, the relative solubility of surfactants in the oil and water phases usually varies with temperature [64]. At a certain temperature, the solubility of the surfactant with oil and aqueous phases is almost the same, this is known as the phase reversal temperature or PIT, as the oil and water surfactant system changes from an O/W emulsion to a W/O emulsion. Because the fill factor is equal to one ($P = 1$). At temperatures higher than PIT, the head assembly gradually dries up and the solubility of the surfactant in water decreases becomes more soluble in oil ($p > 1$) and the formation of a W/O emulsion is preferred. When the temperature decreases ($\approx T < PIT$), the main group of non-ionic surfactants become highly hydrating and tend to be more soluble in water ($p < 1$), which favors the formation of O/W emulsions [3].

Above the PIT, the surfactant molecules are mostly inside the oil droplets, because at this temperature they are more soluble in oil. When this system cools the PIT, the surfactants move rapidly from the oil phase to the aqueous phase, just as the movement of water-miscible solvents in the automatic emulsification method results in the automatic formation of small oil droplets due to the increase in the intermediate zone and generating an intermediate turbulent flow. For this reason, Anton and his colleagues suggested that the PIT method for the formation of NEs has a similar physicochemical basis to the spontaneous method of emulsification [65]. This process is described by the fact that it is easy, prevents degradation of the encapsulated drug during processing, consumes small amounts of energy, and allows for easy industrial overflow.

3.2.2.1.6 Phase Inversion Composition (PIC) Method This method is very similar to the PIT method. The difference is, here the optimal bending of the surfactant is changed by altering the process of formulation, not the temperature [64]. For example, the O/W emulsion can be reversed to W/O emulsion by adding salt, since in this case the filling parameter has increased and become greater than one (p > 1) due to the ability of the salt ions to shield the electric charge on groups of surfactants [66]. Alternatively, by dilution with water, a W/O emulsion containing a high salt concentration can be transformed to an O/W emulsion to minimize ionic strength below the critical level. Another PIC approach for obtaining NEs is to modify the electrical charge and stability of the emulsions by altering the pH. Fatty acid carboxyl groups are not charged at pH values lower than pKa (pH < pKa) and have significantly higher oil solubility (p > 1) so that they can stabilize W/O emulsions but are ionized at higher pH so that they become more sensitive to water solubility (p < 1) and O/W emulsion stabilization. By increasing the pH of a mixture of fatty acids, oils, and water from the bottom to the pKa value, nanoemulsification is enhanced [16, 67]. Various literature has reported the successful formulation of NEs using the PIC method of nanoemulsification (Table 3.3).

3.2.2.1.7 Emulsion Inversion Point This method involves changing the system configuration at a constant temperature. The droplet is created by gradually diluting them with water or oil to establish thermodynamically stable NEs. Moving from W/O to O/W in EIP methods or vice versa requires destructive phase inversion rather than intermittent phase inversion, as is the case with PIC or PIT methods [71, 72]. Transient phase reversal occurs when the properties of the surfactant change by modifying one of the compositional variables, such as temperature, pH, or ionic strength. The catastrophic phase reversal occurs by changing the oil/water phase ratio while the surfactant properties remain constant.

Emulsifiers used to reverse catastrophic phase transformation are generally limited to small molecular surfactants that can stabilize W/O emulsions and O/W emulsions [71, 3]. McClements and Rao have demonstrated in practice that increases the amount of water

TABLE 3.3 Nanoemulsion Prepared Using Phase Inversion Composition (PIC), a Low-Energy Emulsification Technique

Active Nutrients	Emulsifiers	Formulation Outcomes	References
n-tetradecane	POP(D230-2OA)	Smaller droplet sizes 37 nm	[68]
Oleic acid	Oleylamine chloride, potassium oleate, oleic acid	Surfactant concentrations produce smaller droplet and high stabilization	[66]
Reference oil phase	Span 20, Tween 20, C12E4, C18E10	Instability of nanoemulsions due to Ostwald ripening	[69]
Phenytoin-loaded alkyd NEs		Possible for topical wound healing application	[45]
Formation of NEs containing ibuprofen		Effective topical application	[70]

in the W/O emulsion, consisting of water droplets dispersed in the oil with constant stirring, can lead to the formation of additional water droplets in the oil phase with low amounts of added water. Even though the melting rate of the water droplets exceeds the melting rate of the oil droplets until the critical water content is surpassed, and there is thus a phase transition from the W/O to the O/W system. Thus, phase inversion by the catastrophic method is usually stimulated either by increasing or decreasing the volume portion of the external phase above (or below) a certain critical level. The value of the critical level at which the phase transition takes place, and the size of the resulting oil droplets depend on process variables such as the velocity of stirring, the quantity of addition of water, and the emulsion concentration [3, 72].

3.2.2.2 High-Energy Emulsification

In addition to the amount of energy applied, the formulation of NEs by highly energetic approaches depends to a great extent on the surfactants used and on the functional compounds [1]. NEs produced by high-energy approaches, therefore, establish a natural method for the preservation of NEs against modification of formulation, such as monomer, surfactant, cosurfactant addition [42]. Various high-energy approaches are listed in the coming section and represented in Figure 3.4.

High-energy approaches utilize mechanical devices to create powerful forces that can mix and disrupt the phases of oil and water, resulting in the production of tiny oil droplets, such instruments are homogenizers for high-pressure valves, microfluidics, and ultrasonication [35, 36]. High energies are introduced to stabilize the droplets in irregular shape to generate destructive forces that overtake the restored forces, and the Laplace Pressure will quantify these restored forces using the formula: $P = \gamma/2r$, which increases by reducing droplet radius (r) and increasing interfacial tension (γ) [72]. The droplet size produced by high-energy techniques is generally regulated by a balance between two opposing processes within the homogenizer, which are the fragmentation of the droplet and agglomeration of the droplet [73]. It may obtain smaller droplets by increasing the strength or length of homogenization, higher concentration of the emulsifying agent being used adjusting the viscosity ratio [74]. The smallest droplet size that can be obtained with a specific high power device is controlled by the flow profiles and strength of the homogenizer, operating conditions such as energy density and duration of the process, environmental conditions, which means the applied temperature, composition of the sample, which includes the type of oil, the type of emulsion and its concentrations and the characteristics physical and chemical phases, which mean stress and viscosity between them. In simpler terms, the droplet size decreases as the energy increases or length rises, the interfacial tension decreases, the adsorption rate of the emulsifier increases, and the viscosity ratio of the disperse-to-continuous process falls within a certain range (0.05 < $\eta D/\eta C$ < 5) [21]. The formation of small droplets determines the amount of the $\eta D/\eta C$ range, and the nature of the disruptive power produced by the specific homogenizer used, i.e. simple shear versus extensional flow. Thus, the smaller the radius of the droplet, the harder it is to further break them down.

High-energy approaches are most suitable to produce nutritional NEs, as they can be applied to different types of oils such as glycerol-triacyl oils and essential oils. Material. However, the particle size strongly depends on the properties of the oil and emulsion used. For example, it is easiest to get small-sized drops when using essential oils or alkanes as an oil phase, as they have low viscosity and/or pressure between surfaces [3]. Now we present the most commonly used devices in high-energy approaches.

3.2.2.2.1 High-Pressure Homogenization The process of emulsion production is called homogenization, which is achieved by applying enough energy to the oil/water interface, which divides the bulk oil into smaller droplets. In general, the smaller the volume to be achieved, the greater the input power required [75]. As in the case of macroemulsion, simple high-speed stirring can create droplets of a few micron. To reduce the droplet size to a range below a micron, more energy is needed, which can be provided by various dissociation mechanisms, including rotor-stator, high pressure, diaphragm, and ultrasonic systems [76].

In the homogenization of a high-pressure valve, very high pressure is applied to the mixture first and then squeezed through a restrictive valve (Figure 3.5). The very fine droplets of emulsion are produced by very high shear stress [3, 77]. The combination of rigorous disruptive forces, such as shear stress, cavitation, and turbulent flow conditions can break the macro-size oil globules into micro or nano sizes [78]. High-pressure valve homogenizers are frequently used to develop conventional emulsions of small droplet sizes in the food processing industry [78]. β-carotene, thyme oil, and curcumin NEs are among the food NEs prepared by the high-pressure valve homogenization technique [79].

These devices are more useful for minimizing the size of oil droplets in pre-existing coarse emulsions than for directly generating emulsions from two different liquids [3]. The coarse emulsion is generated by the high-shear mixer to describe the rationale in the high-pressure valve homogenizer and is then transferred into yet another chamber by the pump through the inlet of the high-pressure valve homogenizer and then forced on its forward stroke through a narrow valve at the end of the chamber. As it passes through the valve, the coarse emulsion particles are broken down into smaller ones by a combination of strong disrupting forces. Different nozzle models are available to enhance the productivity of disrupting droplets [73].

As the number and homogenization pressure increase, the droplet size generated using a high-pressure valve homogenizer typically decreases. It also relies on homogenizing the viscosity ratio of the two phases. As stated before, only when the disperse-to-continuous phase viscosity ratio falls within a certain range, small droplets normally be formed ($0.05 < \eta D/\eta C < 5$) [3, 76]. Moreover, a sufficient emulsifier is required to cover

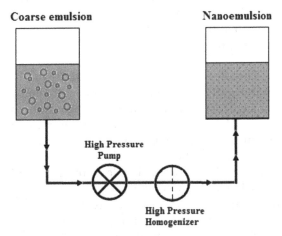

FIGURE 3.5 Layout summarizing the high-pressure homogenization, a type of high-energy nanoemulsification technique.

TABLE 3.4   Nanoemulsion Prepared Using High-Pressure Homogenizer, a High-Energy Emulsification Technique

Active Nutrients	Emulsifiers	Formulation Outcomes	References
Oil of *Nigella sativa*	Tween 20 at three ratios (5%, 10%, and 15%)	Nanoemulsion with highest stability and zeta potential at 22,000 psi and 5 cycles	[80]
β-carotene	Tween 80, 60, 40, 20	Smallest particle sizes and narrowest size distribution	[81]
	SDS, Tween 20, lactoglobulin, sodium caseinate	Smaller droplet sizes (d ~ 60 nm)	[82]
Range of different oil	SDS, Tween 20, sodium caseinate	In multiple cycles, homogenization is more effective than microfluidization	[83]
Lemongrass oil	Tween 80, sodium alginate	Droplet size and PDI inversely depend on treatment time and amplitude	[84]
Red palm oil	Tween 80, Span glycerol as a co-solvent	Homogenization pressure 500 bar produced lower droplet size 119.49 nm and PDI 0.286	[85]

the surfaces of the new droplet formed during homogenization, and the emulsifier should be rapidly adsorbed on the droplet surfaces to prevent recoalescence [73].

To summarize, we need to work at extremely high pressures and to use multiple passes through the homogenizer to obtain the necessary droplet size in NEs (Table 3.4). To obtain droplets less than 100 nm in radius, high emulsifier levels, low interfacial tensions, and adequate viscosity ranges are necessary [16].

3.2.2.2.2 Microfluidizer   This device is similar in nature to the homogenizer of high-pressure valves in that it uses high pressure to squeeze the emulsion premix through a narrow orifice to enhance droplet disruption but changes only in the channels where the emulsion flows (Figure 3.6). The emulsion streaming in the microfluidizer is broken into different streams through a pipe, each of which passes through a separate microchannel and then redirects the two streams to the reaction chamber. Here, it is subjected to extreme breaking forces that lead to the droplets being destroyed very effectively [3, 76]. The droplet size formed can be substantially decreased by increasing the homogenization pressure, many passes, and emulsifier concentration. McClements and Rao have demonstrated in practice that the logarithm of the mean droplet diameter decreases linearly as the logarithm of the homogenization pressure increases for both the ionic surfactant and the globular protein (β-lactoglobulin). However, it can be seen that this bond is sharper for the surfactant than the protein, and this can be explained by the slow rate of protein absorption on the droplet

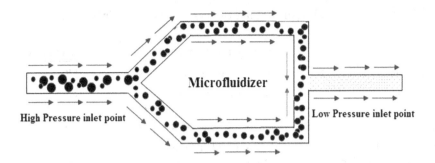

FIGURE 3.6 Layout summarizing the microfluidizer, a typical high-energy nanoemulsification technique.

surfaces, with the formation of an elastic viscous layer, preventing further disruption of the droplets [16].

Also, there is an ideal range of viscosity of the dispersion phase to continuous, which facilitates the formation of small droplets. However, this bond is significantly influenced by the surfactant used, as the average ionic surfactant decreases as the viscosity decreases. On the other hand, the average droplet size is not affected by the viscosity ratio when the globular protein is used as an emulsifier, which can be attributed to the relatively slow absorption. From protein and its ability to form a coating that prevents further disruption of the drops [3, 76].

Microfluidizers are widely used in the preparation of pharmaceutical products, such as food emulsifiers, foods, and beverages such as homogeneous milk, as well as in the production of aromatic emulsifiers. NEs were prepared using the microfluidization technique of various bioactive compounds such as β-carotene and lemongrass essential oil [76, 85].

3.2.2.2.3 Ultrasound Homogenizer When two non-miscible liquids in the presence of a surfactant are exposed to high-frequency sound waves (frequency > 20 kHz) using probes containing piezoelectric quartz crystals that in response to AC voltage, expand and contract, this causes strong shocks. The waves are produced in the surrounding fluid by the tip of the sound wave placed in the fluid. Mechanical vibrations lead to the high-speed creation of liquid jets, the collapse of cavitation-formed micro-bubbles generates extreme disrupting forces leading to droplets breaking and the formation of nano-sized emulsion droplets (70 nm). Within the region where droplet disruption occurs, the emulsion should spend enough process to realize successful and uniform homogenization [3, 86]. In practice, as the strength of ultrasonic waves, sonic time, power level, and emulsifier concentration increase, the droplet size decreases. The form and quantity of the emulsifier used, as well as the viscosity of the oil and water phases, affect the efficiency of homogenization, as shown in Table 3.5 [76, 87]. To generate NEs of the appropriate droplet size, all the specifications mention above should first be optimized. It is worth mentioning that sonication techniques during homogenization can lead to protein denaturation, depolymerization of polysaccharides, or lipid oxidation. Therefore, for industrial-scale applications, this technology has not yet been proven successful [3, 86].

3.2.2.2.4 High-Shear Mixers In energy-intensive processes such as homogenization, dispersion, emulsification, grinding, decomposition, high-shear mixers are commonly used

TABLE 3.5　Nanoemulsion Prepared Using Ultrasonication High-Energy Emulsification Techniques

Active Nutrients	Formulation Outcome	References
Basil oil (*Ocimum basilicum*)	Average droplet sizes 29.3 nm	[88]
Eucalyptus oil	Smaller droplet sizes	[89]
D-limonene	Smaller droplet sizes < 100 nm	[90]
Lemon myrtle and anise myrtle essential oil in water NEs	NEs showed enhanced antibacterial activity and stability	[91]
NEs of coenzyme Q10	Can be used in topical application	[92]
NEs encapsulating curcumin	High nanoencapsulation of curcumin	[93]
Formulation of oil-in-water (O/W) NEs of wheat bran oil	Suitable for use in the food industry	[94]
Neem oil (*Azadirachta indica*) NEs	Effective larvicidal agent	[95]

and cell destruction in the agricultural, food, and chemical reactions industries. Mixers, also known as rotary static mixers, high shear reactors, and high shear homogenizers, have high rotor terminal speeds (ranging from 10 to 50 m/s), high local energy dissipation rates in proximity to the mixing head, and very high-shear rates (range 20000–100000 sec^{-1}) and comparatively higher power consumption compared to conventional mechanically stirred vessels, which is caused by centrifugal forces resulting from the relative motion between the rotor and stator fitted with a narrow spacing (ranging from 100 to 3000 nm) [96]. High-shear mixers are available in both batch and built-in configurations. Batch blocks will have radial or coaxial hollow types, while built-in high-shear mixers have either a vane screen or a rotor-stator tooth configuration. High-shear mixers can be used in conjunction with built-in high-shear mixers as the built-in unit operates in a round ring under the tank equipped with the batch unit. This will enhance the quality of products and reducing operational time. It is made up of a high-speed rotor within a stationary stator. The material or emulsion is constantly drawn into the blending system as the turbine spins and is ejected at supersonic speeds through the stationary stator. As the hydraulic shear blends the emulsion more easily, the size of the droplets will be decreased [65].

Compared with the high-energy procedures, rotor/stator equipment like Ultra-Turrax does not have a substantial reduction in terms of droplet sizes. The effectiveness of such devices when calculated was 0.1%, where 99.9% is dissipated as heat during the homogenization process, so the energy provided mostly being dissipated, generating heat [21, 34, 76].

3.3 CONCLUSION AND FUTURE PERSPECTIVES

NEs have gained great interest and popularity over the last decade due to their exceptional properties, such as high surface area, translucent appearance, strong stability, and adjustable rheology. High-energy methods include high-pressure homogenization, microfluidization, and ultrasonication, as well as low-energy methods, such as spontaneous emulsification, phase inversion, phase reversal temperature, and emulsion reversal point, are the effective methods for preparing NEs. The future industrial importance of many of these methods is little known, as the physics of NE formation remains almost experimental and the rationale has not been widely studied. There is an increasing interest in NE preparation and applications, but only a few of the many discussed methods have commercial applications.

To increase the solubility, bioavailability, and functionality of non-polar bioactive compounds, NEs are one of the most promising systems. To produce new food products, the food and pharmaceutical industry aim to use these systems to incorporate lipophilic bioactive compounds for a better outcome. While NE has substantial benefits over conventional emulsions, such as elegance for food and beverages due to transparent in nature whereas high acceptability for drug industry due to improved bioavailability and physical stability. However, to qualify for the mainstream use of NEs, there is quite a range of regulatory issues that need to be considered first.

First of all, the ingredients used in the formulation of NEs in low-energy or high-energy methods are not acceptable as it uses surfactants, synthetic polymers, synthetic oils or organic solvents for common use in the food industry. Therefore, in the formulation of NEs for food products, food ingredients such as aromatic oils, triglyceride oils, proteins, and polysaccharides should be used as these ingredients are legally appropriate, suitable for labeling, and economically feasible. The second issue, to produce food NEs on an industrial scale, appropriate processes must be used to obtain economical and high stable products. Therefore, many of the methods that have been identified and developed in research laboratories are not feasible for industrial production, especially low-intensity methods that cannot be tested in industrial production. Currently, only high-intensity methods are used to obtain NEs in the food industry.

Finally, there are some safety concerns associated with the use of very small drops of fat in foods. For example, the absorption pathway, bioavailability, or potential toxicity of a lipophilic component encapsulated in nanometer-sized lipid droplets is significantly different from that dispersed in the total lipid phase.

NEs have great potential in a wide range of industries, including the food, pharmaceutical, and cosmetics sectors, due to their unique characteristics and better stability compared to conventional emulsifiers. It is also, the most promising system for improving the solubility, bioavailability, and functionality of non-polar bioactive molecules. Despite their great potential, NEs are not completely stable. Therefore, exhaustive studies are needed to get stable NEs and acceptable for industrialization. The interactions between the bioactive compound and the components of the NE require further research to better understand the effect of the components of NEs on the release of the bioactive molecule and its absorption in different physiological pathways. Therefore, the formulation of NE with biocompatible materials and FDA approval is a prerequisite for future use not only in the pharmaceutical but also in the medical, food, and cosmetics sectors.

CONFLICT OF INTEREST

The authors declare no conflict of interest.

LIST OF ABBREVIATIONS

DAG	Diacylglycerols
DPE	D-phase emulsification
EIP	Emulsion-inversion point
Er	Enhancement ratio
FDA	U.S. Food and Drug Administration
FFA	Free fatty acids
GRAS	Ingredients considered safe by FDA
HLB	Hydrophilic balance system
Jss	Steady state flux
Kp	Permeability coefficient
MAG	Monoacylglycerols
MCT	Medium-chain triglycerides
ME	Membrane emulsification
mN/m	milli-Newton/meter
NE	Nanoemulsions
O/W	Oil in water emulsions
PDI	Polydispersity index
PIC	Phase-inversion composition
PIT	Phase-inversion temperature
SDS	Sodium dodecyl sulfate
SEDDS	Self-emulsifying drug delivery systems
SNEDDS	Self-nano-emulsifying drug delivery systems
TAG	Triacylglycerols
W/O	water in oil emulsions

REFERENCES

1. Silva HD, Cerqueira MA, Vicente AA. Nanoemulsions for food applications: Development and characterization. Food and Bioprocess Technology. 2012; 5(3):854–867.

2. Gupta A, Eral HB, Hatton TA, Doyle PS. Nanoemulsions: Formation, properties and applications. Soft Matter. 2016;12(11):2826–2841.

3. McClements DJ. Edible nanoemulsions: Fabrication, properties, and functional performance. Soft Matter. 2011;7(6):2297–2316.

4. Talegaonkar S, Mustafa G, Akhter S, Iqbal ZI. Design and development of oral oil-in water nanoemulsion formulation bearing atorvastatin: In vitro assessment. Journal of Dispersion Science and Technology. 2009;30:1–12.

5. Solans C, Solé I. Nano-emulsions: Formation by low-energy methods. Current Opinion in Colloid & Interface Science. 2012;17(5):246–254.

6. de Oca-Ávalos JMM, Candal RJ, Herrera ML. Nanoemulsions: Stability and physical properties. Current Opinion in Food Science. 2017;16:1–6. doi.org/10.1016/j.cis.2010.06.010.

7. Shakeel F, Ramadan W, Rizwan M, Faiyazuddin M, Mustafa G, Shafiq S et al. Transdermal and topical delivery of anti-inflammatory agents using nanoemulsion/microemulsion: An updated review. Current Nanoscience. 2010;6:184–198.

8. Mustafa G, Khan ZI, Bansal T, Talegaonkar S. Preparation and characterization of oil in water nano-reservoir systems for improved oral delivery of atorvastatin. Current Nanoscience. 2009;5:428–440.

9. Karthik P, Ezhilarasi PN, Anandharamakrishnan C. Challenges associated in stability of food grade nanoemulsions. Critical Reviews in Food Science and Nutrition. 2017;57(7):1435–1450.

10. Faiyazuddin M, Akhtar N, Akhter J, Shakeel F, Shafiq S, Mustafa G. Production, characterization, in vitro and ex vivo studies of babchi oil-encapsulated nanostructured solid lipid carriers produced by a hot aqueous titration method. Pharmazie. 2010;65:347–354.

11. Mustafa G, Baboota S, Ahuja A, Ali J. Formulation development of chitosan coated intra nasal ropinirole nanoemulsion for better management option of Parkinson: An in vitro ex vivo evaluation. Current Nanoscience. 2012;8:348–360.

12. Alotaibi FO, Mustafa G, Ahuja A. Study of enhanced anti-inflammatory potential of nigella sativa in topical nanoformulation. International Journal of Pharmacy and Pharmaceutical Sciences. 2018;10(7):41–51. doi:10.22159/ijpps.2018v10i7.22966.

13. Baboota S, Abdullah GM, Sahni JK, Ali J. Mechanistic approach for the development of ultrafine oil-water emulsions using monoglyceride and blends of medium and long chain triglycerides: Enhancement of the solubility and bioavailability of Perphenazine. Journal of Excipients & Food Chemicals. 2013;4(1):12–24.

14. Pangeni R, Sharma S, Mustafa G, Ali J, Baboota S. Vitamin E loaded resveratrol nanoemulsion for brain targeting for the treatment of Parkinson's disease by reducing oxidative stress. Nanotechnology. 2014;25(48). doi:10.1088/0957-4484/25/48/485102.

15. Troncoso E, Aguilera JM, McClements DJ. Fabrication, characterization and lipase digestibility of food-grade nanoemulsions. Food Hydrocolloid. 2012;27(2):355–363.

16. McClements DJ, Rao J. Food-grade nanoemulsions: Formulation, fabrication, properties, performance, biological fate, and potential toxicity. Critical Reviews in Food Science and Nutrition. 2011;51(4):285–330.

17. Gordon EM, Cornelio GH, Lorenzo CC, Levy JP, Reed RA, Liu L, Hall FL. First clinical experience using a "pathotropic" injectable retroviral vector (Rexin-G) as intervention for stage IV pancreatic cancer. International Journal of Oncology. 2004;24:177–185.

18. Macedo JPF, Fernandes LL, Formiga FR, Reis MF, Nagashima Júnior T, Soares LAL, Egito EST. Micro-emultocrit technique: A valuable tool for determination of critical HLB value of emulsions. AAPS PharmaSciTech PharmSci Tech. 2006;7:E146–E152.

19. Webster AJ, Cates ME. Osmotic stabilization of concentrated emulsions and foams. Langmuir. 2001;17:595.

20. Tlusty T, Safran SA. Microemulsion networks: The onset of bicontinuity. Journal of Physics: Condensed Matter. 2000;12:A253.

21. Tadros T, Izquierdo P, Esquena J, Solans C. Formation and stability of nano-emulsions. Advances in Colloid and Interface Science. 2004;108–109:303–318.

22. Kumar M, Misra A, Babbar AK et al. Intranasal nanoemulsion based brain targeting drug delivery system of risperidone. International Journal of Pharmaceutics. 2008;358:285–291.

23. Komaiko J, McClements DJ. Optimization of isothermal low-energy nanoemulsion formation: hydrocarbon oil, non-ionic surfactant, and water systems. Journal of Colloid and Interface Science. 2014;425:59–66.

24. Kwon SS, Kong BJ, Cho WG, Park SN. Formation of stable hydrocarbon oil-in-water nanoemulsions by phase inversion composition method at elevated temperature. Korean Journal of Chemical Engineering. 2015;32:540–546.

25. Park H, Han DW, Kim JW. Highly stable phase change material emulsions fabricated by interfacial assembly of amphiphilic block copolymers during phase inversion. Langmuir 2015;31:2649–2654.

26. Nandita D, Ranjan S, Mundra S et al. Fabrication of food grade vitamin E nanoemulsion by low energy approach, characterization and its application. International Journal of Food Properties. 2016;19:700–708.

27. Zahi MR, Wan P, Liang H, Yuan Q. Formation and stability of d-Limonene organogel-based nanoemulsion prepared by a high-pressure homogenizer. Journal of Agricultural and Food Chemistry. 2014;62:12563–12569.

28. Saberi AH, Fang Y, McClements DJ. Fabrication of vitamin E-enriched nanoemulsions: Factors affecting particle size using spontaneous emulsification. Journal of Colloid and Interface Science. 2013;391:95–102.

29. Gulotta A, Saberi AH, Nicoli MC, McClements DJ. Nanoemulsion-based delivery systems for polyunsaturated (ω-3) oils: Formation using a spontaneous emulsification method. Journal of Agricultural and Food Chemistry. 2014;62:1720–1725.

30. Yang J, Jiang W, Guan B et al Preparation of D-limonene oil-in-water nanoemulsion from an optimum formulation. Journal of Oleo Science. 2014;63:1133–1140.

31. Pascual-Pineda LA, Flores-Andrade E, Jiménez-Fernández M, Beristain CI. Kinetic and thermodynamic stability of paprika nanoemulsions. International Journal of Food Science and Technology. 2015;50:1174–1181.

32. Fotticchia I, Fotticchia T, Mattia CA et al. Thermodynamic signature of secondary nano-emulsion formation by isothermal titration calorimetry. Langmuir 2014;30:14427–14433.

33. Zeeb B, Herz E, McClements DJ, Weiss J. Impact of alcohols on the formation and stability of protein-stabilized nanoemulsions. Journal of Colloid and Interface Science. 2014;433:196–203.

34. Aboofazeli R. Nanometric-scaled emulsions (nanoemulsions). Iranian Journal of Pharmaceutical Research. 2010;9(4):325–326.

35. Wooster TJ, Golding M, Sanguansri P. Impact of oil type on nanoemulsion formation and Ostwald ripening stability. Langmuir. 2008;24(22):12758–12765.

36. Leong T, Wooster T, Kentish S, Ashokkumar M. Minimising oil droplet size using ultrasonic emulsification. Ultrasonics Sonochemistry. 2009;16(6):721–727.

37. Freitas S, Merkle HP, Gander B. Microencapsulation by solvent extraction/evaporation: Reviewing the state of the art of microsphere preparation process technology. Journal of Controlled Release. 2005;102(2):313–332.

38. Bouchemal K, Briançon S, Perrier E, Fessi H. Nano-emulsion formulation using spontaneous emulsification: Solvent, oil and surfactant optimisation. International Journal of Pharmaceutics. 2004;280(1–2):241–251.

39. Yin L-J, Chu B-S, Kobayashi I, Nakajima M. Performance of selected emulsifiers and their combinations in the preparation of β-carotene nanodispersions. Food Hydrocolloids. 2009;23(6):1617–1622.

40. Morales D, Gutiérrez J, Garcia-Celma M, Solans Y. A study of the relation between bicontinuous microemulsions and oil/water nano-emulsion formation. Langmuir. 2003;19(18):7196–7200.

41. Uson N, Garcia M, Solans C. Formation of water-in-oil (W/O) nano-emulsions in a water/mixed non-ionic surfactant/oil systems prepared by a low-energy emulsification method. Colloids and Surfaces A: Physicochemical and Engineering Aspects. 2004;**250**(1–3):415–421.

42. Anton N, Benoit J-P, Saulnier P. Design and production of nanoparticles formulated from nano-emulsion templates – A review. Journal of Controlled Release. 2008;**128**(3):185–199.

43. Horn D, Rieger J. Organic nanoparticles in the aqueous phase-theory, experiment, and use. Angewandte Chemie (International Ed. in English). 2001;**40**(23):4330–4361.

44. Wang Z, Neves MA, Isoda H, Nakajima M. Preparation and Characterization of Micro/Nano-emulsions Containing Functional Food Components. Japan Journal of Food Engineering. 2015;**16**:263–276.

45. Teo BSX, Basri M, Zakaria MRS, Salleh AB, Rahman RNZRA, Rahman AMB. A potential tocopherol acetate loaded palm oil esters-in-water nanoemulsions for nanocosmeceuticals. J Nanobiotech. 2010;**8**:1–12.

46. Bansal T, Mustafa G, Khan ZI, Ahmad FJ, Khar RK, Talegaonkar S. Solid self-nano-emulsifying delivery system, a platform technology for formulation of poorly soluble drug. Critical Reviews in Therapeutic Drug Carrier Systems. 2008;**25**(1):63–116.

47. Khan AW, Kotta S, Ansari SH, Sharma RK, Ali J. Self-nanoemulsifying drug delivery system (SNEDDS) of the poorly water-soluble grapefruit flavonoid naringenin: Design, characterization, *in vitro* and *in vivo* evaluation. Drug Delivery. 2015;**22**:552–561.

48. Pouton CW, Porter CJH. Formulation of lipid-based delivery systems for oral administration: materials, methods and strategies. Advanced Drug Delivery Reviews. 2008;**60**:625–637.

49. Agrawal S, Giri TK, Tripathi DK, Ajazuddin, AA. A review on novel therapeutic strategies for the enhancement of solubility for hydrophobic drugs through lipid and surfactant based self micro emulsifying drug delivery system: A novel approach. American Journal of Drug Discovery and Development. 2012;**2**:143–183.

50. Patel G, Shelat P, Lalwani A. Statistical modeling, optimization and characterization of solid self-nanoemulsifying drug delivery system of lopinavir using design of experiment. Drug Delivery. 2016;**23**:3027–3042.

51. Kohli K, Chopra S, Dhar D, Arora S, Khar RK. Self-emulsifying drug delivery systems: An approach to enhance oral bioavailability. Drug Discovery Today. 2010;**15**:958–965.

52. Nakashima T, Shimizu M, Kukizaki M. Particle control of emulsion by membrane emulsification and its applications. Advanced Drug Delivery Reviews. 2000;**45**(1):47–56.

53. Sanguansri P, Augustin MA. Nanoscale materials development – A food industry perspective. Trends in Food Science & Technology. 2006;**17**(10):547–556.

54. Holdich RG, Dragosavac MM, Vladisavljević GT, Kosvintsev SR. Membrane emulsification with oscillating and stationary membranes. Industrial & Engineering Chemistry Research. 2010;**49**:3810–3817.

55. Nazir A, Schroën K, Boom R. Premix emulsification: A review. Journal of Membrane Science. 2010;**362**:1–11.

56. Dragosavac MM, Sovilj MN, Kosvintsev SR, Holdich RG, Vladisavljević GT. Controlled production of oil-in-water emulsions containing unrefined pumpkin seed oil using stirred cell membrane emulsification. Membrane Science. 2008;**322**:178–188.

57. Vladisavljević GTR, Williams A. Manufacture of large uniform droplets using rotating membrane emulsification. Journal of Colloid and Interface Science. 2006;**299**:396–402.

58. Hancocks RD, Spyropoulos F, Norton IT. Comparisons between membranes for use in cross flow membrane emulsification. Journal of Food Engineering. 2013;**116**:382–389.

59. Kosvintsev SR, Gasparini G, Holdich RG, Cumming IW, Stillwell MT. Liquid-liquid membrane dispersion in a stirred cell with and without controlled shear. Industrial & Engineering Chemistry Research. 2005;**44**:9323–9330.

60. Chu B-S, Ichikawa S, Kanafusa S, Nakajima M. Preparation and characterization of β-carotene nanodispersions prepared by solvent displacement technique. Journal of Agricultural and Food Chemistry. 2007;**55**(16):6754–6760.

61. Shinoda K, Saito H. The effect of temperature on the phase equilibria and the types of dispersions of the ternary system composed of water, cyclohexane, and nonionic surfactant. Journal of Colloid and Interface Science. 1968;**26**(1):70–74.

62. Shinoda K, Saito H. The stability of O/W type emulsions as functions of temperature and the HLB of emulsifiers: The emulsification by PIT-method. Journal of Colloid and Interface Science. 1969;**30**(2):258–263.

63. Sadurní N, Solans C, Azemar N, García-Celma MJ. Studies on the formation of O/W nano-emulsions, by low-energy emulsification methods, suitable for pharmaceutical applications. European Journal of Pharmaceutical Sciences. 2005;**26**(5):438–445.

64. Anton N, Vandamme TF. The universality of low-energy nano-emulsification. International Journal of Pharmaceutics. 2009;**377**(1–2):142–147.

65. Anton N, Gayet P, Benoit J-P, Saulnier P. Nano-emulsions and nanocapsules by the PIT method: An investigation on the role of the temperature cycling on the emulsion phase inversion. International Journal of Pharmaceutics. 2007;**344**(1–2):44–52.

66. Maestro A, Solè I, González C, Solans C, Gutiérrez JM. Influence of the phase behavior on the properties of ionic nanoemulsions prepared by the phase inversion composition method. Journal of Colloid and Interface Science. 2008;**327**(2):433–439.

67. Sole I, Maestro A, Pey C, González C, Solans C, Gutiérrez J. Nano-emulsions preparation by low energy methods in an ionic surfactant system. Colloids and Surfaces A: Physicochemical and Engineering Aspects. 2006;**288**(1–3):138–143.

68. Ren G, Sun Z, Wang Z, Zheng X, Xu Z, Sun D. Nanoemulsion formation by the phase inversion temperature method using polyoxypropylene surfactants. Journal of Colloid and Interface Science. 2019;**540**:177–184.

69. Sole I, Pey CM, Maestro A, González C, Porras M, Solans C, Gutiérrez JM. Nano-emulsions prepared by the phase inversion composition method: Preparation variables and scale up. Journal of Colloid and Interface Science. 2010;**344**:417–423.

70. Salim N, Jose García-Celma M, Escribano E, Nolla J, Llinàs M, Basri M, Solans C, Esquena J, Tadros TF. Formation of nanoemulsion containing ibuprofen by PIC method for topical delivery. Materials Today: Proceedings. 2018;**5**:S172–S179.

71. Thakur RK, Villette C, Aubry J, Delaplace G. Dynamic emulsification and catastrophic phase inversion of lecithin-based emulsions. Colloids and Surfaces A: Physicochemical and Engineering Aspects. 2008;**315**(1–3):285–293.

72. Schubert H, Engel R. Product and formulation engineering of emulsions. Chemical Engineering Research and Design. 2004;**82**(9):1137–1143,

73. Jafari SM, Assadpoor E, He Y, Bhandari B. Re-coalescence of emulsion droplets during high-energy emulsification. Food Hydrocolloids. 2008;**22**(7):1191–1202.

74. Mohd-Setapar SH, Nian-Yian L, Kamarudin WNW, Idham Z, Norfahana A-T. Omega-3 emulsion of rubber (*Hevea brasiliensis*) seed oil. Agricultural Sciences. 2013;4(5):84.

75. McClements DJ, Li Y. Structured emulsion-based delivery systems: Controlling the digestion and release of lipophilic food components. Advances in Colloid and Interface Science. 2010;159:213–228.

76. Schultz S, Wagner G, Urban K, Ulrich J. High-pressure homogenization as a process for emulsion formation. Chemical Engineering and Technology. 2004;27:361–368.

77. Quintanilla-Carvajal MX, Camacho-Díaz BH, Meraz-Torres LS, Chanona-Pérez JJ, Alamilla-Beltrán L, Jimenéz-Aparicio A, Gutiérrez-López GF. Nanoencapsulation: A new trend in food engineering processing. Food Engineering Reviews. 2010;2(1):39–50.

78. Stang M, Schuchmann H, Schubert H. Emulsification in high-pressure homogenizers. Engineering in Life Sciences. 2001;1(4):151–157.

79. Ziani K, Chang Y, McLandsborough L, McClements DJ. Influence of surfactant charge on antimicrobial efficacy of surfactant-stabilized thyme oil nanoemulsions. Journal of Agricultural and Food Chemistry. 2011;59(11):6247–6255.

80. Mohammed NK, Muhialdin BJ, Meor Hussin AS. Characterization of nanoemulsion of *Nigella sativa* oil and its application in ice cream. Food Science and Nutrition. 2020;8(6):2608–2618.

81. Yuan Y, Gao Y. Characterization and stability evaluation of b-carotene nanoemulsions prepared by high pressure homogenization under various emulsifying conditions. Food Research International. 2008;41:61–68.

82. Qian C, Mcclements DJ. Formation of nanoemulsions stabilized by model food-grade emulsifiers using high-pressure homogenization: Factors acting particle size. Food Hydrocoll. 2011;25:1000–1008.

83. Lee L, Norton IT. Comparing droplet breakup for a high-pressure valve homogeniser and a microfluidizer for the potential production of food-grade nanoemulsions. Journal of Food Engineering. 2013;114:158–163.

84. Salvia-Trujillo L, Rojas-Graü MA, Soliva-Fortuny R, Martín-Belloso O. Effect of processing parameters on physicochemical characteristics of microfluidized lemongrass essential oil-alginate nanoemulsions. Food Hydrocolloids. 2013;30:401–407.

85. Chong W-T, Tan C-P, Cheah Y-K, Lajis AFB, Habi Mat Dian NL, Kanagaratnam S, Lai O-M. Optimization of process parameters in preparation of tocotrienol-rich red palm oil-based nanoemulsion stabilized by Tween80-Span 80 using response surface methodology. PLoS One. 2018;13(8):e0202771.

86. Ricaurte L, de Perea-Flores MJ, Martinez A, Quintanilla-Carvajal MX. Production of high-oleic palm oil nanoemulsions by high-shear homogenization (microfluidization). Innovative Food Science and Emerging Technologies. 2016;35:75. https://doi.org/10.1016/j.ifset.2016.04.004.

87. Maa Y-F, Hsu CC. Performance of sonication and microfluidization for liquid-liquid emulsification. Pharmaceutical Development and Technology. 1999;4(2):233–240.

88. Ghosh V, Mukherjee A, Chandrasekaran N. Formulation and characterization of plant essential oil based nanoemulsion: Evaluation of its larvicidal activity against Aedes aegypti. Asian Journal of Chemistry. 2013;25:18–20.

89. Sugumar S, Ghosh V, Nirmala MJ et al. Ultrasonic emulsification of eucalyptus oil nanoemulsion: Antibacterial activity against Staphylococcus Aureus and wound healing activity in Wistar rats. Ultrasonics Sonochemistry. 2014;21:1044–1049.

90. Li P, Chiang B. Process optimization and stability of D-limonene-in-water nano-emulsions prepared by ultrasonic emulsification using response surface methodology. Ultrasonics Sonochemistry. 2012;19:192–197.

91. Nirmal NP, Mereddy R, Li L, Sultanbawa Y. Formulation, characterisation and antibacterial activity of lemon myrtle and anise myrtle essential oil in water nano-emulsion. Food Chemistry. 2018;254:1–7.

92. Kaci M, Belhaffef A, Meziane S, Dostert G, Menu P, Velot É, Desobry S, Arab-Tehrany, E. Nanoemulsions and topical creams for the safe and effective delivery of lipophilic antioxidant coenzyme Q10. Colloids Surfaces B Biointerfaces. 2018;167:165–175.

93. Sari TP, Mann B, Kumar R, Singh RRB, Sharma R, Bhardwaj M, Athira S. Preparation and characterization of nanoemulsion encapsulating curcumin. Food Hydrocolloids. 2015;43:540–546.

94. Rebolleda S, Sanz MT, Benito JM, Beltrán S, Escudero I, González San-José ML. Formulation and characterisation of wheat bran oil-in-water nanoemulsions. Food Chemistry. 2015;167:16–23.

95. Anjali C, Sharma Y, Mukherjee A, Chandrasekaran N. Neem oil (Azadirachta indica) nanoemulsion-a potent larvicidal agent against Culex quinquefasciatus. Pest Management Science. 2012;68:158–163.

96. Israelachvili JN. Intermolecular and Surface Forces. United States: Elsevier, Academic Press; 2011.

CHAPTER 4

Stability Perspectives of Nanoemulsions

Md. Shoaib Alam,[1] Ayesha Akhtar,[2]
Javed Ahmad,[3] Leo M.L. Nollet[4]

[1]Research and Development, Jamjoom Pharmaceuticals,
Jeddah, Kingdom of Saudi Arabia
[2]Quality Control, Med City Pharma, Jeddah, Kingdom of Saudi Arabia
[3]College of Pharmacy, Najran University, Najran, Kingdom of Saudi Arabia
[4]University College, Ghent, Belgium

CONTENTS

DOI: 10.1201/9781003121121-4

4.1 INTRODUCTION

Nanoemulsion is a non-equilibrium, transparent, and translucent emulsion system which comprises monodisperse droplets of size range varying between 20 and 200 nm. Nanoemulsions exhibit tremendous potential to enhance the absorption of bioactive molecules [1]. Despite thermodynamical instability, these nanoemulsions are kinetically stable and favor physical stability for the long term [2]. Nanoemulsions are thermodynamically unstable which is shown in Figure 4.1 and comparing the free energy of the colloidal dispersions and separated phases as shown in Figure 4.1.

Consumption of a balanced diet with high nutritious value is the prime way to sustain a good and healthy life with minimal health risk. Nowadays due to the lack of adequate nutrition approximately 600 million men fall ill and around 4.2 million men die every year [3]. Food safety and its nutritional values are intricately interlinked. To overcome these issues and constraints associated with food, numerous studies have been done. Implementing nanoemulsion in probiotic health food and other functional food is among one of the approaches to improve the nutritive value, digestibility, and oral bioavailability of bioactive compounds [4–5]. In general, nano-encapsulation facilitates dispersion and absorption of lipophilic bioactive compounds such as vitamins, flavors, and nutraceuticals in tiny vesicles with nano diameters [6]. The potential applications of nanoemulsion in the food industry has depicted in Figure 4.2.

The bioactive compounds usually exist in two forms that is water-soluble compounds and water-insoluble compounds. Hence, depending upon the characteristics of the bioactive compound, the formation of nanoemulsion is of two different types: O/W type (oil-in-water) (oil is dispersed in aqueous phase) and W/O (water-in-oil) type (water is dispersed in oil phase). The oil phase in a nanoemulsion is composed of different types of acylglycerols (tri, di, mono) and free fatty acids. Non-polar essential

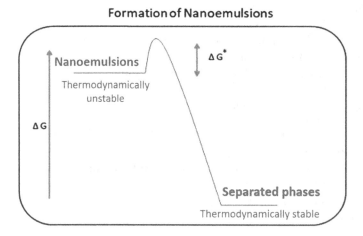

Formation of Nanoemulsions

FIGURE 4.1 Illustration showing the comparison of free energy (ΔG) of nanoemulsion systems compared to the phase separated state. The two states are separated by an activation energy ΔG^*.

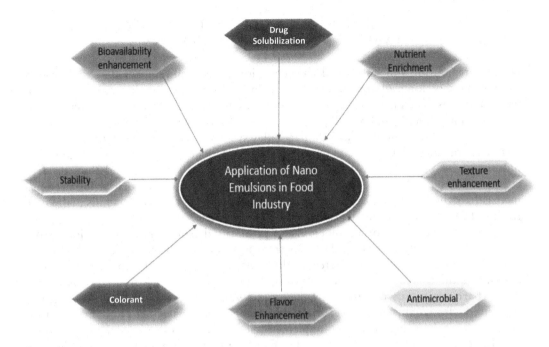

FIGURE 4.2 A brief illustration of nanoemulsion applications in the food industry.

oils, lipids, waxes, oil-soluble vitamins, and various lipophilic components are also used as oil phase [7–10].

Nanoemulsions have been enriched with unique characteristics such as small size, stability to chemical and physical changes, kinetically stability, and a controlled structure that facilitate to be an ideal platform in food application [11–13]. Following factors should be taken into consideration while formulating nanoemulsion systems for food science applications [14–15].

- Phase selection must be appropriate to avoid Ostwald ripening. The two immiscible phases of the nanoemulsion system (dispersed phase and dispersed medium) are stabilized with the aid of emulsifying agent/surfactant.
- The surfactant must be selected carefully such that an ultralow interfacial tension is achieved which is a primary requirement to produce nanoemulsion.
- The concentration of surfactant must be high enough to stabilize the micro droplets to produce a nanoemulsion.
- The surfactant must be flexible or fluid enough to promote the formation of nanoemulsion.

The various advantages of a nanoemulsion system as a delivery vehicle include: (1) Elevated rate of absorption, (2) mitigated variability subsequent to absorption, (3) Providing protection shield from oxidation and hydrolysis via O/W nanoemulsions, (4) Increased solubility and availability of lipophilic bioactive molecule, (5) Increased bioavailability for various bioactive molecules, (6) facilitating numerous modes of delivery of bioactive molecules, for instance, oral, topical and intravenous, (7) Fast and efficient penetration of the bioactive moiety, (8) Encapsulation and delivery of both lipophilic and hydrophilic bioactive molecules by same nanoemulsions [16–18].

4.2 FACTORS TO BE CONSIDERED FOR STABILITY EVALUATION OF NANOEMULSIONS

For the preparation and stability of nanoemulsions, it is required to understand exactly the behavior of force of interaction toward influencing functionality and stability of nanoemulsions. The primary mechanism associated with the stability of nanoemulsions include conjugation, complex interplay of van der Waals forces and steric interactions.

The thermodynamic stability of a particular system is accomplished by the change of its free energy and an appropriate reference state. A nanoemulsion is thermodynamically unstable, which means that the free energy of a nanoemulsion is higher than the free energy of the separate phases (oil and water). The formation of a nanoemulsion becomes increasingly thermodynamically unfavorable as the radius of the droplets falls and where the interfacial tension is similar to that at a planar surface.

The kinetic stability of a nanoemulsion primarily depends upon two major factors which include energy barriers and a mass transport system. The activation energy is the required amount of energy to segregate different states (initial and final), moreover, it determines the rate of conversion from one state to another. The threshold peak of this energy barrier is mostly controlled by the forces, which lie between two droplets such as repulsive hydrodynamic and steric or electrostatic interactions. Kinetic stability of nanoemulsions can be achieved by the implementation of large activation energy (approximately >20 kT) between the two states. On the other hand, for a microemulsion (heating up and mechanical agitation), activation energy is required to reach the thermodynamically stable state when the components come into contact. Mass transport phenomena are responsible for "Ostwald ripening" which changes the homogeneous structure of a nanoemulsion over a long duration time.

4.3 STABILITY ASSESSMENT OF NANOEMULSIONS

Free energy is the amount of energy required to segregate the oil and water phase, usually lesser than required for emulsification. The thermodynamic instability of nanoemulsions is undesirable and responsible for various issues over a longer period of storage [19]. The major instability mechanisms of nanoemulsion are flocculation, creaming, sedimentation, coalescence, Ostwald ripening, and phase separation as shown in Figure 4.3. [20]. All major instability mechanisms of nanoemulsion studies have been described in Table 4.1.

4.3.1 Ostwald Ripening

This is a physiochemical process, where the formation of large droplets occurs by aggregation of small droplets resulting from the molecular diffusion of oil between droplets through the continuous phase. The mechanism behind Ostwald ripening is based on the Kelvin effect [2]. Kelvin effect is summarized as follows: oil solubility of small emulsions droplets has greater value than larger emulsions droplets due to difference in Laplace pressures [21]. The major factor that influences the Ostwald ripening is oil solubility in the continuous phase. In addition to that, these phenomena occur mostly in beverage emulsions that comprise water-soluble flavor oils, short-chain triglycerides, and essential oils [22]. However, it is not observed in dairy-based emulsions, which contain water-insoluble

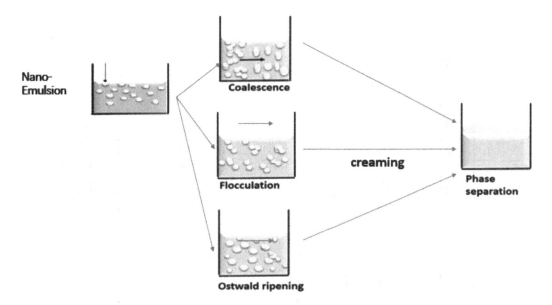

Nano-Emulsion

Coalescence

Flocculation

creaming

Phase separation

Ostwald ripening

FIGURE 4.3   Representation of commonly occurring physical instability in food emulsion.

triglyceride oils. Hence, Ostwald ripening can be minimized by the inclusion of a ripening inhibitor (mostly hydrophobic) [23]. Ostwald ripening is principally modulated by the solubility of oil in the continuous phase and molecular volume of oil (Vm). Moreover, the solubility of a low molar Vm (small radius of curvature) in water is higher, hence it has a greater tendency for re-precipitation by Ostwald ripening. On the other hand, when a larger molar Vm (900 cm^3 mol^{-1}) is insoluble in water and it does not allow re-precipitation then as a consequence Ostwald ripening is prevented from happening.

Additionally, Ostwald ripening mainly occurs in various types of dispersions and the change in particle size distribution are described by the LSW theory (Lifshits, Slezov, and Wagner). However, this theory does not apply if particles are extremely close to each other.

Kabalnov finds that the rate of Ostwald ripening is directly proportional to the interfacial tension [24]. The change in mean droplet diameter (d) with time (t), owing to Ostwald ripening of a single component emulsified lipid in the steady-state system is specified by the following equation (4.1)

$$d^3 - d_0^3 = \bar{\omega} = \frac{32}{9}\alpha cDt, \qquad (4.1)$$

where, ω ($=2\gamma V_m/RT$), γ is the interfacial tension, α represents the Ostwald ripening rate, V_m is the molar volume of the lipid, R is the gas constant and T is the absolute temperature, c is the water solubility of the lipid and D is the translational diffusion coefficient of the lipid in the water.

In general, Ostwald ripening frequently happens in beverages emulsions, which contain an exorbitant amount of hydrophilic flavor oils for instance, essential oils, low molecular weight triglycerides, etc. However, dairy-based emulsions are mostly comprised of water-insoluble triglyceride oils, hence it does not lead to Ostwald ripening.

Md. Shoaib Alam et al.

TABLE 4.1 Different Stability Studies of Food-Grade Nanoemulsion

Food Bioactive Components	Types of Instability Studies	Emulsifiers and Oil Phase	Emulsification Techniques	Droplet Size	Factors Influencing	References
Thyme oil	Ostwald ripening; phase separation; creaming	Tween-80; lauric arginate; sodium dodecyl sulfate	High-speed blending High-pressure homogenization	7–196 nm	Droplet size and solubility	[25]
Olive oil	Ostwald ripening	Sucrose monoesters	Mixing	Less than 200 nm	Composition and droplet size	[28]
Lemon oil	Ostwald ripening; creaming	Tween-80	High-speed stirring High-pressure homogenization	112–296 nm	Droplet growth and composition	[8]
Curcumin	Ostwald ripening; phase separation; creaming; sedimentation	B lactoglobulin oil phase: long-chain triacylglycerols (LCT), medium-chain triacylglycerols (MCT), and short-chain triacylglycerols (SCT)	High speed blending High-pressure homogenization	Less than 200 nm	Solubility	[29]
Peanut oil	Ostwald ripening	Sodium dodecyl sulphate (SDS); Tween-80; polyethylene glycol	Silverson rotor-stator mixing Microfluidization	40 nm	Solubility and droplet size	[37]
Soybean oil	Ostwald ripening	PEG stearates Myrj s40, Myrj s50, and Myrj s100	Ultrasonication	150 nm	Emulsification ion process, temperature, type of emulsifiers, and concentration	[26]

Consequently, Ostwald ripening can be effectively reduced by the addition of non-polar molecules in form of ripening inhibitors. The ripening inhibitors, which are non-polar molecules, are soluble in the oil phase and insoluble in the water phase. By the addition of ripening inhibitor into the nanoemulsion, the instability in nanoemulsion can be prevented. It was observed that long-chain triglyceride (peanut oil) containing nanoemulsion having size 120 nm was physically stable up to 3 months of storage with no Ostwald ripening were found. The large molar volume of long-chain triglyceride oil acts as a kinetic barrier to Ostwald ripening without causing any change in particle size.

Similarly, Ziani et al. has reported that, thyme oil containing emulsions with mean particle diameter >7000 nm were highly unstable with clearly observable creaming during storage period of 3 days. The thyme oil based nanoemulsions were at high risk of Ostwald ripening due to the moderately high water solubility, rapid creaming, and low density of thyme oil. Further, more than 75% of corn oil (hydrophobic material with low water solubility) developed into the nanoemulsion prohibited the phenomenon of Ostwald ripening [25].

Similarly, Delmas et al. reported the Ostwald ripening process in the nanoemulsion with a mixture of oil compounds (soybean oil or Labrafac). The nanoemulsion was kept at different temperatures i.e., 20, 40, 60, and 70°C to observe the size evolution and Ostwald ripening rate. The Ostwald ripening explained a linear relationship between droplet radius and time that designates them as the main destabilization mechanism. It was observed that Ostwald ripening is affected by temperature and there is an exponential increase in Ostwald ripening with increase in temperature during storage. Ostwald ripening plots were observed linear at short time (t < 20 h) storage of nanoemulsion, whereas nonlinearity was observed at long time storage, and by addition of insoluble surfactant, the rate of droplet growth was reduced. It was recommended that the incorporation of insoluble surfactant in an inadequate amount in a nanoemulsion stabilizes this against Ostwald ripening. Further, the stability of nanoemulsions was also attributed to kinetic inhibition of the Ostwald ripening and coalescence processes [26].

Rao and McClements has selected different folds concentration of lemon oil (1X, 3X, 5X, and 10X) which was depending on the degree of distillation of oils. Furthermore, also studied the impact of lemon oil fold on the formation and stability of the nanoemulsion. During storage, nanoemulsions with lower fold oils (1X, 3X, and 5X) were found to be highly unstable. However, the higher oil fold (10X) nanoemulsion was stable against Ostwald ripening due to very low water solubility that inhibited the droplet growth [27].

Recently, Eid et al. have shown the impact of various surfactants such as sucrose laureate, glycerol, and different concentrations of oil on the stability of an olive oil nanoemulsion. Nanoemulsion with 36% oil concentration exhibited a larger droplet size with less uniformity as compared to 50% oil concentration. Even so, in respect of surfactant concentration, a nanoemulsion that comprises 50% of oil and 15% surfactant concentration respectively showed greater droplet size and uniformity compared to 20% and 25% surfactant concentration. The consequences of these studies were that with a specific formulation of nanoemulsion (50% of oil and 15% surfactant concentration respectively) there are lesser susceptibility to Ostwald ripening [28].

On the other side, Ahmed et al. found that in a curcumin nanoemulsion (<200nm, with the composition of an equal portion of long-chain triacylglycerols and short-chain triacylglycerols) there was no observation of Ostwald ripening. It happened because of, mixing of high water solubility (SCT) oil into relatively low water solubility (LCT) oil during preparation of emulsion. Moreover, this type of nanoemulsion was also resistant to particle growth and gravitational separation [29].

The nanoemulsion comprising insoluble oil with higher molar volume or high-distilled oil is more resistant against Ostwald ripening as they represent a kinetic barrier. Additionally, these nanoemulsions are resistant to Ostwald ripening due to low water solubility. However, nanoemulsions with high oil concentration and lesser surfactant concentration are more susceptible to Ostwald ripening.

4.3.2 Coalescence

The emulsification process involves two steps, firstly deformation and disruption of droplets and secondly stabilization of a new interface by emulsifier to avert re- coalescence of the newly formed droplets. On the other hand, final emulsion droplet size (EDS) can be explained as an equilibrium between droplet disruption and re-coalescence [30]. Coalescence is a phenomenon of the collision that happened between two droplets and its conglomeration, which destabilize the emulsions. However, microsized emulsions have more tendencies to coalescence than nanoemulsions. There are two ways to reduce re-coalescence soon after droplet disruption, firstly by the addition of surfactant molecules and secondly by hydrodynamic effects [31–32]. Furthermore, the stability of the emulsions to droplet coalescence would decrease by increasing the storage temperature [33].

Britten and Giroux have proposed to estimate coalescence stability of protein-stabilized emulsions and proven that the rate of coalescence is slower than creaming and flocculation. Casein can produce more stable nanoemulsion and resistance toward coalescence instability due to requirement of high homogenization pressure. The numerous factors are known to influence the occurrence of coalescence in nanoemulsion. Based on the microscopic examination, it was identified that the formation of large oil droplets is due to over agitation of the emulsion which lead to formation of coalescence nuclei [34].

Jafari et al. produced an O/W nanoemulsion through its microfluidization technique and additionally, evaluated the impact of sonication and microfluidization technique on the stability of a nanoemulsion. It was concluded that, if the collision rate is higher during microfluidization, it increases the droplet size and produces larger coalescence which is known as the "over-processing" phenomenon. Moreover, the coalescence was also noticed in the ultra-sonicated emulsion which exhibiting a bimodal droplet size distribution [35].

In 2007, Jafari et al. developed a D-limonene nanoemulsion by microfludization and ultra-sonication techniques and evaluated its stability. It was prepared with various dispersed-phase volume fractions of d-limonene (5 to 15%) by microfluidization at 21, 42, and 63 MPa (2 cycles). It was concluded that microfludization could be considered as an efficient technique for producing small emulsion droplets with narrow distributions. Nevertheless, "over-processing" can occur during microfludization and this leads to re-coalescence of emulsion droplets. To minimize "over-processing" and inhibit coalescence of new droplets, sonication time should have been improved to increase the energy input [36].

In 2008 Kentish and associates have prepared flaxseed oil/water food-grade emulsions (135 mm) to investigate the phenomenon of droplet coalescence and cavitation bubble cloud formation. They concluded that an increased value of residence time could be responsible for reducing droplet size up to a certain extent and low surfactant (Tween-40) contents could also result in droplet coalescence. Moreover, they suggested that apart from the operating factors of emulsion preparation techniques, the occurrence of coalescence can be minimized by managing the nature of surfactants and their concentration [37].

Takegami and co-workers have prepared a novel lipid nanoemulsion (50 nm), which contains two co-surfactants i.e., sodium palmitate (PA) for droplet size reduction and sucrose palmitate to increase its stability. This novel lipid nanoemulsion was prepared from soybean oil and phosphatidylcholine (PC). Phosphatidylcholine was responsible for stability enhancement of the nanoemulsion. This emulsion showed increased mean droplet size with broadening of the droplet size after 12-month storage as a result of coalescence. In addition to that, the saline solution is accountable for swift rate of coalescence that almost increases the particle diameter up to 150 nm within 0.5 h. To inhibit the effect of coalescence, sodium and sucrose palmitate was introduced and the result was examined. The particle diameter was found to be 50 nm for 12 months with no change in mean droplet size [38].

Similarly, Wulff-Perez and co-workers have prepared o/w nanoemulsions (<500 nm) with an organic phase such as natural oils (soybean, olive, and sesame oil) and Pluronic F68 surfactant. Simultaneously these authors did characterization and stability studies. The stability of emulsions was analyzed using Turbiscan (backscattering) and was found to be stable. However, the nanoemulsion stability was starting to decline after crossing the specific threshold concentration of Pluronic F68 and then the system got destabilized due to coalescence [39].

Similarly, Mao and associates developed a β-carotene nanoemulsion having a droplet size in the range of 115 to 303 nm using different surfactants and studied the effect of different surfactants on the stability. The octenyl succinate starch (OSS) nanoemulsion had a better stability than that stabilized by whey protein isolate (WPI) and Tween-20. At the same time decaglycerol monolaurate employed nanoemulsion was lacking stability. The large molecule emulsifiers (e.g., OSS and WPI) were able to form a mechanically strong interfacial layers and cause steric hindrance to prevent droplet coalescence. Furthermore, WPI emulsion droplets carry a large number of ions having zeta potential –17.3 mV, and OSS was imparting a high viscosity to solution attributing improved stabilizing properties [40].

Likewise, Henry et al. studied the phenomenon of droplet breakup and coalescence in a decane nanoemulsion and examined the effect of different surfactants and its concentrations on the processes. In his study phosphatidylcholine, phosphatidylglycerol, whey protein, and β-lactoglobulin were used as a surfactant. Moreover, at a higher surfactant concentration no coalescence effect was observed except for phosphatidylcholine [41].

Anjali et al. in 2012 studied the nanoemulsion stability at various concentrations, which were prepared from neem oil and non-ionic surfactant such as Tween-20 (31.03–251.43 nm). The larvicidal effect of the above nanoemulsion was quantified against *Culex quinquefasciatus*. They concluded that the preparation with a higher surfactant concentration (1 oil: 3 surfactants) with lesser droplet size (31.03 nm) was more stable [42].

Similarly, Rao and McClements in 2011 studied food-grade nanoemulsions, which were composed of sucrose monopalmitate (SMP) as surfactant and lemon oil. This study aimed to analyze the conditional temperature and concentration of surfactant and lemon oil to get stable nanoemulsions. Moreover, the effect of mixing, heat treatment, and homogenization at various colloidal systems was also examined. It was concluded that nanoemulsions at lower temperature 5°C and 23°C were stable however, at a higher temperature (40°C) they exhibited coalescence within 3 days of storage [8].

Moreover, Qian et al. in 2018 has studied beta carotene enriched nanoemulsion and observed the influence of temperature, pH, ionic strength, and emulsifier type on the physical and chemical stability of nanoemulsion which was formulated with various carrier oils. The objective of this study was to select suitable oils (palm oil, coconut oil, fish oil, and corn oil) as a carrier to form stable nanoemulsion and to enhance the bioaccessibility of beta-carotene. The nanoemulsion was designed with 90% v/v aqueous solution,

2% whey protein isolate (WPI) w/v and 10% v/v dispersed oil. Based on the study it was found that WPI are more effective in inhibiting the degradation and thus stabilizing beta-carotene nanoemulsion. [43].

The above-mentioned studies demonstrate that if the amount of input energy increased beyond the threshold value during the nanoemulsification process, it would lead to droplet coalescence due to over-processing. Furthermore, the surfactant concentrations and their nature, the ratio of oil and surfactant concentration showed a tremendous impact on nanoemulsion stability (coalescence). If the surfactant concentration is lower, it leads to coalescence because of the insufficiency of surfactant to stabilize the nanoemulsion. Few known specific emulsifiers, which imparted stability to nanoemulsion against coalescence, are OSS, WPI, and sucrose palmitate. Furthermore, the emulsion, with higher emulsifier concentration, can decline the interfacial free energy and facilitate a mechanical barrier to coalescence.

4.3.3 Flocculation

Flocculation instability of nanoemulsions is associated with long-term storage conditions. It is a process where more than two droplets conglomerate. Flocculation is the result of interaction between surfactant and biopolymer hence, a higher surfactant concentration and increased ionic strength lead to flocculation of nanoemulsions. Therefore, surfactant concentration and ionic strength should be optimized to prevent flocculation [44–45].

Jafari et al. in 2007 stated that a D-limonene nanoemulsion can be stabilized by incorporation of Tween-20 surfactant into Hi-cap (biopolymer) to reduce the emulsion size during a micro fluidization process. On the other hand, surfactants and biopolymers, which are present in the emulsions, have the potency to approach each other to adsorb the oil-water interface and decrease the interfacial tension. Hence, surfactant concentration and surfactant types have a major impact on the flocculation process [36].

Wulff-Perez et al. in 2009 studied the preparation of an oil-water nanoemulsion with natural oils and Pluronic F68 and its characterization and stability. They have clearly shown that the colloidal stability of the nanoemulsion formulation can be promptly managed by the chemical structure of the interface. The nature of the surfactant, which has been used, is an amphiphilic moiety with uncharged tri-block copolymer Pluronic F68 (3.1×10^{-3} mol/l). They observed that phase separation occurred without coalescence due to the depletion-flocculation mechanism. Destabilization of nanoemulsion happened above certain surfactant concentration, which causes attraction between drops by a depletion mechanism and formation of non-adsorbing micelles. However, they cannot agglomerate due to their surfactant layers, but they are close enough to cause flocculation and creaming [39].

Relatively, Mao et al. in 2009 have not reported any flocculation while preparation of nanoemulsions (β-carotene) due to the low concentration of Tween-80 [40].

Qian et al. 2012 studied the β-carotene O/W nanoemulsions system and its stability by using a globular protein (β-Lacto globulin) for encapsulation. They used a larger particle size of β-lactoglobulin (70–80 nm) to stabilize the nanoemulsions for 15 days of storage [43].

Other than the nature of surfactant and its concentration, pH, emulsifier type and its ionic strength can also have a great impact on the physical and chemical stability of the nanoemulsion i.e., occurrence of flocculation. Salt could possibly reduce the electrostatic repulsion and slightly escalate the hydrophobic attraction among the droplets while storage leading to flocculation.

Takegami et al. in 2008 explored the impact of co-surfactants (sodium palmitate and sucrose palmitate) on a lipid nanoemulsion. This novel lipid nanoemulsion was prepared using soybean oil and phosphatidylcholine (PC) with a mean droplet size of 50 nm. Phosphatidylcholine was responsible for stability enhancement of the nanoemulsion. However; addition of phosphatidylcholine has shown increase in mean droplet size with broadening of the droplet size after 12-month storage as a result of flocculation. Contrastingly, there was no sign of flocculation in sucrose palmitate-lipid nanoemulsion after 12 months of storage. It was concluded that sucrose palmitate has the potential to enhance the storage stability of a nanoemulsion by preventing the flocculation occurrence [38].

Recently, Rao et al. studied the influence of the ratio of digestible to indigestible oil and carrier oil composition on the physical stability of a β-carotene nanoemulsion having a size less than 150 nm in the simulated gastrointestinal conditions. When the β-carotene nanoemulsion was incubated in artificial saliva (pH 7), the particle dimension of the β-carotene nanoemulsion increased (approx. d43 > 1 μm). This happened due to excessive flocculation along with a significant decrease in the negative charge of the nanoemulsions. The alteration in droplet aggregation and charge inside the mouth phase has been accredited to the availability of salts and mucin within the saliva. The salt present in saliva can reduce the zeta potential of droplets due to an electrostatic screening whereas mucin adsorbed on the droplet surfaces and may change the droplet charge. As a result, electrostatic repulsion between the droplets reduces flocculation [27].

The interaction between surfactant and biopolymer shows a flocculation of droplets in the emulsion. Furthermore, the higher concentration of surfactant results in the flocculation of nanoemulsion. Hence, the flocculation can be prevented by optimizing the concentration of surfactant during the preparation of an emulsion. The flocculation is the result of an increase in ionic strength, which is due to the higher salt concentration that reduces the electrostatic repulsion between nanoemulsion droplets. Due to the presence of salts and mucin in the saliva, nanoemulsions became susceptible to flocculation in the mouth phase.

4.3.4 Phase Separation

Phase separation is defined as the separation of two distinct and clear phases from a single homogeneous mixture of two immiscible liquids such as oil and water (nanoemulsions). Phase separation occurs when the driving force for phase separation surpasses the driving force for mixing. Phase separation is considered an unfavorable instability property of nanoemulsions, which can be visually observed. In nanoemulsions two immiscible liquids are compelled to interact with different interfaces; hence, they are immiscible liquids and their interaction energy indulges phase separation. Phase separation reveals unfavorable spontaneous separation of two distinct layers which introduces a specific morphology into the solution that leads to an immiscible blend with astonishing rheological properties. The kinetics which are associated with phase separation would be determined by the rheological behavior of interconnected oil droplets which induces the strength of depletion attraction and gel-like droplet network formation. Additions of polysaccharides at various concentrations are preferred to overcome the phase separation by the formation of a droplet network. Consequently, the types of emulsifiers, the storage period of nanoemulsions and temperature are the major factor which influences the phase separation mechanism in the emulsion.

In 2011, Lee et al. performed a comparative study between conventional emulsions and the nanoemulsions which comprise triacylglycerol oil (corn or fish oil) and WPI as an emulsifier. It was noticed that the mean particle diameter of the emulsifier has risen at the isoelectric point (PI) and enhances phase separation of a different pattern. At pH 4.5 to 5.5 the distinct cream layer was also noticed in the conventional emulsion however, no cream layer was observed in the case of nanoemulsions [46]. That same year Ziani et al., reported the phase separation in thyme oil nanoemulsion with an increase of the mean particle diameter (>7000 nm) after 3 days of storage [25].

A year later in 2012 Ahmed et al., studied the impact of triacylglycerols with various chain lengths on the stability of curcumin nanoemulsions. Triacylglycerols with medium-chain have more resistance toward phase separation at 4°C over 10 days. However, nano-emulsions containing triacylglycerols with short-chain were more susceptible toward phase separation within a few hours. This could be diminished by the addition of 50% of long-chain triacylglycerols into the lipid phase [29].

In the following year 2013, Ghosh et al. studied the impact of the emulsification time and different surfactant concentrations over phase separation stability of a nanoemulsion composed of basil oil. In his experimental design, it was clearly shown that at the same ratio of oil and surfactant (1:1, 1:2, 1:3, and 1:4) with elevated sonication time (15 mins) reduces the droplet diameter and increased the nanoemulsions stability for one month [47]. On the other hand, Kim et al. studied the stability of lemongrass oil nanoemul-sion composed of different concentrations of lemon oil (0.5% to 4.0%) and surfactant (Tween-80). If the oil concentration increased from 0.5 to 4%, the mean droplet size of nanoemulsions also increased from 56 nm to 86 nm which enhances a phase separation at 4% concentration of lemon oil in emulsion. It might have occurred due to a higher concentration of oil and increased value of mean droplet diameter. Consequently, an elevated concentration of oil leads to phase separation and responsible for the instability of nano-emulsions [48]. However, Sugumar et al. reported no phase separation during storage and centrifugation in a eucalyptus oil nanoemulsion (17 nm) due to smaller droplet size [49]. Likewise, Anjali et al. also reported no signs of phase separation in various formulations of neem oil nanoemulsion [42]. All the formulated nanoemulsions were found to be physically stable at room temperature due to the significant role of the steric effect.

Similarly, in 2013, Liang et al. noticed no phase separation of modified starch (β-carotene) nanoemulsions over 30 days of storage. At storage condition of 25°C, the average diameter of nanoemulsions was increased; however, there was no significant change in diameter size at 4°C.This enhanced physical stability of nanoemulsions was attributed to the lower storage temperature and higher dispersed molecular density of modified starch that yielded a thicker and denser layer around the oil droplets [50].

Nanoemulsions, which are prepared with a higher water solubility oil (SCT), are more susceptible to phase separation due to an increased mean droplet size, although it can be diminished by the addition of lower water solubility oil (LCT) into the nano-emulsions mixture. Therefore, various factors have their impact on phase separation and stability of nanoemulsions such as process variables, technique applied, pH, oil and sur-factant ration, sonication time, and storage condition.

4.3.5 Sedimentation

The formation of crystalline lipids and tiny oil droplets are responsible for sedimentation in oil-water or water-oil nanoemulsions due to its wetting properties. The emulsion droplets have more density value than the rest of the liquid molecules and it leads to downward movements of particles and sedimentation. Conventionally, droplets of oil-water emulsions are more inclined toward upward movement such as cream, while on the contrary droplets of water-oil are more inclined toward downwards movement such as sedimentation. Moreover, both are considered as major constrain to the stability of nanoemulsions, which are associated with gravitational force. Emulsions with small droplet sizes have more resistance toward gravitational force with a higher value of Brownian motion and are able to overcome the sedimentation during storage. Apart from this, external factors are also responsible for sedimentation in nanoemulsions such as the composition of emulsion, emulsification techniques, and storage conditions.

In 2010, Teo et al. Investigated the storage stability of nanoemulsions composed of vitamin-E loaded palm oil esters (80–200 nm) along with the different concentrations of Pluronic F68. In two different experimental designs, it was concluded that a formulation with a higher emulsifier concentration was more stable (2.4%) during a storage duration of 4 weeks at 45°C. On the other hand, the same formulation with storage conditions of 45°C for 5 weeks exhibited creaming and sedimentation. However, at storage condition 4°C to 25°C for 4 weeks, none of the formulations has exhibited phase separation [51].

In 2013, Ng et al. Performed a comparative stability study between tocotrienol-loaded palm oil esters containing nanoemulsions and curcumin containing long-chain triacylglycerols nanoemulsions. In the case of tocotrienol nanoemulsions, it exhibited sedimentation. On the other hand, curcumin nanoemulsions do not exhibit any sedimentation because of the low water solubility effect of LCT and MCT as discussed earlier [52].

Those nanoemulsions composed of low water-soluble oil were more stable against sedimentation during storage. Additionally, the rate of sedimentation of a nanoemulsion enhances when the concentration of oil increases.

4.3.6 Creaming

Creaming in nanoemulsion is the gravity-driven phase separation due to the density difference between the two liquids. The creaming process can be quantified in terms of cream layer thickness.

$$C = 100\frac{(V_t - V_s)}{V_t},$$

where Vt (mL) represents the total volume of the sample and Vs (mL) is the volume of the lower phase layer (serum). Therefore, the C value of 100 indicates a stable emulsion.

Ziani et al. observed creaming in thyme oil nanoemulsions due to their larger droplet size (d > 5000 nm). Physical stability of grape marc polyphenols encapsulated nanoemulsion was examined for creaming volume percentage at different storage conditions and it was observed that creaming volume percentage increases with an increase in storage temperature [47].

Similarly, Rao and McClements studied the stability of lemon oil nanoemulsion which was stored at different temperature conditions for 12 hours. It was concluded that when using lower oil concentration then nanoemulsion was highly unstable due to droplet growth and increase in participle size. The nanoemulsion was observed stable at a higher oil concentration with a slight change in particle size. Furthermore, the creamy layer becomes visible at the top with a lower oil concentration while no changes with a higher oil concentration [27].

In the same way, Ahmed et al. investigated stability of curcumin nanoemulsion and concluded that there were no signs of creaming observed with LCT and MCT nanoemulsion while stored up to 24 h. The curcumin nanoemulsion remained homogeneous and yellowish in appearance. Nanoemulsion prepared using SCT were unstable because of Ostwald ripening effects [29].

However, in 2011, Donsi et al. has shown the impact of various emulsifiers on the storage stability of nanoemulsions (74–366 nm) which were based on the essential oil. They concluded that no observation of any instability phenomena occurred such as creaming and phase separation up to four weeks of storage [53]. Furthermore, in another experimental design by Li et al., for a polymethoxyflavones based nanoemulsion same result were observed due to smaller droplet size distributions [46]. Similarly, in 2012, Dev et al. reported the same stability phenomena in EPA and DHA-rich fish oil-based nanoemulsions. Due to its kinetic stability and emulsifier property and storage conditions, creaming has not been observed in fish oil-based nanoemulsions and β-carotene. Consequently, the creaming stability of nanoemulsions can be diminished by increasing the storage temperature concerning the composition of nanoemulsions. Moreover, the stability of nanoemulsions can also be enhanced by varying degree of distillation oil [54].

4.4 PERSPECTIVES OF NANOEMULSIONS STABILITY IN FOOD INDUSTRY

Nanoemulsions possess greater potential against flocculation and gravitational separation rather than conventional emulsions to increase the shelf life of food. A major group of researchers concluded that the stability of a formulation of nanoemulsion, is attributed to small particles size with kinetic stability. The kinetic stability of nanoemulsions can be enhanced by, either managing their particle-size distributions or by the addition of stabilizing agents. Nanoemulsions are optically clear and transparent at lower particle size (<50 nm) hence, used in products such as fortified water and soft drinks. Under conditional circumstances, nanoemulsions can be used in food and beverage products to create novel texture, due to their more suitable particle size and greater number of particles to particle interactions [55]. To increase bioavailability, nanoemulsions can be encapsulated in bioactive agents for instance lipids, vitamins, acidulates, preservatives, colorings, antioxidants, and nutraceuticals [56]. Most of the bioactive compounds are classified under lipophilic, therefore, oil in water mode of nanoemulsion preparation is preferred. In consequence to enhance its solubility, the dispersibility of bioactive agent (lipophilic) substance in aqueous media is increased. Moreover, it also intensifies stability, taste (reduce bitterness and astringency), the texture of food products, and absorption and bioavailability [57]. The mechanism behind the increased bioavailability of bioactive compounds through nanoemulsion system is due to the retention of nanodroplets by mucus layer as well as active and passive transport of nanodroplets through transcellular/paracellular route [58].

4.5 CORRELATIVE MECHANISM OF INSTABILITY

The prevailing opinion in this literature suggested that the instability mechanism of nanoemulsions such as coalescence, flocculation and subsequent creaming are correlated with each other and may occur simultaneously. The instability occurs as a result of increasing in droplet size and eventually leads to coalescence and Ostwald ripening. Oil-water nanoemulsions are stabilized by Brij 30, Tween-80, and Span 80 and with liquid paraffin as a dispersed phase. Similarly, Pluronic F68 and other oils are generally utilized for pharmaceutical and cosmetic applications.

4.6 STRATEGIES TO MINIMIZE THE NANOEMULSION INSTABILITY

For commercial perspective, it is important that nanoemulsion based formulations/ products always physiochemically stable in any conditions like temperature, mechanical forces, and ionic strength throughout their production. The addition of suitable stabilizers, including emulsifiers, weighting agents, texture modifiers and ripening inhibitors can improve the physical stability of nanoemulsions.

4.6.1 Nanoemulsion Stabilizer

A nanoemulsion stabilizer is known as one of the most essential components during the preparation to reward its commercial requirement. A nanoemulsion stabilizer is primarily designed to overcome various challenges such as improved kinetic stability and appropriate composition of oil and water phase. Widely used nanoemulsion stabilizers in the food industry are emulsifiers, weighing agents, texture modifiers, and ripening inhibitors. The physicochemical properties of nanoemulsions are responsible for their stability are their composition, size, physical state, rheology, interfacial composition, charges (electrical), etc. Nanoemulsion stabilizer examples were summarized in Table 4.2.

TABLE 4.2   Summarizes the Stabilizers with an Example to Enhance the Long-Term Stability of Nanoemulsions

Stabilizers	Types	Examples
Emulsifiers	Low-molecular-weight surfactant (LMWS)	Tween series, Span series, phospholipid, glycolipid
	High-molecular-weight emulsifier (HMWE)	Protein and polysaccharide (chitosan, pectin, Pereskia aculeata Miller, gum Arabic (GA), sugar beet pectin (SBP), etc.)
	Mixture of protein and polysaccharide	Sodium caseinate (sc), GA and whey protein hydrolysate (WPH)
Ripening retarders	Long-chain triglycerides	Sunflower oil, grape seed oil, corn oil, palm oil
Texture modifiers	Soluble polymers displaying extended structures (water-soluble polysaccharides and proteins)	Proteins (whey protein isolate, gelatin or soy protein isolate), sugars (high-fructose corn syrup or sucrose), polysaccharides (carrageenan, xanthan, pectin, alginate)
Weighting agents	Hydrophobic substances with a higher density than water liquids nanoemulsions	Ester gum, brominated vegetable oil, sucrose acetate isobutyrate, rosin gum isobutyrate

4.6.2 Choice of Suitable Emulsifier

In simple terms emulsifiers are amphiphilic molecules used to form and stabilize nanoemulsions or any type of emulsions by decreasing their interfacial pressure. Low interfacial pressure facilitates disruption of droplets and formation of a defensive interfacial layer which would not allow droplet accumulations. The impact of emulsion droplet size over its stability, optical properties applicability and quality of emulsion are well known. Emulsifiers are used in almost all products of the food industry such as bakery, ice-cream, confectionery, etc.

4.6.3 Ripening Retarders

These are long-chain triglycerides hydrophobic molecules with extremely low solubility in water and acting as stability enhancers. The ripening retarder's substances are added at the dispersion phase of a nanoemulsion to prevent droplet expansion as a result of Ostwald ripening. Ostwald ripening is a known instability phenomenon where large droplets are co agglomerating due to molecular diffusion of oil between droplets through the continuous phase. Examples of ripening retarders include a long-chain triglyceride, grape seed oil, mineral oil, ester gum sunflower oil, corn oil, palm oil, etc. [54].

4.6.4 Texture Modifiers

Nanoemulsions are additionally used to enhance the texture of food products due to their various characterizations over conventional emulsions. Texture modifiers are either thickening or gelling agent used to attain modified rheological properties. The texture modifier usually inhibits the droplet movement to overcome gravitational separation instability. Water-soluble polysaccharides and proteins are often considered suitable for texture modifiers, for instance, proteins (whey protein isolate, gelatin or soy protein isolate), sugars (high-fructose corn syrup or sucrose) polysaccharides (carrageenan, xanthan, pectin, alginate, etc.) [2, 59–61].

4.6.5 Addition of Weighting Agents

The weighting agent are hydrophobic substances which are commonly employed in the formation O/W nanoemulsions. Weighting agent increases the density until their levels become comparable to that of the aqueous phase. Weighting agents used in the food industry are considered to be highly hydrophobic substances to decrease the possibility of gravitational separation and creaming effect. For instance, Ester gum, brominated vegetable oil, sucrose acetate isobutyrate, or rosin gum isobutyrate [62–63].

4.7 STABILITY RECOMMENDATION/GUIDELINES AS PER INTERNATIONAL CONFERENCE ON HARMONIZATION (ICH)

The integrity and quality of bioactive molecule preparation changes under the influence of certain environmental factors such as light, humidity, and temperature. World Health Organization has been addressing the salient issues as it relates to bioactive stability and storage with particular attention to the developing countries most of which are located in the tropical climate zones where bioactive molecule stability poses more serious problems that may lead to changes in the shelf-life of the food product. Nanoemulsions are isotropic, thermodynamically stable, transparent system of oil, water, and surfactants with a droplet size range of 20–200 nm. Nanoemulsions may become unstable due to the effect of environmental factors on the globular size leading to coalescence, flocculation, and Ostwald ripening.

Accelerated stability studies are carried out to check the effect of the above factors on the overall integrity and quality of the nanoemulsion formulation. Change in viscosity can equally be used as a study tool in determining the stability of nanoemulsions. Any variation in globular size or number, or in the orientation or migration of emulsifier over a period of time may be detected by a change in apparent viscosity. To compare the relative stabilities of a range of similar products, it is often necessary to speed up the process of creaming and coalescence. This can be achieved by employing exaggerating temperature fluctuations to which any product is subjected under normal storage conditions and this is the basis of accelerated stability studies. Change in the droplet size, as well as droplet count, is measured by Transmission Electron Microscope or Scanning Electron Microscope, while the use of Photon Correlation Microscope with Zetta-sizer is essential for Zeta-potential determination.

Accelerated stability studies to be performed on optimized nanoemulsions are in accord with International Conference on Harmonization (ICH) protocols. Replicates of the prepared samples are kept at refrigerator temperature ($4 \pm 0.7°C$), room temperature ($25 \pm 0.5°C/60 \pm 5\%$ RH.) and $40 \pm 0.5°C/75 \pm 5\%$ RH. Samples are taken at 0, 30, 60, and 90 days. The samples are assessed for droplet size, viscosity, pH, conductivity, and refractive index.

Three batches of the optimized nanoemulsion are equally taken in a glass vial and kept at accelerated temperatures of 30, 40, 50, and 60°C at ambient humidity. Samples of it are taken and analyzed.

Shakeel et al. completed the stability of therapeutic products and saw the physical and chemical stability of the dosage unit which are suitable and capable to maintain protection against microbiological contamination. At the start and throughout the intended shelf-life period, the product must be characterized physically, chemically, and microbiologically. Hence, the nanoemulsion formulation was characterized for droplet size, viscosity, and RI up to three months period. Throughout stability studies droplet size, viscosity and RI were evaluated at 4 and 25°C at 0, 1, 2, and 3 months [64].

Affandi et al. studied nanoemulsions of astaxanthin by using high-pressure homogenization under different emulsifying conditions, where particle size was from 120 to 180 nm and astaxanthin stability was achieved during 3 months at 4 and 25°C [65].

4.8 CONCLUSIONS AND FUTURE TRENDS

Nanoemulsions are enriched with unique characteristics such as small size, stability to chemical and physical changes, kinetical stability. This controlled structure facilitates an ideal platform in food applications. The different instability mechanisms of nanoemulsions like flocculation, creaming, sedimentation, coalescence, Ostwald ripening, and phase separation are stated to be influenced by several factors like ionic strength, temperature, solubilization, particle size distribution, particle charge, pH strength, and heat treatment. The primary mechanism associated with the stability of nanoemulsion is by conjugation and complex interplay of van der Waals forces and steric interactions.

Free energy is the amount of energy required to segregating the oil and water phase and is commonly lesser than required for emulsification hence, nanoemulsions are thermodynamically unstable. Nanoemulsion stabilizers are primarily designs to overcome its various challenges such as improved kinetic stability, the appropriate composition of oil and water phase. Widely used nanoemulsion stabilizers in the food industry are emulsifiers, weighing agents, texture modifiers, and ripening inhibitors. The physicochemical properties of nanoemulsions responsible for their stability are composition, size, physical state, rheology, interfacial composition, and charges (electrical). Optimizing the circumstances to enhance the stability of nanoemulsions facilitates the development of high-throughput production and the extensive application in food, beverages, and pharmaceutical.

CONFLICT OF INTEREST

The authors declare no conflict of interest.

REFERENCES

1. Nakajima H, Tomomossa S, Okabe M. First Emulsion Conference 1993. Paris, France.
2. Tadros T, Izquierdo P, Esquena J, Solans C. Formation and stability of nano-emulsions. Advances in Colloid and Interface Science. 2004 May 20;108:303–18.
3. World Health Organization. Assuring food safety and quality: Guidelines for strengthening national food control systems. In Assuring food safety and quality: guidelines for strengthening national food control systems 2003 (pp. 73–73).
4. Salvia-Trujillo L, Martín-Belloso O, McClements DJ. Excipient nanoemulsions for improving oral bioavailability of bioactives. Nanomaterials. 2016 Jan;6(1):17.
5. Bouwmeester H, Dekkers S, Noordam MY, Hagens WI, Bulder AS, De Heer C, Ten Voorde SE, Wijnhoven SW, Marvin HJ, Sips AJ. Review of health safety aspects of nanotechnologies in food production. Regulatory Toxicology and Pharmacology. 2009 Feb 1;53(1):52–62.
6. Abbas S, Hayat K, Karangwa E, Bashari M, Zhang X. An overview of ultrasound-assisted food-grade nanoemulsions. Food Engineering Reviews. 2013 Sep;5(3):139–57.
7. Wooster TJ, Golding M, Sanguansri P. Impact of oil type on nanoemulsion formation and Ostwald ripening stability. Langmuir. 2008 Nov 18;24(22):12758–65.
8. McClements DJ, Rao J. Food-grade nanoemulsions: formulation, fabrication, properties, performance, biological fate, and potential toxicity. Critical Reviews in Food Science and Nutrition. 2011 Mar 25;51(4):285–330.
9. Gurpreet K, Singh SK. Review of nanoemulsion formulation and characterization techniques. Indian Journal of Pharmaceutical Sciences. 2018 Aug 31;80(5):781–9.
10. Shafiq S, Shakeel F, Talegaonkar S, Ahmad FJ, Khar RK, Ali M. Development and bioavailability assessment of ramipril nanoemulsion formulation. European Journal of Pharmaceutics and Biopharmaceutics. 2007 May 1;66(2):227–43.
11. Gupta A, Eral HB, Hatton TA, Doyle PS. Nanoemulsions: formation, properties and applications. Soft Matter. 2016;12(11):2826–41.
12. Aswathanarayan JB, Vittal RR. Nanoemulsions and their potential applications in food industry. Frontiers in Sustainable Food Systems. 2019 Nov 13;3:95.

13. Yuan Y, Gao Y, Mao L, Zhao J. Optimisation of conditions for the preparation of β-carotene nanoemulsions using response surface methodology. Food Chemistry. 2008 Apr 1;107(3):1300–6.
14. Devarajan V, Ravichandran V. Nanoemulsions: as modified drug delivery tool. International Journal of Comprehensive Pharmacy. 2011;4(1):1–6.
15. Bhatt P, Madhav S. A detailed review on nanoemulsion drug delivery system. International Journal of Pharmaceutical Sciences and Research. 2011 Oct 1;2(10):2482.
16. Bouchemal K, Briançon S, Perrier E, Fessi H. Nano-emulsion formulation using spontaneous emulsification: solvent, oil and surfactant optimisation. International Journal of Pharmaceutics. 2004 Aug 6;280(1–2):241–51.
17. Baboota S, Shakeel F, Ahuja A, Ali J, Shafiq S. Design, development and evaluation of novel nanoemulsion formulations for transdermal potential of celecoxib. Acta Pharmaceutica. 2007 Sep 1;57(3):315–32.
18. Wagner JG, Gerard ES, Kaiser DG. The effect of the dosage form on serum levels of indoxole. Clinical Pharmacology & Therapeutics. 1966 Sep;7(5):610–9.
19. Wang L, Li X, Zhang G, Dong J, Eastoe J. Oil-in-water nanoemulsions for pesticide formulations. Journal of Colloid and Interface Science. 2007 Oct 1;314(1):230–5.
20. Wang L, Mutch KJ, Eastoe J, Heenan RK, Dong J. Nanoemulsions prepared by a two-step low-energy process. Langmuir. 2008 Jun 17;24(12):6092–9.
21. Taylor P. Ostwald ripening in emulsions: estimation of solution thermodynamics of the disperse phase. Advances in Colloid and Interface Science. 2003 Dec 1;106(1–3):261–85.
22. Pham BT, Zondanos H, Such CH, Warr GG, Hawkett BS. Miniemulsion polymerization with arrested Ostwald ripening stabilized by amphiphilic RAFT copolymers. Macromolecules. 2010 Oct 12;43(19):7950–7.
23. Durand A, Marie E, Rotureau E, Leonard M, Dellacherie E. Amphiphilic polysaccharides: useful tools for the preparation of nanoparticles with controlled surface characteristics. Langmuir. 2004 Aug 3;20(16):6956–63.
24. Kabalnov A. Ostwald ripening and related phenomena. Journal of Dispersion Science and Technology. 2001 Feb 4;22(1):1–2.
25. Ziani K, Chang Y, McLandsborough L, McClements DJ. Influence of surfactant charge on antimicrobial efficacy of surfactant-stabilized thyme oil nanoemulsions. Journal of Agricultural and Food Chemistry. 2011 Jun 8;59(11):6247–55.
26. Delmas T, Piraux H, Couffin AC, Texier I, Vinet F, Poulin P, Cates ME, Bibette J. How to prepare and stabilize very small nanoemulsions. Langmuir. 2011 Mar 1;27(5):1683–92.
27. Rao J, Decker EA, Xiao H, McClements DJ. Nutraceutical nanoemulsions: influence of carrier oil composition (digestible versus indigestible oil) on β-carotene bioavailability. Journal of the Science of Food and Agriculture. 2013 Oct;93(13):3175–83.
28. Eid AM, Elmarzugi NA, El-Enshasy HA. Preparation and evaluation of olive oil nanoemulsion using sucrose monoester. International Journal of Pharmacy and Pharmaceutical Sciences. 2013;5(Suppl 3):434–40.
29. Ahmed K, Li Y, McClements DJ, Xiao H. Nanoemulsion- and emulsion-based delivery systems for curcumin: encapsulation and release properties. Food Chemistry. 2012 May 15;132(2):799–807.
30. Mohan S, Narsimhan G. Coalescence of protein-stabilized emulsions in a high-pressure homogenizer. Journal of Colloid and Interface Science. 1997 Aug 1;192(1):1–5.
31. Floury J, Desrumaux A, Legrand J. Effect of ultra-high-pressure homogenization on structure and on rheological properties of soy protein-stabilized emulsions. Journal of Food Science. 2002 Nov;67(9):3388–95.

32. Forgiarini A, Esquena J, Gonzalez C, Solans C. Formation of nano-emulsions by low-energy emulsification methods at constant temperature. Langmuir. 2001 Apr 3;17(7):2076–83.

33. Olson DW, White CH, Richter RL. Effect of pressure and fat content on particle sizes in microfluidized milk. Journal of Dairy Science. 2004 Oct 1;87(10):3217–23.

34. Britten M, Giroux HJ. Coalescence index of protein-stabilized emulsions. Journal of Food Science. 1991 May;56(3):792–5.

35. Mahdi Jafari S, He Y, Bhandari B. Nano-emulsion production by sonication and microfluidization—a comparison. International Journal of Food Properties. 2006 Sep 1;9(3):475–85.

36. Jafari SM, He Y, Bhandari B. Optimization of nano-emulsions production by microfluidization. European Food Research and Technology. 2007 Sep;225(5):733–41.

37. Kentish S, Wooster TJ, Ashokkumar M, Balachandran S, Mawson R, Simons L. The use of ultrasonics for nanoemulsion preparation. Innovative Food Science & Emerging Technologies. 2008 Apr 1;9(2):170–5.

38. Takegami S, Kitamura K, Kawada H, Matsumoto Y, Kitade T, Ishida H, Nagata C. Preparation and characterization of a new lipid nano-emulsion containing two cosurfactants, sodium palmitate for droplet size reduction and sucrose palmitate for stability enhancement. Chemical and Pharmaceutical Bulletin. 2008 Aug 1;56(8):1097–102.

39. Wulff-Pérez M, Torcello-Gómez A, Gálvez-Ruíz MJ, Martín-Rodríguez A. Stability of emulsions for parenteral feeding: Preparation and characterization of o/w nanoemulsions with natural oils and Pluronic f68 as surfactant. Food Hydrocolloids. 2009 Jun 1;23(4):1096–102.

40. Mao L, Xu D, Yang J, Yuan F, Gao Y, Zhao J. Effects of small and large molecule emulsifiers on the characteristics of β-carotene nanoemulsions prepared by high pressure homogenization. Food Technology and Biotechnology. 2009 Aug 7;47(3):336–42.

41. Henry JV, Fryer PJ, Frith WJ, Norton IT. The influence of phospholipids and food proteins on the size and stability of model sub-micron emulsions. Food Hydrocolloids. 2010 Jan 1;24(1):66–71.

42. Anjali CH, Sharma Y, Mukherjee A, Chandrasekaran N. Neem oil (Azadirachta indica) nanoemulsion—a potent larvicidal agent against Culex quinquefasciatus. Pest Management Science. 2012 Feb;68(2):158–63.

43. Qian C, Decker EA, Xiao H, McClements DJ. Physical and chemical stability of β-carotene-enriched nanoemulsions: influence of pH, ionic strength, temperature, and emulsifier type. Food Chemistry. 2012 Jun 1;132(3):1221–9.

44. Lips A, Smart C, Willis E. Light scattering studies on a coagulating polystyrene latex. Transactions of the Faraday Society. 1971;67:2979–88.

45. Reerink H, Overbeek JT. The rate of coagulation as a measure of the stability of silver iodide sols. Discussions of the Faraday Society. 1954;18:74–84.

46. Lee SJ, Choi SJ, Li Y, Decker EA, McClements DJ. Protein-stabilized nanoemulsions and emulsions: comparison of physicochemical stability, lipid oxidation, and lipase digestibility. Journal of Agricultural and Food Chemistry. 2011 Jan 12;59(1):415–27.

47. Ghosh V, Mukherjee A, Chandrasekaran N. Ultrasonic emulsification of food-grade nanoemulsion formulation and evaluation of its bactericidal activity. Ultrasonics Sonochemistry. 2013 Jan 1;20(1):338–44.

48. Kim IH, Lee H, Kim JE, Song KB, Lee YS, Chung DS, Min SC. Plum coatings of lemongrass oil-incorporating carnauba wax-based nanoemulsion. Journal of Food Science. 2013 Oct;78(10):E1551–9.

49. Sugumar S, Nirmala J, Ghosh V, Anjali H, Mukherjee A, Chandrasekaran N. Bio-based nanoemulsion formulation, characterization and antibacterial activity against food-borne pathogens. Journal of Basic Microbiology. 2013 Aug;53(8):677–85.

50. Liang R, Shoemaker CF, Yang X, Zhong F, Huang Q. Stability and bioaccessibility of β-carotene in nanoemulsions stabilized by modified starches. Journal of Agricultural and Food Chemistry. 2013 Feb 13;61(6):1249–57.

51. Teo BS, Basri M, Zakaria MR, Salleh AB, Rahman RN, Rahman MB. A potential tocopherol acetate loaded palm oil esters-in-water nanoemulsions for nanocosmeceuticals. Journal of Nanobiotechnology. 2010 Dec;8(1):1–1.

52. Ng SH, Woi PM, Basri M, Ismail Z. Characterization of structural stability of palm oil esters-based nanocosmeceuticals loaded with tocotrienol. Journal of Nanobiotechnology. 2013 Dec;11(1):1–7.

53. Donsì F, Annunziata M, Sessa M, Ferrari G. Nanoencapsulation of essential oils to enhance their antimicrobial activity in foods. LWT—Food Science and Technology. 2011 Nov 1;44(9):1908–14.

54. Silva HD, Cerqueira MÂ, Vicente AA. Nanoemulsions for food applications: development and characterization. Food and Bioprocess Technology. 2012 Apr;5(3):854–67.

55. Borrin TR, Georges EL, Moraes IC, Pinho SC. Curcumin-loaded nanoemulsions produced by the emulsion inversion point (EIP) method: an evaluation of process parameters and physico-chemical stability. Journal of Food Engineering. 2016 Jan 1; 169:1–9.

56. Jain A, Ranjan S, Dasgupta N, Ramalingam C. Nanomaterials in food and agriculture: an overview on their safety concerns and regulatory issues. Critical Reviews in Food Science and Nutrition. 2018 Jan 22;58(2):297–317.

57. Maier C, Zeeb B, Weiss J. Investigations into aggregate formation with oppositely charged oil-in-water emulsions at different pH values. Colloids and Surfaces B: Biointerfaces. 2014 May 1;117:368–75.

58. Odriozola-Serrano I, Oms-Oliu G, Martín-Belloso O. Nanoemulsion-based delivery systems to improve functionality of lipophilic components. Frontiers in Nutrition. 2014 Dec 5;1:24.

59. Walstra P. Principles of emulsion formation. Chemical Engineering Science. 1993 Jan 1;48(2):333–49.

60. Esmaeili A, Gholami M. Optimization and preparation of nanocapsules for food applications using two methodologies. Food Chemistry. 2015 Jul 15;179:26–34.

61. Aggarwal G, Nagpal M. Pharmaceutical polymer gels in drug delivery. In Polymer Gels 2018 (pp. 249–284). Springer, Singapore.

62. Artiga-Artigas M, Acevedo-Fani A, Martín-Belloso O. Effect of sodium alginate incorporation procedure on the physicochemical properties of nanoemulsions. Food Hydrocolloids. 2017 Sep 1;70:191–200.

63. Thomas L, Zakir F, Mirza MA, Anwer MK, Ahmad FJ, Iqbal Z. Development of Curcumin loaded chitosan polymer based nanoemulsion gel: in vitro, ex vivo evaluation and in vivo wound healing studies. International Journal of Biological Macromolecules. 2017 Aug 1;101:569–79.

64. Shakeel F, Baboota S, Ahuja A, Ali J, Shafiq S. Accelerated stability testing of celecoxib nanoemulsion containing cremophor-EL. African Journal of Pharmacy and Pharmacology. 2008 Oct 31;2(8):179–83.

65. Affandi MM, Julianto T, Majeed A. Development and stability evaluation of astaxanthin nanoemulsion. Asian Journal of Pharmaceutical and Clinical Research. 2011;4(1):142–8.

CHAPTER 5

Fate of Nanoemulsions in Biological Systems

Musarrat Husain Warsi,[1] Usama Ahmad,[2]
Abdul Qadir,[3] Abdul Muheem,[3] Javed Ahmad[4]

[1]Department of Pharmaceutics and Industrial Pharmacy,
College of Pharmacy, Taif University, Taif-Al-Haweiah,
Kingdom of Saudi Arabia
[2]Department of Pharmaceutics, Faculty of Pharmacy,
Integral University, Lucknow, India
[3]Department of Pharmaceutics, School of Pharmaceutical
Education & Research, Jamia Hamdard, New Delhi, India
[4]Department of Pharmaceutics, College of Pharmacy,
Najran University, Najran, Kingdom of Saudi Arabia

CONTENTS

5.1 INTRODUCTION

Nanoemulsions have advantages over emulsions due to their wide usage inside the food and beverage industry. The advantages are mainly due to better stability of nanoemulsions especially in terms of particle aggregation and settling due to gravitational influence [1].

DOI: 10.1201/9781003121121-5

The droplets in nanoemulsions have a weak tendency to scatter light and due to this, they are frequently used in commodities such as fortified soft drinks and water that have to be optically clear or somewhat turbid, some soup categories, sauces, and dips [2–4]. Nanoemulsions also have rheological benefits over emulsions like the formation of exceedingly viscous or gel type systems make them more suitable to be utilized in the food industry [1, 2, 5]. One more interesting and widely used application of nanoemulsions in the nutraceutical industry is the enhancement of the bioavailability of lipophilic bioactive components/nutraceuticals [6].

Nanoemulsions are broadly classified into two broad categories based on the spatial arrangement of their two major components, namely oil and water. While the dispersion of oil droplets in water droplets is referred to as oil in water (O/W) nanoemulsion and when there is the dispersion of the water droplets inside the oil droplets it is referred to as water in oil (W/O) nanoemulsion. These two liquid phases are also referred to as the external phase (one which is dispersed) and the internal phase (one which is continuous and surrounding dispersed phase) [7]. Nanoemulsions like conventional emulsions contain two immiscible liquids stabilized by a surfactant molecule [8–11]. However, the mean droplet diameters of these two systems are different. An emulsion usually has mean radii of 100 nm to 100 mm and due to its larger droplet size, they are thermodynamically unstable systems [12].

The stability of nanoemulsions is attributed to their smaller droplet size and spherical shape which produces high interfacial tension leading to an increase in Laplace pressure which facilitates the limitation of the oil-water interfacial surface. Nanoemulsions must not be confused with microemulsions as these have different molecular geometry and are spherical, ellipsoid, or worm-like in silhouettes that produce lower interfacial tension. The distinction between these two systems is very important to proceed clearly with the best approach in optimization of nutraceutical characterization parameters, stability assessment, and overall performance of the system [12]. In the remaining section of the present chapter, the main emphasis will be on nanoemulsions, but the provided content could also be appropriate for conventional type of emulsions and microemulsions.

The stability profile of nanoemulsions can be enhanced by controlling the components used in their formation that is oil and water phases and surfactants as these govern the microstructure of nanoemulsion. Surface active agents or surfactants have a crucial role in stabilizing the emulsions. These are also stated as emulsifiers, wetting agents, or ripening retarders [9, 13, 14]. The government of different countries has laid down certain regulations on the use of surfactants in the food industry correspondingly. Apart from government regulations, the selection of stabilizers is based on cost concerns and practical aspects like easiness during use, the trustworthiness of resource, and matrix compilation. The other important component which plays a vital role in the formation and stability of nanoemulsion is the selection of the oil phase. The different aspects of oil phases like their polarity, interfacial tension, hydrophilicity, and flow behavior will determine the formulation formation region and its overall stability. The method used to fabricate the formulation also plays a key part in establishing the stability of nanoemulsion as some of them produce very fine droplets leading to better stability profiles [9, 12].

Research studies on monitoring the fate of nanoemulsions within the human body has gained considerable momentum nowadays [15–17]. The information provided in this chapter could be utilized for getting insights of bioavailability enhancement and targeted delivery of various lipophilic nutraceuticals, or vitamins [18–20]. Numerous approaches could be employed to fabricate functional food products that are designed to reduce calories, or control hormonal responses and improve overall human health [9, 20].

5.2 FORMULATION AND CHARACTERISTIC FEATURES OF NANOEMULSION

Nanoemulsions can be manufactured by employing various techniques, classically categorized as high-energy or low-energy approach [21–24]. High-energy strategies employ mechanical equipment proficient of creating strong unruly forces that blend and interrupt the phases of water and oil foremost to the formation of tiny droplets of oil [12, 23, 25]. When the different circumstances are modified, e.g., temperature or composition, low energy attempts depend on the instinctive creation of minuscule oily droplets surrounded by mixed oil–water-emulsifier schemes. In accordance with the methodology applied, the functional characteristics, configuration of the system and the size of the droplet can be modulated. A schematic diagram of different methods is shown in Figure 5.1.

The physicochemical attributes required for a specific nanoemulsion delivery system differ for the specific requirements of the final application. Nanoemulsions can be constructed with distinct optical, rheological, and stability characteristics by regulating their compositions.

5.2.1 Optical Properties

In some commercial applications, it is essential to control the visual appearance of a nanoemulsion based carrier system [26–28]. Such as, the food industry would like to integrate lipophilic bioactive components (like nutraceuticals and vitamins) into optically translucent beverages (e.g., fortified waters and soft drinks). Therefore, it is vital

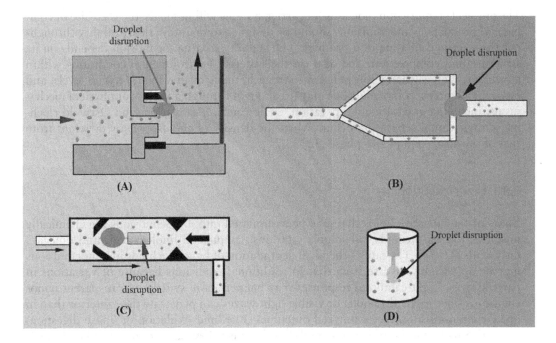

FIGURE 5.1   Schematic diagram of different techniques used in formation of nanoemulsion: (A) high pressure valve homogenizer, (B) microfluidizer, (C) ultrasonic jet homogenizer, and (D) ultrasonic probe homogenizer.

for carrier systems that they do not increase turbidity [26, 28]. On the other side, if the bioactive elements are to be incorporated into an optically opaque product (such as a fortified yogurt or cream) then this attribute is less important. Nanoemulsions that appear transparent, turbid, or opaque depending on their compositions and structures may be created. The optical properties of nanoemulsions are governed by three main characteristics: droplet size; droplet concentration; and refractive index contrast [26–28]. When the droplet radius is below 40 nm, nanoemulsions appear to be transparent, translucent when the droplet radius is between around 40 and 60 nm, and turbid or obscure when the droplet size is above 60 nm.

5.2.2 Rheological Properties

The rheological properties of nanoemulsion based delivery systems are very crucial in various commercial tenders. The nanoemulsions used to deliver lipophilic bioactive components in fortified waters or soft drinks should not greatly upsurge the viscosity of the final product. On the other hand, it may be suitable to have an extremely viscous or gel-like substance for some efficient food products (such as fortified yogurts, desserts, dressings, and sauces) and creams used for topical applications of bioactive components. The rheology of nanoemulsions may vary from low viscosity fluids to gel-like materials varying on their formation and structure [28–31].

5.2.3 Physical Stability

The physical stability of nanoemulsion carrier systems will influence the shelf-life of commercial products. Nanoemulsions should be designed to maintain their stability throughout the expected lifetime of a product. The stability of a nanoemulsion depends on its structure and composition, and also on the different environmental conditions within a product, such as changes in pH, ionic strength, temperature, light, oxygen levels, and mechanical forces. Nanoemulsions may breakdown through a variety of physical mechanisms, including gravitational separation, coalescence, flocculation, and Ostwald ripening. Gravitational separation occurs when the density of the droplets is different from that of the adjacent aqueous phase [32, 33].

5.2.4 Chemical Stability

Many of the bioactive lipids that may be assimilated into nanoemulsions are chemically unstable, like carotenoids, oil-soluble vitamins, conjugated linoleic acid, and omega-3 fatty acids [32–34]. The rate of chemical degradation of bioactive lipids in nanoemulsions may be appreciably diverse from that in traditional emulsions because of variations in particle sizes. Light-catalyzed responses may happen more swiftly in transparent nanoemulsions since more ultraviolet or visible light waves can penetrate their interior than in opaque emulsions. Surface-catalyzed outcomes (like lipid oxidation or lipase digestion) may arise more promptly in nanoemulsions because they have higher interfacial areas than conventional emulsions. It is therefore important to carefully design nanoemulsion-based delivery systems to hinder any prospective chemical degradation reactions that might occur. Different approaches have been established to retard the degradation of bioactive lipophilic components, including reducing oxygen and light levels, maintaining low storage temperatures, incorporating appropriate antioxidants, chelating transition metal catalysts, and controlling droplet interfacial properties [33, 35, 36].

5.3 FATE OF NANOEMULSIONS IN IN VIVO SYSTEM

It is essential to be aware of the biological fate of the administered encapsulated nutraceuticals in nanoemulsions. The physicochemical properties of materials change as we move toward the nanometer scale and this will affect the pharmacokinetic profile, i.e., the biological fate of the component and delivery system as a whole. The fate of nanoemulsions within the biological systems usually depend upon their physicochemical properties which include particle concentration, size, shape, electrical charge, and interfacial characteristics. However, many more investigations are needed to identify the release pattern of encapsulated components within the biological system [37].

After administration in the body, nanoemulsions are subjected to the diverse physiochemical physiological system after ingestion as they pass through the gastrointestinal tract (GIT) [38, 39]. A basic understanding of all the possible environments throughout the GIT is essential to address the biological fate of nanoemulsion [38, 39]. Throughout GIT the pH, ionic compositions, enzyme functions, biopolymer levels, biosurfactant levels, and flow/stress patterns, the distinct GIT regions vary. All these environments can greatly cause significant changes in the physicochemical properties of nanoemulsions. This must be taken into consideration while designing an effective delivery system [20].

5.3.1 Mouth

The majority of the drugs/nutraceuticals are administered orally and the first area that comes in contact is the mouth or oral cavity [38–40]. It becomes necessary to understand this area as the droplet of the formulation administered might be altered depending upon the time it remains in the oral cavity which might range from few seconds to minutes. Due to the process of mastication, i.e., chewing and mixing with saliva, both structural and compositional changes may occur in nanoemulsion which might affect the fate of formulation. Human saliva contains multiple components in its composition. It possesses different biopolymers (such as mucin), salts (such as sodium and calcium), enzymes (such as amylase), and has a reasonably neutral pH [41–43]. Due to flocculation mucin may encourage the accumulation of nanoemulsion within the oral cavity. This structural modification of nanoemulsion might alter the overall fate and efficacy of the administered nutraceutical. The composition of the bolus coming to the stomach from the mouth might play an important role during lipid digestion.

5.3.2 Stomach

The masticated food passes to the esophagus and reaches into the gastric site. The nanoemulsion is further treated in the stomach and is converted into such a form that it may be absorbed efficiently in the small intestine. Inside the stomach, the gastric fluids have about 1–3 pH with an ionic strength of about 100 mM under fasting conditions and involve multiple digestive enzymes such as gastric proteases, lipases, and surface-active materials like proteins and secretory phospholipids [41]. Immediately after the food reaches the inside stomach, the pH of the stomach increases dramatically and then gradually decreases as the food moves toward the small intestine. The pH-time pattern of the stomach alters the nature of food ingested [41].

Apart from the pH of the stomach, digestive enzymes also impart a vital part in determining the fate of nanoemulsions. Gastric pepsin and lipase are the most significant digestive enzymes found inside the stomach [44]. Protein hydrolysis by pepsin can modify the fate of both free as well as entrapped lipid droplets. If these droplets are protein-coated in a nanoemulsion, then pepsin hydrolysis of the adsorbed protein layer will cause instability as the droplets are no longer controlled from accumulation. Although if the lipid globules are stuck inside a protein microgel, hydrolysis of the protein matrix will result in the release of the lipid droplets. The intensity and extent of protein hydrolysis of pepsin under gastric conditions depends on the protein type, with flexible open proteins (like casein) digesting more quickly than compact globular proteins (like innate β-lactoglobulin) [45].

The stability of nanoemulsions inside the stomach is governed by many more factors. The existence of polysaccharides such as chitosan, pectin, methylcellulose, and fucoidan can stimulate the accumulation of protein-coated lipids inside the stomach and modify their stability. Due to polysaccharides capability to encourage bridge or depletion flocculation droplets may become unstable or they may become more stable if polysaccharide forms a protective layer around them [46, 47]. Therefore, to assess the overall fate of nanoemulsion in GIT the amount of polysaccharide must be carefully taken into account.

5.3.3 Small Intestine

The processed food (from the stomach) which is now in the fluid form (chyme) passes across the pylorus sphincter and move into the small intestine where maximum digestion and absorption of macro and micronutrients takes place [39]. In this region, multiple digestive enzymes and salts are at work which promotes digestion, haulage and assimilation of nutrients. Phospholipids, bile salts, digestive enzymes, and coenzymes (e.g., lipases, phospholipases, colipases, amylases, proteases), bicarbonate salts, and other minerals are present in the fluids in the small intestine. Maximum lipid digestion takes place inside the duodenum, the proximal area of the small intestine. The acidic chyme combines with basic intestinal secretions and makes the complex's pH to neutral. Pancreatic enzymes work best at this pH [20].

In the small intestine, there are two types of factors that alter digestion, solubilization, and absorption of lipid globules. These include (i) the quality of the lipid ingested and (ii) the characteristics of the intestinal fluids. Here the focus will remain mainly on the effect of lipid digestion as they affect solubilization of free fatty acids (FFAs) and monoacylglycerol (MAGs) by enabling lipase to access the corresponding triacylglycerol (TAGs) [20]. In comparison, high multivalent cation levels can hinder lipid ingestion by facilitating droplet flocculation, thus plummeting lipase accessibility to the surfaces of the lipidic droplet [20, 47]. The gelation of certain forms of ionic biopolymers (such as alginate) that form hydrogels and capturing of lipid droplets is facilitated, multivalent ions thereby obstructing lipid digestion. The lipase is unable to have access to these captured lipid droplets. The fate of nanoemulsion inside the small intestine greatly depends upon the quality of food ingested and the overall health status of an individual [48].

5.3.4 Colon

The ultimate destination of any administered food in the GIT is the large intestine. This region can be divided into cecum, colon, and rectum. The colon has a significant role in inflaming the swallowed dietary fibers that have traveled throughout the GIT [49]. Most of the nanoemulsions components are consumed throughout the GIT before reaching the large intestine. It is necessary to ensure that the components making up to colon does not cause any harmful effect in the region. Surplus maltodextrin, for instance, exhibits inhibition of cellular antibacterial activities and disrupt intestinal antimicrobial protection processes in the colon. Further, unabsorbed iron may rise pathogen accumulation in infants and trigger inflammation in the intestinal. [50, 51].

5.3.5 Intestinal Lymphatic Transport

Lipidic carrier system and the existence of food collectively augment the oral absorption of nutraceuticals that are lipophilic in nature. The physiological process for the digestion and absorption of such lipid via the lymphatic system is very pertinent for this augmented nutraceutical bioavailability. Such lipidic carriers may enhance the lymphatic haulage of hydrophobic molecules by promoting the fabrication of the chylomicron.

The intestinal lymphatic transport system has been established as a promising nutraceutical delivery pathway. The anatomy of the mesenteric lymphatic system is unique and allows the transport of ingested bioactives bypassing the first-pass effect. Subsequent to administration of lipids and several lipophilic bioactives orally, they get distributed inside the intestinal enterocyte, and are associated with secreted enterocyte lipoproteins. Such an intestinal lymphatic transport is recognized as an absorptive pathway. The bioactive compound is conjugated to chylomicron and is released into the lymphatic circulation, rather than site circulation, from the enterocyte. Thus, the metabolically active liver is avoided but inevitably reverts to the systemic circulation. This advantage enables the lymphatic system to play a key role in the bioavailability enhancement of certain bioactive compounds. For this mode of transport, the biological responses of lipid digestion and absorption are important. Lipid vehicles by stimulating the development of chylomicrons, improve the lymphatic transport of lipophilic materials. In combination with the triglyceride center of chylomicrons, lipophilic nutraceuticals join the lymphatic system. Besides lipid digestion, a bioactive compounds' physicochemical characteristics also determine its in-vivo fate. The bioactive compound loading per chylomicron is affected by its partition coefficient and solubility in the triglyceride. In 1986, Charman and Stella suggested that bioactive molecules for lymphatic transport should have a log P > 5, given the discrepancy between intestinal blood vs lymph blood flow (500:1) and the reality that chylomicron makes up about 1% of lymph [51, 52]. Liu et al. demonstrated the significance of the physiology of lipid digestion in promoting lymphatic transportation. After administering soybean emulsion stabilized by milk fat globule membrane, 19.2% of total vitamin D3 dose reached into the lymph. While, by administering this emulsion system with pancreatic lipase and bile salt, the amount of lymphatic transport was increased to 20.4% [53]. The real impact of the dispersed phase of the vehicle on the degree of lymphatic delivery of the bioactive molecule was deeply investigated [54]. The systems used were developed to predict the physicochemical features indicative of the end phase of lipid digestion and included lipid solution (oleic acid mixture and glycerol monooleate), stabilized emulsion polysorbate-80, and mixed micellar system polysorbate-80. After intraduodenal administration, an increased lymphatic transport was observed, which buoyed the theory that such nutraceutical formulations (dispersed systems) may intensify lymphatic transport. Such improved lymphatic transport has been advocated as a prospective mechanism of increased bioavailability [55].

5.4 PRACTICAL APPLICATIONS OF NANOEMULSIONS IN FOOD TECHNOLOGY

Research into the production and use of food-grade nanoemulsions has recently increased due to the growing demand for effective delivery systems for the packaging, protection, and release of active food components. Indeed, a recent report ("Food Encapsulation: A Global Strategic Business Report," Global Industry Analysts, Inc., San Jose, CA) estimated that the global food supply market would be close to $40 billion by 2015. For the rest of this section, we discuss an outline of the prospective of nanoemulsions in the food and beverage industries.

Numerous studies recommend that nanoemulsions may be an effective way to synthesize and deliver antimicrobial agents [56, 57]. Antimicrobial nanoemulsions are designed to eliminate contamination of food packaging and use in various food environments [56]. Nanoemulsions based on non-ionic surfactants, soybean oil, and tributyl phosphate, have been reported to be very effective in fighting various foodborne illnesses, including bacteria, viruses, fungi, and seeds [57]. However, they are often more effective in fighting Gram-positive bacteria than Gram-negative bacteria, which are said to be influenced by the cell-wall lipopolysaccharide surrounding Gram-negative bacteria. The antimicrobial nanoemulsions methods that inhibit bacterial growth depend on the nature of the antimicrobial compounds (e.g., essential oils, proteins, surfactants), and the type of nanoemulsion droplets itself (e.g., size, charge, composition). Nanoemulsions can be synthesized that contain various types of antimicrobial agents that can work in harmony. Nanoemulsion particles may be designed to deliver one or more antimicrobial compounds after contact with bacterial sites, the interaction between droplets and microorganisms can be enhanced by regulating colloidal contact in the system [7, 58].

5.5 CONCLUSION AND FUTURE PERSPECTIVE

Nanoemulsions have been widely used in the solubility and bioavailability enhancement of numerous bioactive compounds and now their advantages have been extended to nutraceuticals. The food sector is expected to attract more attention in the immediate future. Nanoemulsions have numerous advantages over emulsions and are utilized for encapsulating bioactive lipophilic components due to their good physical stability, high optical clarity, and aptitude to enhance the bioavailability of captured components. These advantages make them a suitable choice for industries to encapsulate nutraceuticals for commercial application.

Most of the studies on the fate of nanoemulsions are being carried out using simulated gastrointestinal models but it needs to be extended to animal and human studies. The impact of particle size, charge, and composition on the bioaccessibility, absorption, and metabolism of encapsulated lipophilic components are comprehensively required to be investigated for establishing the clinical relevance and safety profile of encapsulated nutraceuticals [28, 58].

CONFLICT OF INTEREST

The authors declare no conflict of interest.

REFERENCES

1. Tadros, T., Izquierdo, R., Esquena, J., and Solans, C. (2004). Formation and stability of nano-emulsions. Advances in Colloid and Interface Science, 108–109:303–318.
2. Mason, T.G., Wilking, J.N., Meleson, K., Chang, C.B., and Graves, S.M. (2006). Nanoemulsions: Formation, structure, and physical properties. Journal of Physics-Condensed Matter, 18(41):R635–R666.
3. Velikov, K.P., and Pelan, E. (2008). Colloidal delivery systems for micronutrients and nutraceuticals. Soft Matter, 4(10):1964–1980.
4. Wooster, T., Golding, M., and Sanguansri, P. (2008). Impact of oil type on nano-emulsion formation and Ostwald ripening stability. Langmuir, 24(22):12758–12765.
5. Sonneville-Aubrun, O., Simonnet, J.T., and F. L'Alloret (2004). Nanoemulsions: A new vehicle for skincare products. Advances in Colloid and Interface Science, 108:145–149.
6. Acosta, E. (2009). Bioavailability of nanoparticles in nutrient and nutraceutical delivery. Current Opinion in Colloid & Interface Science, 14(1):3–15.
7. McClements, D. J., and Rao, J. (2011). Food-grade nanoemulsions: formulation, fabrication, properties, performance, biological fate, and potential toxicity. Critical Reviews in Food Science and Nutrition, 51(4):285–330, DOI: 10.1080/10408398.2011.559558.
8. Friberg, S., Larsson, K., and Sjoblom, J. (2004). Food Emulsions. Marcel Dekker, New York.
9. McClements, D. J. (2005). Food Emulsions: Principles, Practice, and Techniques. CRC Press, Boca Raton.
10. Dickinson, E. (1992). Introduction to Food Colloids. Royal Society of Chemistry, Cambridge, UK.
11. Dickinson, E., and Stainsby, G. (1982). Colloids in Foods. London: Applied Science.
12. McClements, D. J. (2011). Edible nanoemulsions: fabrication, properties, and functional performance. The Royal Society of Chemistry, 7:2297–2316. DOI: 10.1039/c0sm00549e.
13. Kabalnov, A. S., and Shchukin, E. D. (1992). Advances in Colloid & Interface Science, 38:69–97.
14. Capek, I. (2004). Advances in Colloid & Interface Science, 107:125–155.
15. McClements, D.J., and Li, Y. (2010). Review of in vitro digestion models for rapid screening of emulsion-based systems. Journal of Functional Food, 1:32–59.
16. Cerqueira, M.A., Pinheiro, A.C., Silva, H.D., Ramos, P.E., Azevedo, M.A., Flores-Lopez, M.L., Rivera, M.C., Bourbon, A.I., Ramos, O.L., and Vicente, A.A. (2014). Design of bio-nanosystems for oral delivery of functional compounds. Food Engineering Reviews, 6:1–19.
17. Mei, L., Zhang, Z.P., Zhao, L.Y., Huang, L.Q., Yang, X.L., Tang, J.T., and Feng, S.S. (2013). Pharmaceutical nanotechnology for oral delivery of anticancer drugs. Advanced Drug Delivery Reviews, 65:880–890.
18. van Aken, G.A. (2010). Relating food emulsion structure and composition to the way it is processed in the gastrointestinal tract and physiological responses: what are the opportunities? Food Biophysics, 5:258–283.

19. McClements, D.J. (2012). Advances in fabrication of emulsions with enhanced functionality using structural design principles. Current Opinion in Colloid & Interface Science, 17:235–245.
20. Zhang, R., and McClements, D. J. (2018). Characterization of gastrointestinal fate of nanoemulsions. Nanoemulsions, 577–602, https://doi.org/10.1016/B978-0-12-811838-2.00018-7.
21. Acosta, E. (2009). Current Opinion in Colloid & Interface Science, 14:3–15.
22. Tadros, T., Izquierdo, R., Esquena, J., and Solans, C. (2004). Advances in Colloid & Interface Science, 108–109:303–318.
23. Leong, T. S. H., Wooster, T. J., Kentish, S. E., and Ashokkumar, M. (2009). Ultrasonic Sonochemistry, 16:721–727.
24. Pouton, C. W., and Porter, C. J. H. (2008). Formulation of lipid-based delivery systems for oral administration: Materials, methods and strategies. Advanced Drug Delivery Review, 60 (6):625–637.
25. Gutierrez, J. M., Gonzalez, C., Maestro, A., Sole, I., Pey, C. M., and Nolla, J. (2008). Nano-emulsions: New applications and optimization of their preparation, Current Opinion in Colloid & Interface Science, 13 (4):245–251.
26. Velikov, K. P., and Pelan, E. (2008). Colloidal delivery systems for micronutrients and nutraceuticals. Soft Matter, 4(10):1964–1980.
27. McClements, D. J. (2002). Theoretical prediction of emulsion color. Advances in Colloid & Interface Science, 97(1–3):63–89.
28. McClements, D. J. (2013). Nanoemulsion-based oral delivery systems for lipophilic bioactive components: nutraceuticals and pharmaceuticals. Therapeutic Delivery, 4(7):841–857.
29. Mason, T. G., Wilking, J. N., Meleson, K., Chang, C. B., and Graves, S. M. (2006). Nanoemulsions: formation, structure, and physical properties. Journal of Physics: Condensed Matter, 18(41):R635–R666.
30. Tadros, T., Izquierdo, P., Esquena, J., and Solans, C. (2004). Formation and stability of nano-emulsions. Advances in Colloid & Interface Science, 108–109(0):303–318.
31. McClements, D. J. (2011). Edible nanoemulsions: fabrication, properties, and functional performance. Soft Matter, 7(6):2297–2316.
32. Lee, S. J., Choi, S. J., Li, Y., Decker, E. A., and McClements, D. J. (2011). Protein-stabilized nanoemulsions and emulsions: comparison of physicochemical stability, lipid oxidation, and lipase digestibility. Journal of Agriculture and Food Chemistry, 59(1):415–427.
33. Qian, C., Decker, E. A., Xiao, H., and McClements, D. J. (2012). Physical and chemical stability of betacarotene-enriched nanoemulsions: influence of pH, ionic strength, temperature, and emulsifier type. Food Chemistry, 132(3): 1221–1229.
34. Belhaj, N., Arab-Tehrany, E., and Linder, M. (2010). Oxidative kinetics of salmon oil in bulk and in nanoemulsion stabilized by marine lecithin. Process Biochemistry, 45(2):187–195.
35. McClements, D. J., and Decker, E. A. (2000). Lipid oxidation in oil-in-water emulsions: impact of molecular environment on chemical reactions in heterogeneous food systems. Journal of Food Science, 65(8):1270–1282.
36. Boon, C. S., McClements, D. J., Weiss, J., and Decker, E. A. (2010). Factors influencing the chemical stability of carotenoids in foods. Critical Reviews in Food Science and Nutrition, 50(6):515–532.
37. McClements, D. J. and Xiao, H. (2012). Potential biological fate of ingested nano-emulsions: influence of particle characteristics. Journal of Functional Foods, 3:202–220.

38. McClements, D. J., Decker, E. A., Park, Y., and Weiss, J. (2009). Structural design principles for delivery of bioactive components in nutraceuticals and functional foods. Critical Reviews in Food Science and Nutrition, 49:577–606.

39. Singh, H., Ye, A., and Horne, D. (2009). Structuring food emulsions in the gastrointestinal tract to modify lipid digestion. Progress in Lipid Research, 48:92–100.

40. McClements, D. J., Decker, E. A., and Park, Y. (2008). Controlling lipid bioavailability through physicochemical and structural approaches. Critical Reviews in Food Science and Nutrition. 49:48–67.

41. Kalantzi, L., Goumas, K., Kalioras, V., Abrahamsson, B., Dressman, J. B., and Reppas, C. (2006). Characterization of the human upper gastrointestinal contents under conditions simulating bioavailability/bioequivalence studies. Pharm. Res. 23: 165–176.

42. Amado, F. M. L., Vitorino, R. M. P., Domingues, P. M. D. N., Lobo, M. J. C., and Duarte, J. A. R. (2005). Analysis of the human saliva proteome. Expert Reviews in Proteomics, 2:521–539.

43. Vitorino, R., Lobo, M. J. C., Ferrer-Correira, A. J., Dubin, J. R., Tomer, K. B., Domingues, P. M., and Amado, F. M. L. (2004b). Identification of human whole saliva protein components using proteomics. Proteomics, 4:1109–1115.

44. Sams, L., Paume, J., Giallo, J., and Carriere, F. (2016). Relevant pH and lipase for in vitro models of gastric digestion. Journal of Functional Food, 7:30–45.

45. Mandalari, G., Adel-Patient, K., Barkholt, V., Baro, C., Bennett, L., Bublin, M., Gaier, S., Graser, G., Ladics, G. S., and Mierzejewska, D. (2009). In vitro digestibility of B-casein and B-lactoglobulin under simulated human gastric and duodenal conditions: a multi-laboratory evaluation. Regulatory, Toxicology and Pharmacology, 55:372–381.

46. Zhang, R., Zhang, Z., Zhang, H., Decker, E. A., and McClements, D. J. (2015a). Influence of emulsifier type on gastrointestinal fate of oil-in-water emulsions containing anionic dietary fiber (pectin). Food Hydrocolloids, 45:175–185.

47. McClements, D.J. (2015). Food Emulsions: Principles, Practices, And Techniques. CRC Press, Boca Raton, FL.

48. Tso, P., and Crissinger, K. (20000. Overview of digestion and absorption. In: Biochemical and Physiological Aspects of Human Nutrition. W.B. Saunders, Philadelphia, PA, pp. 75–90.

49. Basit, A.W. (2005). Advances in colonic drug delivery. Drugs, 65:1991–2007.

50. Nickerson, K. P., Homer, C. R., Kessler, S. P., Dixon, L. J., Kabi, A., Gordon, I. O., Johnson, E. E., Carol, A., and Mcdonald, C. (2014). The dietary polysaccharide maltodextrin promotes salmonella survival and mucosal colonization in mice. PLoS One, 9 (7): e101789.

51. Nickerson, K. P., Chanin, R., and Mcdonald, C. (2015). Deregulation of intestinal anti-microbial defense by the dietary additive, maltodextrin. Gut Microbes, 6:78–83.

52. Charman, W. N. A., and Stella, V. J. (1986). Estimating the maximal potential for intestinal lymphatic transport of lipophilic drug molecules. International Journal of Pharmaceutics, 34(1-2):175–178.

53. Liu, H.-X., Adachi, I., Horikoshi, I., and Ueno, M. (1995). Mechanism of promotion of lymphatic drug absorption by milk fat globule membrane. International Journal of Pharmacy, 118:55–64.

54. Porter, C. J. H., Charman, S. A., and Charman, W. N. (1996a). Lymphatic transport of halofantrine in the triple-cannulated anesthetized rat model: effect of lipid vehicle dispersion. Journal of Pharmacy Science, 85:351–356.

55. Kommuru, T. R., Gurley, B., Khan, M. A., and Reddy, I.K. (2001). Self-emulsifying drug delivery systems (SEDDS) of coenzyme Q_{10}: formulation development and bio-availability assessment. International Journal of Pharmacy, 212:133–246.
56. Hamouda, T., Hayes, M. M., Cao, Z. Y., Tonda, R., Johnson, K., Wright, D. C., Brisker, J., and Baker, J.R. (1999). A novel surfactant nanoemulsion with broad-spectrum sporicidal activity against bacillus species. Journal of Infectious Diseases, 180(6):1939–1949.
57. Hamouda, T., Myc, A., Donovan, B., Shih, A. Y., Reuter, J. D., and Baker, J. R. (2001). A novel surfactant nanoemulsion with a unique non-irritant topical anti-microbial activity against bacteria, enveloped viruses and fungi. Microbiological Research, 156(1):1–7.
58. Sekhon, B. (2010). Food nanotechnology – an overview. Nanotechnology, Science and Applications, 3:1–15.

SECTION ❚❚

Characterization

CHAPTER 6

Interfacial Characterization of Nanoemulsions

Deepak Kumar, Sheetal Yadav,
Rewati Raman Ujjwal, Keerti Jain

Department of Pharmaceutics, National Institute of Pharmaceutical
Education and Research (NIPER), Raebareli, India

CONTENTS

DOI: 10.1201/9781003121121-6

6.1 INTRODUCTION

Emulsification is the fractionation of an oil phase into small globules resulting in an enormous increase in the surface area. This phenomenon is facilitated by mitigated surface tensions and rapid adsorption means. Therefore, provided certain energy to the system, tinier globules are created with surfactants efficient enough to adsorb and decrease the interfacial tension. The stability of emulsions is regulated by various interfacial phenomenons influenced by the bulk and adsorption properties of surfactants, both at equilibrium and in dynamic conditions. These include creaming or sedimentation (which is the gravity-driven phase separation due to the density difference between the two liquids); Ostwald ripening (net mass transfer from small to big droplets) and droplet coalescence (merging of two droplets into a single larger one) (Schmitt et al., 2004). To overcome these destabilizations, specifically droplet coalescence, several interfacial aspects including repulsive interaction between adsorbed layers, the interfacial coverage, the steric effects and the high dilational viscoelasticity of the interfacial layers needs to be addressed (Boos et al., 2013; Llamas et al., 2018). The repulsive interactions are pertinent in case of ionic surfactants where even a small fraction of adsorbed molecules at the drop surface is enough to avert coalescence (Llamas et al., 2018) and to keep intact the stability of emulsions. The interfacial coverage is the comparative area of the interface inhabited by the surfactant. The higher is the coverage area the lesser is the coalescence and stable is the emulsion. Then steric hindrance is related to composite surface layers with large surfactant molecules at high adsorption coverage which play an important role in stabilizing the emulsion.

The dilational viscoelasticity, E, is the dynamic response of the interfacial tension, γ, to extensional perturbations of the surface area, A. For small amplitude harmonic perturbations, E is a frequency dependent complex quantity, defined as:

$$E = \Delta\gamma \, \Delta A / A0 \, e i \phi,$$

where $\Delta\gamma$ and ΔA are the amplitudes of the oscillating surface tension and surface area, respectively, A0 is the reference area and ϕ is the phase shift between the oscillating surface tension and surface area. According to its definition, high values of the dilational viscoelasticity make the liquid films between drops in emulsions more stable and hamper their local thinning for Marangoni effects.

The deep understanding of the interfacial phenomenon is crucial to deal with fundamental stability issues of nanoemulsions (NEs) (Ravera et al., 2021). The structure of NE is depicted in Figure 6.1. As surfactants play pivotal role in addressing these stability issues, the profound characterization and understanding of the interfacial properties and surfactants used for stabilizing NE at liquid-liquid interface is important to get insights of stabilizing mechanisms. The present chapter deals with the characterization of interfacial properties of NE and it also highlights the applicability of interfacial characterization in preparation and stabilization of food NE.

6.1.1 Influence of Preparation Methods on Interfacial Properties and Stability of NE

NE formation requires high energy from mechanical equipments. However, it can also be prepared by low energy methods. The preparation methods of NE can be divided into high-energy and low energy approaches (Figure 6.2).The preparation method influences the stability of the liquid–liquid interface to a great extent. That's why it is worth mentioning of the preparation method due to its relevancy with interfacial stability.

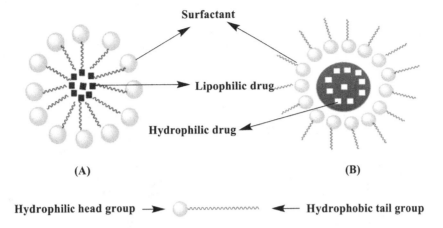

FIGURE 6.1   Illustration of NE (A) o/w emulsion, (B) w/o emulsion.

6.1.1.1 *High Energy Approach*

In the case of the high-energy method, high pressure is provided by the homogenizer, sonication and/or microfluidizer (Figure 6.3) to break the macro droplets of crude emulsion into smaller ones and further adsorption of the surfactant on to the interface ensuing steric stabilization of the resulting nanodispersion (Marzuki et al., 2019). The high-pressure homogenization (HPH) requires high pressure (500–5000 psi) by forcing the crude emulsion mixture through a small orifice. This technique produces very smaller droplets of less than 1μm (Sharma et al., 2020). The HPH method shows its suitability toward oil in water NEs (<20% oil phase) only. Iqbal and co-workersworked out a high energy HPH

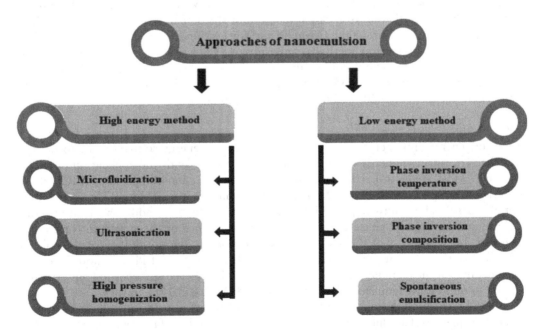

FIGURE 6.2   Different types of approach for the preparation of food NE.

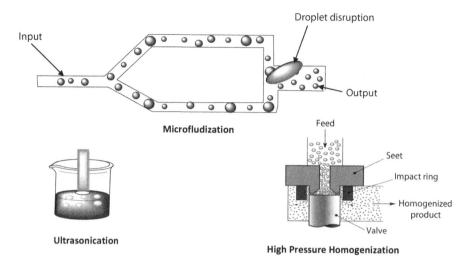

FIGURE 6.3   High-energy approach for the preparation of the NE.

method to prepare food grade NE of quercetin (QT) and curcumin using modified starch (HI-Cap-100) as emulsifier (Iqbal et al., 2020).

In the microfluidization method, the NE formation requires a microfluidizer to execute the process. The displacement pump of the equipment forces the product from the interaction chamber to the microchannel. Further, in the impingement area, the smaller particles get converted into the submicron range (Azmi et al., 2019). Mixing of both oil and aqueous phases in the inline homogenizer is responsible for a stable emulsion formation (Sharma et al., 2013). Ultrasonication breaks coarse droplets into NE *via* generating an ultrasonic agitation by sound waves with more than 20 kHz frequency (Kumar et al, 2019). Mahmood and Anwar utilized the ultrasonication method for the preparation of the vitamin D NE using a mixture of the surfactant Tween 80 and soy lecithin (Mehmood & Ahmed, 2020). Therefore, these high energy methods cause substantial decrease in interfacial tension and lead to formation of a highly stable transparent NE as demonstrated in different research studies where optically clear food NE were prepared successfully employing the high energy approach (Iqbal et al., 2020; Mehmood & Ahmed, 2020).

6.1.1.2 Low Energy Approach

The low energy approach does not require the external energy source for the formation of the NE. Intrinsic energy of the component forms NE with the application of gentle mixing. Most of the utilized approaches are phase inversion composition (PIC), phase inversion temperature (PIT) and spontaneous emulsification (Montes de Oca-Ávalos et al., 2017). The emulsion phase inversion technique involves addition of an aqueous phase into an organic phase or vice versa in the presence of the surfactant or co-surfactant. In the process of the PIC, NE formation occurs due to altering of the volume fraction of the water at a given temperature while in the case of PIT method, inversion temperature is the main factor. PIT is a temperature having the tendency to convert water in oil to oil in water emulsion (Aswathanarayan & Vittal, 2019).

In PIT, the fundamental is that the positive and negative surfactant curvature appear at the droplet interface due to decrease or increase in solubility of the temperature sensitive surfactant with temperature modulation (Liu et al., 2019). Spontaneous emulsification

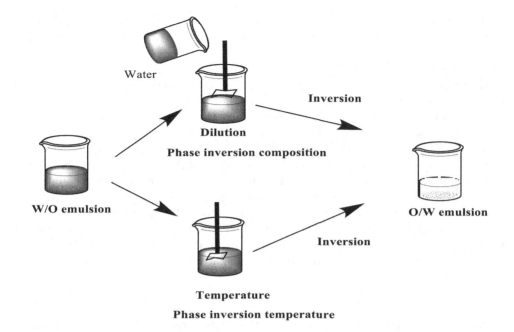

Water

Inversion

Dilution

Phase inversion composition

W/O emulsion

O/W emulsion

Inversion

Temperature

Phase inversion temperature

FIGURE 6.4 Low energy approach for the preparation of NE.

avoids the help of the external aid and spontaneous formation of emulsion takes place *via* continuous stirring in the presence or absence of the surfactant (Komaiko & Mcclements, 2016). Therefore, in the low energy approach, interfacial stabilization leads to formation of fine NE dispersion (Figure 6.4).

6.2 INTERFACIAL CHARACTERIZATION OF NES

6.2.1 Thermodynamic Stability Studies

The NEs are exposed to distinct thermodynamic stability tests to confirm stability of prepared NE under stress conditions. The investigations comprise of successive heating and cooling of NEs where they are subjected to several cycles of cooling (4°C) and heating (45°C). In thermodynamic stability study, NEs are exposed to freeze-thaw cycles and variations in the size, shape, or phase are observed. Kinetic instability, e.g., creaming, settling or phase separation is examined by centrifugation of NEs at 3000–4000 rpm (Singh et al., 2017). The NEs that do not show any sign of creaming and phase separation pass these tests and thermodynamic stability of these NE's is attributed to the interfacial stability.

6.2.2 Determination of Interfacial Tension

Interfacial tension is the surface tension that occurs between two phases. The changes in interfacial tension of a NE is observed with different surfactant concentrations and at variable temperatures. The surfactant is used as a stabilizing agent for the emulsion and it exerts its stabilizing effect by decreasing the interfacial tension at the oil–water interface.

Similarly, film formation avoids the coalescence of oil droplets (Kumar & Mandal, 2018). There are several methods for the determination of interfacial tension as given hereunder.

6.2.2.1 Tracker Automatic Drop Tensiometer

For the determination of interfacial tension, the measurement of the existence of distinct concentrations of surfactants is done through the inverted drop process with a tracker automatic drop tensiometer. A glass cuvette (30 × 30 × 70 mm) is occupied with the surfactant's aqueous medium is used for the process of measurement. The apparatus contains a hook shaped needle in which the sample is loaded at the centre and finally immersed in the aqueous surfactant medium. The drop of the oil is spread on the needle's tip via self-regulating plunging. The first drop is a waste and through the photographic camera of the device (640 × 480 px), the shape of the next five drops is taken. With the help of an algorithm which is based on the Laplace formula, the interfacial surface tension is determined. At the point where the two positions connect which is suitable for the experimental data with the help of segmental linear regression, the calculated values of interfacial tension and surfactant concentration at plateau remain measured (Pavoni et al., 2020).

6.2.2.2 Pendant Drop Method

Pendant drop method is used to determine the dilatant rheology of adsorbed layers at droplet interfaces. This method of determination provides assurance to their potential for measuring the surface layer response and elasticity after oscillation is applied (Figure 6.5). The analysis of the image of the axisymmetric fluid droplet is involved in this approach and fitting of the Young–Laplace equation which perpetuates gravitational deformation of the drop along with the interfacial tension. The light diffusion does not change in this method at spurious and peripheral reflections from the drop interface which evolve from external sources. The images of the droplets which are obtained through the digital camera provide trustworthy results through the homogenous background of the image and lensing effects which in turn are of much assistance in the analysis (Berry et al., 2015; Yakhshi-Tafti et al, 2011).

6.2.2.3 Du Nouy Ring Method

For the determination of the interfacial tension of NE, the Du Nouy ring method is used through monitoring the torsion required to pull the ring from the liquid interfaces continuously. The method comprises of a ring which is attached to the wire at the downward position, a liquid vessel and a device to measure the force (Macy, 1935). Because of the upward and downward movement of vessels, the ring which is positioned in the NE exerts force to the wire. Hence, it is inferred that the amount of force which is required for the detachment of the ring from the vessel is equivalent to the length of ring perimeter multiplied by the surface tension. How much force is required for the ring to be pulled from the liquid interface can be determined with the help of the following equation:

$$F = 4\pi R\gamma B,$$

where F is the maximum force measured by device for detachment of ring, R is the radius of ring, γ is interfacial tension of liquid or NE and B is correction factor.

The correction factor is an essential parameter as the detachment of the ring lifts the additional volume of liquid or NE. It, however, is also dependent on the ring's dimension and the liquid's density. For obtaining the precise value of the interfacial tension, the ring

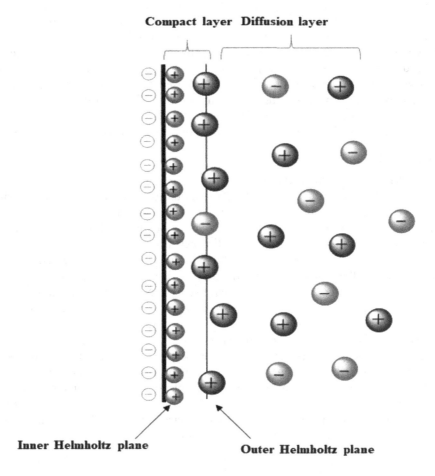

Compact layer Diffusion layer

Inner Helmholtz plane **Outer Helmholtz plane**

FIGURE 6.5 Experimental setup of the pendant drop tensiometer.

is positioned parallel to the liquid surface. Usually for the purpose of preparation of the ring, platinum and platinum-iridium metals are used. The choice of metal is an essential aspect in order to maintain the contact angle between the ring and liquid to zero or near to zero (Korenko & Šimko, 2010).

6.2.3 Determination of the Interfacial Composition of NEs

It is advisable to use similar ingredients however in distinct percentage ratios for the formulation of both the forms of nanocarriers. For example, a small amount of proteins, polysaccharides, and low molecular weight surface-active agents can be used in the formulation of NE (Singh et al., 2020). All in all, oil-in-water (o/w) NEs consist of with a miniscule amount of emulsifier covered lipid droplets dispersed within an aqueous phase. By explanation, the width of the droplets in NEs is determined to be ≤200 nm (Nirale et al., 2020). Usually, the structure and composition of NEs comprise of emulsifier, water and oil, but additional additives and biological active constituents might also be added into these polyphase systems. Two of the major groups of aspectsthat can be controlled when formulating NEs are (a) droplet appearances: Oil globule size, emulsifier nature and the oil phase conformation that can be mixed by a selection of distinct type of dispensing processes and ingredients. (b) Additives: The role of additives with features, e.g., as antioxidants, chelating agents, permeation enhancers, as well as thickener agents can be included in the water, oil, or interfacial areas of NEs. These are the additives that can be added either earlier or after homogenization based on their polarization. Interfacial composition of NE is the major focus of this section. Various other methods ae also available for the purpose of determination of interfacial composition of NE. However the depletion technique is the most useful technique of interfacial composition of NE.

6.2.3.1 Depletion Technique

In order to determine the interfacial composition of NE, the depletion technique is a suitable one if the concentration of surfactant is already known. In case the composition of surfactant is not known in the system, other methods can be used as well. In that situation, it becomes possible to separate the NE droplets from the continuous phase through the technique of centrifugation or filtration. The droplets of NE which are collected are washed to make sure the unnecessary components are removed from the continuous phase. However, the washing of droplets of NE will be completed when the droplets are dispersed in a suitable buffer solution and then centrifuged or filtered repetitively. Once the NE droplets are removed from the continuous phase, addition of highly surface-active agents can be done to the media in order to transfer the original emulsifiers from the o/w interfaces. Hence, it becomes a compulsion to enhance the additional chemical component in order to disrupt any covalent bonds which might be formed among adsorbed emulsifier molecules, e.g., mercaptoethanol could be added to disrupt the disulfide bond among globular protein and facilitate emulsifier displacement. In conclusion, the emulsifier is removed from the droplet of surface, the emulsifier is removed from the continuous phase and the analysis is done by the analytical method to know its concentration in the continuous phase, e.g., mass spectrometry, NMR spectrometry, electrophoresis and chromatography (Monahan et al., 1996).

6.2.4 Determination of Interfacial Structure

Interface's interfacial orientation has an effective influence on the emulsion stability which might be of assistance in order to ascertain the adverse interaction among the molecules. It is also noted that the greatest number of times the conversion in the interface layer occurred because of the instability issues, for example, aggregation process, phase separation and phase inversion. Disparity in the thickness of the interface was observed with proteins and a low molecular weight emulsifier (LMWE). The range of thickness of the interface stabilised with protein and LMWE was 1–15 nm and 1 nm, respectively. There is extreme dependency of the surface load of the interface on the process and formulation variables. The formation of interface layer is also dependent upon the protein concentration or surfactant. Relatively enough, the formation of monolayer interface takes place in the presence of lower protein, however a protein-enriched environment may enhance the chances of the formation of multilayer interface. Hence, determining the interfacial structure through Langmuir trough measurement, microscopy, scattering and spectroscopy is an essential aspect (Berton-Carabin et al., 2018).

6.2.4.1 Optical Microscopy

Optical microscopy is helpful for the measurement of the two-dimensional phase separation of protein and surface-active agent at the interface (Blonk & Aalst, 1993). One of the major drawbacks of the process is that it is not appropriate for the thin interface NE except for a Pickering emulsion. So, electron microscopy is preferred over optical microscopy because of its property of high-resolution. Scanning electron microscopy is utilised for the investigation of surface molecular interaction whereas transmission microscopy is used to measure the internal molecular arrangement of the nanodroplets (Berton-Carabin et al., 2018). To successfully investigate the colloidal interparticle interaction, atomic force microscopy has been employed (Helgeson, 2016).

6.2.4.2 Spectroscopy Techniques

It was investigated that the multiple spectroscopy techniqueslike,infrared (IR), UV-visible, x-ray diffraction spectroscopy and nuclear magnetic resonance (NMR) are all promising for the purpose of measuring the interface structure. The molecules which are present in the NE are capable enough to absorb electromagnetic radiation which can further be correlated with properties of the interfaces. Similarly, NMR spectroscopy can be used to investigate the molecular mobility within the o/w interface. In order to measure the protein mobility in NE, P31 NMR is a suitable method. For the determination of molecular characteristics of NE, vibrational sum-frequency spectroscopy can be used. For example, the molecular properties, i.e., interfacial molecular structure, bonding and oil-water interface behavior of NE formulated using dioctyl sodium sulfosuccinate were investigated (Hensel et al., 2017). One more effective technique for the measurement of the interface structure is electron paramagnetic resonance (EPR) spectroscopy. It combines the knowledge of the necessary interfacial characteristics like packing, molecular arrangement, and mobility. Microwaves are absorbed by the molecules containing the unpaired electrons in their paramagnetic centre. EPR spin spectroscopy was used for the purpose of determining the interfacial properties of the NE loaded with curcumin and vitamin D3 (Demisli et al., 2020). Also Theochari et al. (2020) reported that EPR spectra support the preparation method and composition of the limonene-based o/w NE was not affected by the polarity, rigidity and fluidity of the surfactants' monolayer.

6.2.4.3 Scattering Technique

Numerous methods are used for the determination of structural organization of molecules adsorbed to interfaces including x-ray diffraction spectroscopy, ultrasonic techniques and light scattering. These techniques produce the information required to predict the interfacial thickness of droplet size through detecting the angle of a monochromatic beam of light passing through the sample. These methods are also effective in providing the information related to the interaction and molecular arrangement. These techniques are further used for the purpose of differentiating the interface thickness and arrangement of the particles with different concentrations of emulsifier (Cristofolini, 2014).These techniques are viable in determining the changes of interfacial thickness and molecular arrangement due to alteration of environmental conditions, e.g., temperature, ionic strength and pH. These methods are vital for the prediction of stability of coalescence or emulsion flocculation. For example, in order to determine the interfacial thickness of the emulsifier, which is surrounded by the colloidal globule, the light scattering technique is used as an indirect method. This technique is appropriate for the colloidal particle that is accurately monodisperse and stable without an emulsifier. The droplet of particle radius is a measure of absence and presence of the emulsifier, and the difference between the radius is taken to be equal to the thickness of the interfacial surface (Hu et al., 2017).

6.2.4.4 Langmuir Trough Measurements

Langmuir Trough is the measurement apparatus for interfacial structure of the molecule at air-liquid and liquid-liquid interfaces. Langmuir apparatus comprises of a vessel which holds the liquids to determine the moveable barrier which is capable of changing the area of the liquid-air interface. The technique measures the surface tension at the air-liquid interface generally a wilhemy plate (Zembyla et al., 2019). Based on the solubility of the surface-active agent, dissolved liquid or spread on surface of liquids are analyzed by Langmuir technique. Reduction of surface interfacial area is done in a controlled manner with the help of a motor to drive the rotatable barrier, and the surface pressure is measured as a function of interfacial area. Increase in the concentration of solute material, solute-solute interaction and reduction of interfacial area results in the increase of surface pressure. Hence the movement of solute molecules into the liquid phase is still very easy. If the interfacial area is further reduced, the solute particles become tightly packed together in order to form a strong repulsive force among the molecules because of the increase in the surface pressure. Rise in the surface pressure and reduction in the interfacial area is able to deliver vital information regarding the packing of the surfactant material within the interfacial surface at the saturation instant. The Langmuir trough technique is used in connection with microscopy, reflection, and scattering techniques to provide information related to changes in structural interfaces and interactions of solute particles as the change in interfacial area occurs (Galli & Serabian, 2015; Mottola et al., 2019).

6.2.5 Electrical Properties of the Interface

Electrical properties of NE play a vital role in its stability and physicochemical properties which depend extremely on the type of emulsifier it adsorbs. Electrical characteristics of interfaces vary depending upon the surrounding ionic strength and pH. The electrical interaction may interfere with the magnitude and the interaction within the colloidal droplets. The electrical double layer exists at the interface, which can be understood by two terms: compact Helmholtz layer and outer diffuse layer (Figure 6.6). In the case of Helmholtz layer, the charge on the oil phase is neutralized with the opposite counter-ions held into the water (Lewis et al., 1937). Here, the counterion held on the surface of water might not help in the electrokinetic phenomenon. The existence of a diffuse layer existing is outside of the compact layer; it comprises of counter-ions and co-ions which are relatively affected by the particle surface charge. Usually, the co-ions have similar electric charges which promote the electrostatic repulsion. These co-ions are relatively in lower concentration in the water environment while the counterions are higher in concentration. The surface charge density and electrical surface potential may be considered as the primary parameters to characterize the electrical properties of the o/w interface (Speed, 2016).

The surface charge of the NE is extremely correlated with the stability of the formulations. Electrokinetic potential of the NE is represented by Zeta potential. A lower zeta potential value indicates extensive attraction between the droplets and is responsible for the droplet coalescence and breakdown. In the context of stability, the zeta potential value ± 30 mV represents the NE stability (Zhong & Zhang, 2019). Malvern zeta sizer based on the DLS phenomenon is majorly employed for the determination of the zeta potential (Ding et al., 2018; Jin et al., 2016). Selection of an appropriate emulsifier also contributes to the alteration of zeta potential like anionic surfactants imparts negative zeta potential meanwhile positive zeta potential is observed with cationic surfactants (McClements & Rao, 2011).

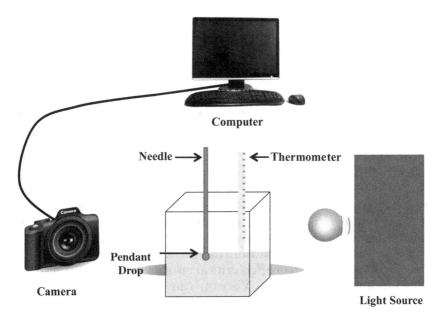

Computer

Needle → ←Thermometer

Pendant Drop

Camera

Light Source

FIGURE 6.6   Schematic diagram of the electrical properties of the interface in NE.

6.2.6 Determination of the Interfacial Rheology

Interfacial rheology establishes a critical relationship between the interfacial stress and interfacial deformation (Murray, 2002). Different types of deformation are involved in the interface including stress, and dilatants/compression. Dilatational rheology investigates the rheological properties that occur due to alteration in the surface area. Furthermore, low molecular surfactant, polymer and protein interfacial dynamic properties can also be measured. The interfacial dilatational viscoelasticity is represented by the given formula

$$\in = \frac{d\gamma}{\ln A} = A\frac{d\gamma}{dA},$$

where γ is deformation and A is surface area.

In the past few decades, various measuring techniques for the estimation of the interfacial rheological factor have been reported. In the case of shear viscosity, the indirect viscometer, i.e., deep-channel surface viscometer, containing two stationary concentric cylinders and a rotating dish, is utilized for the measurement. The channel surface viscometer demonstrated the flow rate through a constricted channel under external pressure (Pelipenko et al., 2012). In dilatational rheology relaxation or dilation of the surface is responsible for the interfacial deformation. The oscillating bubble drop method used for the dilatational rheology measurement involves the compression and dilation of the interface due to a harmonic oscillation. Therefore the investigation of the difference between different types the interfacial tension provides alteration in the dilatational viscoelasticity (Karbaschi et al., 2014; Ravera et al., 2010). Drop based method is classified as drop shape tensiometry (DST) and capillary pressure tensiometry (CPT) method. In DST, the interface property is estimated through the alteration in shape after applied shear while CPT involves the pressure estimation of the bubble inside the capillary to obtain the interfacial property. Another widely used method is Langmuir Trough method that applied the surface pressure.

$$\pi = \sigma 0 - \sigma(\Gamma, T),$$

where, π represent the surface pressure and $\sigma 0$ interfacial tension of the interface (Jaensson & Vermant, 2018).

6.2.7 Determination of Viscosity

Viscosity is the measurement of the resistance of a fluid to deform under applied shear stress. It is a critical parameter for the stability of NEs that depends on the emulsion component (water or oil) concentration and types of surfactant. Brookfield viscometer is the most widely utilized one to measure the viscosity of the NE under applied shear stress. The viscosity of the system offers critical insights to observe the phase inversion phenomenon (Marzuki et al., 2019). Viscosity can be determined from the equation below:

$$F = \mu \; A \frac{u}{y},$$

where F is the stress force, u is the velocity, A is the area and μ is the proportionality factor called dynamic viscosity.

The kinematic viscosity v is related to the dynamic viscosity by dividing by the density of the fluid, $v = \frac{\mu}{v}$ (Yang & Xu, 2016).

6.3 APPLICATIONS OF INTERFACIAL PROPERTIES INFLUENCED FOOD NANOEMULSIONS

Dziza and associates reported a research work in which they investigated interfacial properties of a set of surfactants and elucidated their effects on the characteristics of the respective oil–water emulsions. The surfactants applied in the study included saponin, Tween 80 and citronellol glucoside (CG), whereas the oil was Miglyol 812N which is a medium chain triglyceride (MCT) oil. As all these ingredients are biocompatibility, therefore they are being thoroughly utilized in food, cosmetic or pharmaceutical products. In the study, dynamic and equilibrium interfacial tensions and dilational viscoelasticity of the prepared NE systems were evaluated as a function of the surfactant concentration and analyzed using existing adsorption models. What they concluded from the study findings is that the ingredients concentration in the emulsion matrix differ significantly from the nominal concentration of the solutions before dispersing them, and it is anticipated that this happened because of the huge area of droplets available for surfactant adsorption in the emulsion. From the study findings, the researchers successfully derived a correlation between the observed emulsion behavior and the actual surfactant coverage of the droplet interface (Dziza et al., 2020).

6.4 CONCLUSION

The interfacial characteristics of the NEs exhibit a pivotal role in the formulation process as well as the stability of the NE. NE characterization became an emerging research area because of its utility in therapeutics and nutritionals delivery system. Size and stability of the NE are dependent on the employed preparation method, i.e., high and low energy approach. The characterization of the interfacial rheology and viscosity became necessary for the NE stability because early estimation could avoid the problem of the instability because of coalescence and phase separation. Future research is required to develop evaluation methods for characterization of the NEs particularly its interfacial stability.

ACKNOWLEDGMENTS

We acknowledge Department of Pharmaceuticals (DoP), Ministry of Chemicals and Fertilizers, Government of India, for their support. The NIPER-R communication number for this manuscript is NIPER-R/Communication/185.

REFERENCES

Aswathanarayan, J. B., & Vittal, R. R. (2019). Nanoemulsions and Their Potential Applications in Food Industry. *Frontiers in Sustainable Food Systems*, 3, 1–21. https://doi.org/10.3389/fsufs.2019.00095

Azmi, N. A. N., Elgharbawy, A. A. M., Motlagh, S. R., Samsudin, N., & Salleh, H. M. (2019). Nanoemulsions: Factory for Food, Pharmaceutical and Cosmetics. *Processes*, 7(9). https://doi.org/10.3390/pr7090617

Berry, J. D., Neeson, M. J., Dagastine, R. R., Chan, D. Y. C., & Tabor, R. F. (2015). Surface and Interfacial Tension Using Pendant Drop Tensiometry. *Journal of Colloid and Interface Science*, 454, 226–237. https://doi.org/10.1016/j.jcis.2015.05.012

Berton-Carabin, C. C., Sagis, L., & Schroën, K. (2018). Formation, Structure, and Functionality of Interfacial Layers In Food Emulsions. *Annual Review of Food Science and Technology*, 9, 551–587. https://doi.org/10.1146/annurev-food-030117-012405

Blonk, J. C. G., & Aalst, H. Van. (1993). Confocal Scanning Light Microscopy in Food Research. *Food Research International Journal*, 26, 297–311.

Boos, J.; Preisig, N.; Stubenrauch, C. (2013).Dilational Surface Rheology Studies of N-Dodecyl-B-D-Maltoside, Hexaoxyethylene Dodecyl Ether, and Their 1:1 Mixture. *Advances in Colloid Interface Science*, 197–198, 108–117.

Cristofolini, L. (2014). Synchrotron X-Ray Techniques for the Investigation of Structures and Dynamics in Interfacial Systems. *Current Opinion in Colloid and Interface Science*, 19(3), 228–241. https://doi.org/10.1016/j.cocis.2014.03.006

Demisli, S., Mitsou, E., Pletsa, V., Xenakis, A., & Papadimitriou, V. (2020). Development and Study of Nanoemulsions and Nanoemulsion-Based Hydrogels for the Encapsulation of Lipophilic Compounds. *Nanomaterials*, 10(12), 1–19. https://doi.org/10.3390/nano10122464

Ding, Z., Jiang, Y., & Liu, X. (2018). Nanoemulsions-Based Drug Delivery for Brain Tumors. In *Nanotechnology-Based Targeted Drug Delivery Systems for Brain Tumors*. Elsevier Inc. https://doi.org/10.1016/B978-0-12-812218-1.00012-9

Dziza, K.,Santini, E.,Liggieri, L.,Jarek, E.,Krzan, M.,Fischer, T., & Ravera, F.(2020). Interfacial Properties and Emulsification of Biocompatible Liquid-Liquid Systems. *Coatings*, 10, 397. https://doi.org/10.3390/coatings10040397

Galli, M. C., & Serabian, M. (2015). Regulatory Aspects of Gene Therapy and Cell Therapy Products: A Global Perspective, Springer, Cham, v–vi. https://doi.org/10.1007/978-3-319-18618-4

Helgeson, M. E. (2016). Colloidal Behavior of Nanoemulsions: Interactions, Structure, and Rheology. *Current Opinion in Colloid and Interface Science*, 25, 39–50. https://doi.org/10.1016/j.cocis.2016.06.006

Hensel, J. K., Carpenter, A. P., Ciszewski, R. K., Schabes, B. K., Kittredge, C. T., Moore, F. G., & Richmond, G. L. (2017). Molecular Characterization of Water and Surfactant AOT at Nanoemulsion Surfaces.*Proceedings of the National Academy of Sciences of the United States of America*, 114(51), 13351–13356. https://doi.org/10.1073/pnas.1700099114

Hu, Y. T., Ting, Y., Hu, J. Y., & Hsieh, S. C. (2017). Techniques and Methods to Study Functional Characteristics of Emulsion Systems. *Journal of Food and Drug Analysis*, 25(1), 16–26. https://doi.org/10.1016/j.jfda.2016.10.021

Iqbal, R., Mehmood, Z., Baig, A., & Khalid, N. (2020). Formulation and Characterization of Food Grade O/W Nanoemulsions Encapsulating Quercetin and Curcumin: Insights on Enhancing Solubility Characteristics. *Food and Bioproducts Processing*, 123, 304–311. https://doi.org/10.1016/j.fbp.2020.07.013

Jaensson, N., & Vermant, J. (2018). Tensiometry and Rheology of Complex Interfaces. *Current Opinion in Colloid and Interface Science*, 37, 136–150. https://doi.org/10.1016/j.cocis.2018.09.005

Jin, W., Xu, W., Liang, H., Li, Y., Liu, S., & Li, B. (2016). 1 - Nanoemulsions for Food: Properties, Production, Characterization, and Applications. In *Emulsions*. Elsevier Inc. https://doi.org/10.1016/B978-0-12-804306-6/00001-5

Karbaschi, M., Lotfi, M., Krägel, J., Javadi, A., Bastani, D., & Miller, R. (2014). Rheology of Interfacial Layers. *Current Opinion in Colloid and Interface Science*, 19(6), 514–519. https://doi.org/10.1016/j.cocis.2014.08.003

Komaiko, J. S., & Mcclements, D. J. (2016). Formation of Food-Grade Nanoemulsions Using Low-Energy Preparation Methods: A Review of Available Methods. *Comprehensive Reviews in Food Science and Food Safety*, 15(2), 331–352. https://doi.org/10.1111/1541-4337.12189

Korenko, M., & Ŝimko, F. (2010). Measurement of interfacial Tension in Liquid-Liquid High-Temperature Systems. *Journal of Chemical and Engineering Data*, 55(11), 4561–4573. https://doi.org/10.1021/je1004752

Kumar, M., Bishnoi, R. S., Shukla, A. K., & Jain, C. P. (2019). Techniques for Formulation of Nanoemulsion Drug Delivery System: A Review. *Preventive Nutrition and Food Science*, 24(3), 225–234. https://doi.org/10.3746/pnf.2019.24.3.225

Kumar, N., & Mandal, A. (2018). Surfactant Stabilized Oil-in-Water Nanoemulsion: Stability, Interfacial Tension, and Rheology Study for Enhanced Oil Recovery Application. *Energy and Fuels*, 32(6), 6452–6466. https://doi.org/10.1021/acs.energyfuels.8b00043

Lewis, B. Y. W. C. M., Total, M., Density, C., & Surface, O. (1937). The Electric Charge at an Oil-Water, *Transaction of Faraday Society*, 33, 708–713. https://doi.org/10.3390/TF9373300708

Liu, Q., Huang, H., Chen, H., Lin, J., & Wang, Q. (2019). Food-Grade Nanoemulsions: Preparation, Stability and Application in Encapsulation of Bioactive Compounds. *Molecules*, 24(23), 1–37. https://doi.org/10.3390/molecules24234242

Llamas, S., Santini, E., Liggieri, L., Salerni, F., Orsi, D.; Cristofolini, L., & Ravera, F.(2018). Adsorption of Sodium Dodecyl Sulfate at Water-Dodecane Interface in Relation to the Oil in Water Emulsion Properties. *Langmuir, 34*(21), 5978 5989. https://doi.org/10.1021/acs.langmuir.8b00358

Macy, R. (1935). Surface Tension by the Ring Method: Applicability of the Du Nouy Apparatus. *Journal of Chemical Education, 101*, 573–576. https://doi.org/10.1021/ed012p573

Marzuki, N. H. C., Wahab, R. A., & Hamid, M. A. (2019). An Overview of Nanoemulsion: Concepts of Development and Cosmeceutical Applications.*Biotechnology and Biotechnological Equipment, 33*(1), 779–797. https://doi.org/10.1080/13102818 .2019.1620124

McClements, D. J., & Rao, J. (2011). Food-Grade nanoemulsions: Formulation, fabrication, properties, performance, Biological fate, and Potential Toxicity. *Critical Reviews in Food Science and Nutrition, 51*(4), 285–330. https://doi.org/10.1080/ 10408398.2011.559558

Mehmood, T., & Ahmed, A. (2020). Tween 80 and Soya-Lecithin-Based Food-Grade Nanoemulsions for the Effective Delivery of Vitamin D. *Langmuir, 36*(11), 2886–2892. https://doi.org/10.1021/acs.langmuir.9b03944

Monahan, F. J., McClements, D. J., & German, J. B. (1996). Disulfide-Mediated Polymerization Reactions and Physical Properties of Heated WPI-Stabilized Emulsions. *Journal of Food Science, 61*(3), 504–509. https://doi.org/10.1111/j.1365-2621.1996. tb13143.x

Montes de Oca-Ávalos, J. M., Candal, R. J., & Herrera, M. L. (2017). Nanoemulsions: Stability and Physical Properties.*Current Opinion in Food Science, 16*, 1–6. https:// doi.org/10.1016/j.cofs.2017.06.003

Mottola, M., Caruso, B., & Perillo, M. A. (2019). Langmuir Films at the Oil/Water Interface Revisited. *Scientific Reports, 9*(1), 1–13. https://doi.org/10.1038/s41598-019-38674-9

Murray, B. S. (2002). Interfacial Rheology of Food Emulsifiers and Proteins. *Current Opinion in Colloid and Interface Science, 7*(5–6), 426–431. https://doi.org/10.1016/ S1359-0294(02)00077-8

Nirale, P., Paul, A., & Yadav, K. S. (2020). Nanoemulsions for Targeting the Neurodegenerative Diseases: Alzheimer's, Parkinson's and Prion's. *Life Sciences, 245*, 117394. https://doi.org/10.1016/j.lfs.2020.117394

Pavoni, L., Perinelli, D. R., Ciacciarelli, A., Quassinti, L., Bramucci, M., Miano, A., Casettari, L., Cespi, M., Bonacucina, G., & Palmieri, G. F. (2020). Properties and Stability of Nanoemulsions: How Relevant is the Type of Surfactant? *Journal of Drug Delivery Science and Technology, 58*(April), 101772. https://doi.org/10.1016/j. jddst.2020.101772

Pelipenko, J., Kristl, J., Rošic, R., Baumgartner, S., & Kocbek, P. (2012). Interfacial Rheology: An Overview of Measuring Techniques and Its Role in Dispersions and Electrospinning. *Acta Pharmaceutica, 62*(2), 123–140. https://doi.org/10.2478/ v10007-012-0018-x

Ravera, F., Dziza, K., Santini, E., Cristofolini, L., & Liggieri, L. (2021). Emulsification and Emulsion Stability: The Role of the Interfacial Properties. *Advances in Colloid and Interface Science, 288*, 102344. https://doi.org/10.1016/j.cis.2020.102344

Ravera, F., Loglio, G., & Kovalchuk, V. I. (2010). Current Opinion in Colloid & Interface Science Interfacial Dilational Rheology by Oscillating Bubble/Drop Methods. *Current Opinion in Colloid & Interface Science, 15*(4), 217–228. https:// doi.org/10.1016/j.cocis.2010.04.001

Schmitt, V.,Cattelet, C.,Leal-Calderon, F. (2004). Coarsening of Alkane-in-Water Emulsions Stabilized by Nonionic Poly(Oxyethylene) Surfactants: The Role of Molecular Permeation and Coalescence. *Langmuir*, *20*, 46–52.

Sharma, S., Loach, N., Gupta, S., & Mohan, L. (2020). Phyto-Nanoemulsion: An Emerging Nano-Insecticidal Formulation. *Environmental Nanotechnology, Monitoring and Management*, *14*, 100331. https://doi.org/10.1016/j.enmm.2020.100331

Sharma, N., Mishra, S., Sharma, S., Deshpande, R. D., & Kumar Sharma, R. (2013). Preparation and Optimization of Nanoemulsions for Targeting Drug Delivery. *International Journal of Drug Development and Research*, *5*(4), 37–48.

Singh, M., Bharadwaj, S., Lee, K. E., & Kang, S. G. (2020). Therapeutic Nanoemulsions in Ophthalmic Drug Administration: Concept in Formulations and Characterization Techniques for Ocular Drug Delivery. *Journal of Controlled Release*, *328*, 895–916. https://doi.org/10.1016/j.jconrel.2020.10.025

Singh, Y., Meher, J. G., Raval, K., Khan, F. A., Chaurasia, M., Jain, N. K., & Chourasia, M. K. (2017). Nanoemulsion: Concepts, Development and Applications in Drug Delivery. *Journal of Controlled Release*, *252*, 28–49. https://doi.org/10.1016/j.jconrel.2017.03.008

Speed, D. E. (2016). 10 - Environmental Aspects of Planarization Processes. In *Advances in Chemical Mechanical Planarization (CMP)*. Elsevier Ltd. https://doi.org/10.1016/B978-0-08-100165-3.00010-3

Theochari, I., Demisli, S., Christodoulou, P., Zervou, M., Xenakis, A., & Papadimitriou, V. (2020). Structure, Activity and Dynamics of Extra Virgin Olive Oil-in-Water Nanoemulsions Loaded with Vitamin D3 and Calcium Citrate. *Journal of Molecular Liquids*, *306*, 112908. https://doi.org/10.1016/j.molliq.2020.112908

Yakhshi-Tafti, E., Kumar, R., & Cho, H. J. (2011). Measurement of Surface Interfacial Tension as a Function of Temperature Using Pendant Drop Images. *International Journal of Optomechatronics*, *5*(4), 393–403. https://doi.org/10.1080/15599612.2011.633206

Yang, B., & Xu, J. (2016). Thermophysical Properties and SANS Studies of Nanoemulsion Heat Transfer Fluids. *Neutron Scattering*. https://doi.org/10.5772/62313

Zembyla, M., Lazidis, A., Murray, B. S., & Sarkar, A. (2019). Water-in-Oil Pickering Emulsions Stabilized by Synergistic Particle-Particle Interactions [Research-article]. *Langmuir*, *35*(40), 13078–13089. https://doi.org/10.1021/acs.langmuir.9b02026

Zhong, Q., & Zhang, L. (2019). Nanoparticles Fabricated from Bulk Solid Lipids: Preparation, Properties, and Potential Food Applications. *Advances in Colloid and Interface Science*, *273*, 102033. https://doi.org/10.1016/j.cis.2019.102033

CHAPTER 7

Physical Characterization Technique for Nanoemulsions

Mohammad Zaki Ahmad,[1] Md. Rizwanullah,[2]
Abdul Aleem Mohammad,[1] Javed Ahmad,[1]
Leo M.L. Nollet[3]

[1]Department of Pharmaceutics, College of Pharmacy,
Najran University, Najran, Kingdom of Saudi Arabia
[2]Department of Pharmaceutics, School of Pharmaceutical Education
& Research, Jamia Hamdard, New Delhi, India
[3]University College, Ghent, Belgium

CONTENTS

7.1 INTRODUCTION

Nanoemulsions (NE) are being progressively utilized in a wide range of food and drug industries due to its featured physicochemical properties and functional attributes [1–4]. The functional characteristics of NE-based foodstuffs depend on the choice of the most suitable components (e.g., oil, aqueous phase, surfactant, co-surfactant and active components) [1, 2, 5]. These factors affect the physical properties of NE, including particle size distribution, surface charge, physical condition and interfacial properties [3–5]. Characterization of NE comprises evaluation of physicochemical parameters, such as compatibility of the different components of NE components, isotropicity, assay, content

DOI: 10.1201/9781003121121-7

uniformity, morphological appearance, pH, viscosity, density, conductivity, surface tension, size and size distribution, and zeta potential (ZP) and surface charge concerning the effect on the physical attribute of the composition [3–5]. In this section, we will discuss the different techniques used in physical characterization of NE, i.e., dynamic light scattering (DLS), differential Scanning Calorimetry (DSC), Fourier transform infrared (FT-IR), X-ray diffraction (X-RD), nuclear magnetic resonance (NMR), small-angle X-ray scattering (SAXS) and sensory analysis.

7.2 PHYSICAL CHARACTERIZATION OF NANOEMULSION

7.2.1 Dynamic Light Scattering Analysis

7.2.1.1 Particle Size and Polydispersity Index

DLS is one of the most common techniques used for determining the particle size and polydispersity index (PDI) of NE, also known as photon correlation spectroscopy technique using Malvern Zetasizer. This technique is based on the Brownian motion of the particles and the Stokes-Einstein equation [5, 6].

$$D_f = \frac{k_b T}{6\pi\eta R_h} \tag{5.1}$$

D_f is translational diffusion coefficients of particles or droplets, k_b is the Boltzmann constant, T is absolute temperature, η is the viscosity of the continuous phase and R_h is the hydrodynamic diameter of the particles or globules [5]. In this technique, samples are illuminated with laser beam and intensity fluctuations in the scattered light is analyzed as shown in Figure 7.1, [6] size measurement is typically carried out at a single scattering angle of 90 degrees [5]. The particle size distributions are usually presented as the intensity-weighted mean of the R_h and PDI (Figure 7.1). PDI characterizes the width of the particle-size distribution (homogeneity or heterogeneity) [5, 7]. NE with PDI less than 0.2 indicates a narrow particle or droplet-size distribution (homogeneity) [5, 7], however, PDI value close to 1 shows a wide particle size distribution (heterogeneity) [5, 7, 8]. This technique is capable of measuring the droplets size between the 3 nm and 3 μm [5]. However, one of the major limitations of DLS is the aqueous phase's viscosity, as it assumed to be Newtonian fluid, and this is not always the case of an emulsion [8]. Therefore, samples should be diluted before use to avoid multiple scattering [5].

In recent study by Algahtani et al., DLS was successfully employed in order to determine characterization parameters of NE for curcumin (CUR) formulation [9]. They demonstrated the direct relation between the concentration of oil phase and mean droplet size of NE. At 18% of oil concentration mean droplet size was 10 nm, while at 24% it rises to more than 50 nm [9]. Furthermore, when oil concentration reduced to half the concentration of Smix, mean droplet size reduced to less than 50 nm [9]. In another study, same group have used DLS effectively to investigate the formulation and evaluation of retinyl palmitate NE [1]. At fixed Smix concentration (45%), when oil concentration was increased from 10% to 20%, the mean droplet size of the NE was increased by the factor of 4.5 [1]. Likewise, when the oil concentration was increased from 15% to 20% at constant Smix concentration (50%), the

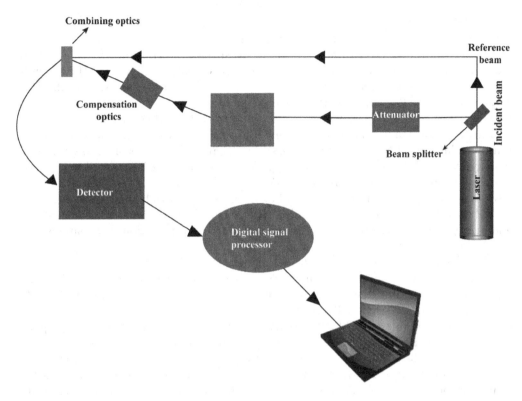

FIGURE 7.1   Working principle of photon correlation spectroscopy (Zetasizer).

mean droplet size increased by a factor of 2.5 [1]. Similarly, Lazzari et al. successfully demonstrated the application of DLS technique in stability of nanoparticles in buffer, simulated gastric, saliva, intestinal and lysosomal fluid [10]. The reported stability test was used in advance to investigate the formulation stability before in vivo application. Khan et al. demonstrated the utilization of DLS technique to investigate the adsorption of protein on gold nanoparticles [11]. Kim et al. investigated the lycopene-NE stability utilizing the DLS technique. Reduced degradation of lycopene was observed with mean droplet size less than 100 nm in comparison to NE with mean droplet size more than 100 nm [12]. Similarly, Maher et al. evaluated the β-Casein Stabilized NE. They reported the direct relation between protein content and mean droplet size of the NE. It was demonstrated that when protein content increased up to 7.5% w/w β-casein, the mean particle size increases from 187 to 199 nm, however, further increase in β-casein to 10% w/w results in decreased particle size (193 nm). It is proposed that increased mean particle size is due to a self-consistent field (SCF) theory in which the protein forms layers around the oil droplets [13]. Recently Hu et al. investigated the effect of mean droplet size of NE on the therapeutic efficacy of perilla oil in rats [14]. They reported that bioavailability and therapeutic efficacy (antioxidant and anti-inflammatory effects) of perilla oil NE was enhanced by decreasing the mean droplet size of the NE [14]. Similarly, Cheong et al. reported that relatively enhanced oral bioavailability of Kenaf (*Hibiscus cannabinus* L.) Seed Oil NE was observed with small nanosized droplets [15].

7.2.1.2 Zeta Potential

ZP is one of the crucial physicochemical parameters known to affect the stability of NE. It determines the electrostatic charge on the surface of NE droplets [6, 16, 17]. In scientific language, it is the electro kinetic potential in colloidal systems [17]. It measures the potential difference between the dispersed phase and dispersion media [17]. Measurement of ZP is currently the simplest and most straightforward way to measure the stability of NE, and it conclusions can be easily made by evaluation the data related to concentration, distribution and adsorption [18]. This is one of the significant characterization techniques of NE to determine the surface charge, that can be used to understand the physical stability of colloids [6, 17, 19]. The magnitude of ZP is related to the short term and long-term physical stability of NE [19]. NE with high ZP (~ ±30 mV) is generally considered to have better physical stability, this is because of high repulsive force that exceed the attractive force [6, 19]. Besides, small magnitude of ZP may leads flocculation and aggregation of particles due to Vander wall forces of attraction that may result in physical instability of NE [20].

ZP of NE is influenced by multiple factors such as surfactant and co-surfactant, ionic strength, morphology and pH of the system [6] For example, Algahtani et al. (2020) reported the ZP of NE that resist the aggregation of droplets [1, 2]. Kim et al. (2014) reported a significantly high ZP of lycopene –NE (–40 mV), which is significantly different from an acceptable range of ZP (±30 mV) for the stable NE [12]. High ZP of prepared NE indicates that NE is effectively covered with emulsifier and act as a barrier to avoid any oxidant on the surface of NE and thus provides the stability of NE [12]. However, several researchers have reported that oxidative stability of NE depends upon the surface charge on oil phase from where oxygen molecules interact with the fatty acids inside the droplet [21–23]. Similarly, Hu et al. reported that ZP with high magnitude leads to reduced agglomeration, thereby attributing high physical stability [14].

7.2.2 Differential Scanning Calorimetry Analysis

DSC is a fundamental tool in thermal analysis. In this technique differential heat requirement to increase the temperature of reference and sample is quantified as a function of temperature and materials heat capacity is evaluated [24–26]. A sample of known mass is heated or cooled and the changes in heat capacity are plotted as a function of heat flow. This allows the detection of transitions such as melts, glass transitions, phase changes, and curing [24–26] (Figure 7.2). This technique can be used to measure the crystallinity of a lipid in NE [27] and stability of emulsion [28]. The impurities present in the oil phase cause nucleation and crystal formation, which in turn affect the stability of NE. The quantity of destabilized oil in the NE can be evaluated using this instruments [27]. Thanasukarn et al. evaluated the crystallization effects of oil and its influence on the stability of NE [29]. Similarly, Uson et al. evaluated the crystallization temperature of a mixture of surfactants (Smix) [30]. Similarly, Maher et al. (2011) utilizes DSC for the determination of glass transition temperature (Tg) in carbohydrate–protein mixtures and NE [13]. They demonstrated that Tg of the maximally freeze-concentrated carbohydrate-protein mixture decreased with increasing protein content [13]. It may be due to the fact that the presence of protein in either a liquid or dehydrated system, affects the Tg of that system [13]. Li et al. investigated the physical properties of lycopene loaded NE using DSC [31]. They observed the sharp peak of lycopene at 169.0°C (melting point of

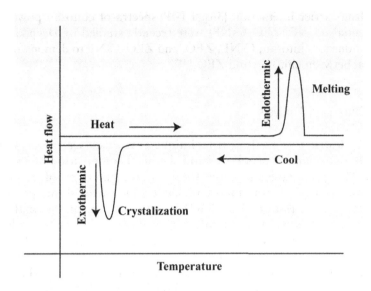

FIGURE 7.2   Thermal analysis of the melting and crystallization of oil droplets in and oil in water nanoemulsion using DSC.

crystalline lycopene) [31]. Though, encapsulated lycopene does not exhibit a sharp melting point, signifying the amorphous nature of encapsulated lycopene [31]. It can be presumed that the configuration of lycopene alters during the process of encapsulation [31].

7.2.3 Fourier Transform Infrared Analysis

FT-IR analysis is used to study the drug excipient interaction, polymerization, cross-linking as well as drug loading in the formulation. This technique is based on the absorption of infrared (IR) radiation when it passes through a sample, i.e., NE. The spectrum resulting from FT-IR represents the molecular absorption and transmission, creating a molecular fingerprint of the sample. The fingerprint from each sample is the characteristic absorption peak that represents the corresponding vibration between bonds of the atoms from each constituent. Since, each material is an unique combination of atoms, no two different constituents produce the exact same FT-IR spectrum [6, 32]. Therefore, outcomes of FT-IR can result in a positive identification of different ingredients. The major advantage of FT-IR includes characterization, identification, and quantification of complex mixtures of constituents, including gases, liquids and solids [32]. Furthermore, it can evaluate the consistency or quality of a sample in very small time [6, 32, 33]. Araujo et al. evaluated the thalidomide encapsulated NE and studied the crystallization behavior through FT-IR analysis [34]. They reported that crystals were in different polymorphic form in NE than in pure drugs [34]. Harwansh et al. developed a betulinic acid loaded NE and studied drug-excipients compatibility through FT-IR analysis [35].

This FTIR spectrum confirms that betulinic acid was compatible with their excipients used in the formulation. Similarly, Ahmad et al. studied the excipients-anticancer moiety interaction of NEs through FT-IR [36]. FT-IR analysis of the NE suggest absence of chemical drug-carrier interaction [36]. FT-IR spectra of chitosan powder, *Zingiber zerumbet* essential oil NE (ZEO-CsNE) were recently studied by Deepika et al. FTIR analyses were done for chitosan, CsNE, ZEO, and ZEO-CsNE to demonstrate the chemical interaction between chitosan and ZEO [37].

7.2.4 X-Ray Diffraction Analysis

XRD is a non-destructive analytical technique that discloses the crystallographic details and physicochemical properties and chemical composition of a sample under observation. This technique measures the scattered intensity of beams of X-ray striking the materials as a function of incident and scattered angles, polarization and wavelength [6, 38] (Figure 7.3). XRD is mainly used for the analysis of crystalline compounds because of their diffraction pattern. However, it has a broad field of applications such as identification of phase behavior of single phase materials, study of crystal structure, structural analysis of materials, identification of amorphous materials in a partially crystalline mixture [6, 39]. Furthermore, it can be used for the identification of structure, phases, crystalline orientation, average grain size, strain and crystal defects of sample [40]. Mulik et al. studied the diffraction pattern of CUR in a lipid nanocarrier system (solid lipid nanoparticles, SLN). The diffraction pattern of the CUR, blank SLN and CUR loaded SLN was analyzed. From the diffraction pattern, it was clear that amorphous state contributes to higher drug loading capacity [41].

XRD-analysis of chitosan, ZEO-CsNE reported by Deepika et al. demonstrated that chitosan (ZEO-CsNE) with lower crystallinity may be utilized as a compatible source of an antifungal agent for direct application in a food system [37].

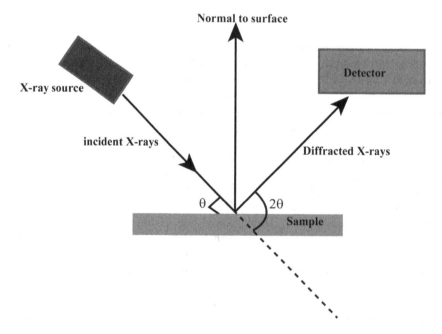

FIGURE 7.3 Schematic representation on the working principles of X-ray diffraction.

7.2.5 Nuclear Magnetic Resonance Analysis

NMR is a analytical tool that analyses the compound in either solid or liquid state [6, 8, 42]. More recently, the use of NMR techniques, which have customarily been used to characterize micellar systems, has gained attention to the characterization of colloidal systems [43]. This is one of the most widely used techniques for measuring the solid contents of food emulsion [8, 42]. This instrument can rapidly analyze the concentrated and optically opaque emulsion, without and further sample preparation [44]. The NMR technique utilizes interactions between radio waves and the nuclei of hydrogen atoms to obtain information about the solid content of a material [8, 42, 44]. It is used as a complementary technique to optical spectroscopy and mass spectrometry that results in detailed information about structure and stereochemistry [6, 42, 45]. The application of NMR for the characterization of nanosystems is attributed to Jenning et al. [46, 47]. They reported the successful incorporation of medium-chain triglycerides oil in a matrix of a solid long-chain glyceride (glyceryl behenate) and the characterization of liquid lipids inside the matrix of SLNs through 1H NMR [46, 47].

The location of the molecules of active ingredients in the microemulsion formulations was investigated by Lv et al. using 1H NMR spectroscopy and DLS measurements [48]. They have shown that molecules of active ingredients have not been solubilized either in the palisade layer or in the oily core of the microemulsion drops, but in the hydrophilic shells of the microemulsion drops, which are made up of many oxyethylene groups [48]. In another study, Casadei et al. uses the 1H NMR technique to determine the quantitative analysis of unloaded active ingredients and entrapment efficiency of active ingredients within lipid nanocarriers [49].

Furthermore, NMR technique is used for the determination of droplet sizes by the means of NMR pulsed gradient method to determine self-diffusion coefficients [50, 51]. Self-diffusion is the random movements of molecules in an isotropic solution without thermal gradients [43]. It is dependent on the molecular size and shape and on the molecule-solvent interactions [43].

Li et al. utilizes the NMR technique to study the effect of loading efficacy of lycopene on the physicochemical behavior of NE [31]. When loading of increasing lycopene concentrations from 0.1 to 0.3% w/w, strengthening in anisotropy of double bond was observed [31]. However, enhanced loading of lycopene results in decreased encapsulation efficiency of oil phase. Besides, it was also demonstrated that lycopene molecules were confined in the hydrophobic core of the O/W NE droplets [31]. Similarly, Hong et al. reported the formulation and evaluation of lycopene NE stabilized by modified octenyl succinic anhydride (OSA) starch [52]. They reported the slightly lower diffusivity of OSA molecules in NE without lycopene in compare to OSA starch solution. It may be due to the adsorption of OSA starch molecules at oil/water interface in NE [52]. Also, low mobility of OSA starch molecules results in a significantly lower self-diffusion coefficient [52], which is further reduced by lycopene loading, suggesting interaction between the lycopene and OSA molecules [3, 52].

7.2.6 Small-Angle X-Ray Scattering Analysis

SAXS is a technique used for the characterization of colloidal dispersion for its shape, size and structure (nanostructure) [53]. SAXS is a non-destructive method of analysis with minimum sample preparation. In this method, elastic scattering of X-rays by a sample of nanometric range, is recorded at very low angles (typically 0.1–10°) [6, 27, 54, 55]. This low angular range gives the information about size and shape, characteristic distance, pore sizes etc. [27]. Zhang et al. used this technique to analyses AOT/water/isooctane/CO_2 NE [56]. Cruz et al. studied the interaction of sorbitan monostearate with caprylic/capric triglyceride, in the core, and the interfacial hydrolysis of indomethacin ethyl ester, in the core of solid–lipid nanoparticles [57].

7.2.7 Sensory Analysis

Sensory analysis or qualitative characterization of appearance of NE are performed by using human panelists [4, 58–60]. Usually, there are three common methods used in sensory analysis, i.e., descriptive, preference, and discrimination test [4, 59–61]. Descriptive test is used to evaluate the quantitative information such as color, color intensity or turbidity of a NE based food products [4, 59, 60]. Discriminant test is used to quantify whether subject can discriminate two or more samples that may have different properties [59]. Preference tests are used to determine the order of preference by consumers [62]. For example, panelists may be instructed to state their preference for a particular sample when they are presented with three different NE based food products [4, 59, 62].

7.3 CONCLUSION

The techniques of physical characterization of NEs are very crucial for evaluating different critical attributes of prepared NE formulations. The useful information about the droplet size, shape, crystallinity, and structural integrity are ultimately going to define the in vivo fate of bioactive ingredient encapsulated in NEs. Nevertheless, there have been so much advancement in the characterization techniques still there is great scope of improvisation. For instance, there are certain loopholes with these techniques that restrict the complete and successful evaluation of prepared NE formulations such as DLS have limited scales of measurement and neglect many important factors such as temperature, viscosity, dilution, refractive index of the samples. Furthermore, the detection limit of crystallinity of DSC and XRD is also very restricted unabling the characterization of sensitive samples. This introduces a new landscape of research field to develop and validate proper physical characterization methods for characterizing NEs in food.

REFERENCES

1. Algahtani MS, Ahmad MZ, Ahmad J. Nanoemulgel for Improved Topical Delivery of Retinyl Palmitate: Formulation Design and Stability Evaluation. Nanomaterials (Basel, Switzerland). 2020;10(5):848. doi: 10.3390/nano10050848.
2. Algahtani MS, Ahmad MZ, Nourein IH, Ahmad J. Co-Delivery of Imiquimod and Curcumin by Nanoemugel for Improved Topical Delivery and Reduced Psoriasis-Like Skin Lesions. Biomolecules. 2020;10(7):968. doi: 10.3390/biom10070968
3. Chime S, Kenechukwu F, Attama A. Nanoemulsions—Advances in Formulation, Characterization and Applications in Drug Delivery. In: Sezer A, editor. Application of Nanotechnology in Drug Delivery. Croatia: InTech; 2014. p. 77–126.

4. Chung C, McClements DJ. Characterization of physicochemical properties of nano-emulsions: appearance, stability, and rheology. In: Seid Mahdi J, David Julian M, editors. Nanoemulsions. London, UK: Elsevier; 2018. p. 547–76.

5. Bourbon AI, Gonçalves RF, Vicente AA, Pinheiro AC. Characterization of particle properties in nanoemulsions. In: Seid Mahdi J, David Julian M, editors. Nanoemulsions. London, UK: Elsevier; 2018. p. 519–46.

6. Silva HD, Cerqueira MÂ, Vicente AA. Nanoemulsions for Food Applications: Development and Characterization. Food Bioprocess Technology. 2012;5(3):854–67.

7. Klang V, Valenta C. Lecithin-Based Nanoemulsions. Journal of Drug Delivery Science Technology. 2011;21(1):55–76.

8. McClements DJ. Characterization of Emulsion Properties. In: McClements DJ, editor. Food Emulsions: Principles, Practices, and Techniques. Boca Raton, FL: CRC Press; 2016. p. 623–75.

9. Alghtani MS, Ahmad MZ, Ahmad J. Nanoemulsion Loaded Polymeric Hydrogel for Topical Delivery of Curcumin in Psoriasis. Journal of Drug Delivery Science and Technology. 2020; 59:101847.

10. Lazzari S, Moscatelli D, Codari F, Salmona M, Morbidelli M, Diomede L. Colloidal Stability of Polymeric Nanoparticles in Biological Fluids. Journal of Nanoparticle Research: An Interdisciplinary Forum for Nanoscale Science and Technology. 2012;14(6):920.

11. Khan S, Gupta A, Verma NC, Nandi CK. Kinetics of Protein Adsorption on Gold Nanoparticle with Variable Protein Structure and Nanoparticle Size. The Journal of Chemical Physics. 2015;143(16):164709.

12. Kim SO, Ha TV, Choi YJ, Ko S. Optimization of Homogenization-Evaporation Process for Lycopene Nanoemulsion Production and its Beverage Applications. Journal of Food Science. 2014;79(8):N1604–10.

13. Maher PG, Fenelon MA, Zhou Y, Kamrul Haque M, Roos YH. Optimization of β-Casein Stabilized Nanoemulsions Using Experimental Mixture Design. Journal of Food Science. 2011;76(8):C1108–17.

14. Hu M, Xie F, Zhang S, Qi B, Li Y. Effect of Nanoemulsion Particle Size on the Bioavailability and Bioactivity of Perilla Oil in Rats. Journal of Food Science. 2021;86(1):206–14.

15. Cheong AM, Tan CP, Nyam KL. Effect of Emulsification Method and Particle Size on the Rate of in vivo Oral Bioavailability of Kenaf (*Hibiscus cannabinus L.*) Seed Oil. Journal of Food Science. 2018;83(7):1964–9.

16. WHO. Coronavirus disease (COVID-19) Pandemic 2020 [Available from: https://www.who.int/emergencies/diseases/novel-coronavirus-2019.]

17. Honary S, Zahir F. Effect of Zeta Potential on the Properties of Nano-Drug Delivery Systems – A Review (Part 2). Tropical Journal of Pharmaceutical Research 2013;12(2):265–73.

18. Rabinovich-Guilatt L, Couvreur P, Lambert G, Goldstein D, Benita S, Dubernet C. Extensive Surface Studies Help to Analyse Zeta Potential Data: The Case of Cationic Emulsions. Chemistry and Physics of Lipids. 2004;131(1):1–13.

19. Joseph E, Singhvi G. Multifunctional Nanocrystals for Cancer Therapy: A Potential Nanocarrier. Nanomaterials for Drug Delivery and Therapy. London, UK: Elsevier; 2019. p. 91–116.

20. Joseph E, Singhvi G. Chapter 4 – Multifunctional Nanocrystals for Cancer Therapy: A Potential Nanocarrier. In: Grumezescu AM, editor. Nanomaterials for Drug Delivery and Therapy. London, UK: William Andrew Publishing; 2019. p. 91–116.

21. Mei LY, McClements DJ, Wu JN, Decker EA. Iron-Catalyzed Lipid Oxidation in Emulsion as Affected by Surfactant, pH and NaCl. Food Chemistry. 1998;61(3):307–12.

22. Yoshida Y, Niki E. Oxidation of Methyl Linoleate in Aqueous Dispersions Induced by Copper And Iron. Archives of Biochemistry and Biophysics. 1992;295(1):107–14.

23. Nguyen HH, Choi KO, Kim DE, Kang WS, Ko S. Improvement of Oxidative Stability of Rice Bran oil Emulsion By Controlling Droplet Size. Journal of Food Process Preservation. 2013;37(2):139–51.

24. Differential Scanning Calorimetry (DSC) 2020 [Available from: https://www.perkinelmer.com/lab-solutions/resources/docs/GDE_DSCBeginnersGuide.pdf.

25. Koshy O, Subramanian L, Thomas S. Chapter 5 – Differential Scanning Calorimetry in Nanoscience and Nanotechnology. In: Thomas S, Thomas R, Zachariah AK, Mishra RK, editors. Thermal and Rheological Measurement Techniques for Nanomaterials Characterization. London, UK: Elsevier; 2017. p. 109–22.

26. Shah MR, Imran M, Ullah S. Chapter 2 – Nanostructured Lipid Carriers. In: Shah MR, Imran M, Ullah S, editors. Lipid-Based Nanocarriers for Drug Delivery and Diagnosis. Oxford, UK: William Andrew Publishing; 2017. p. 37–61.

27. Aswathanarayan JB, Vittal RR. Nanoemulsions and Their Potential Applications in Food Industry. Front Sustain Food Systems. 2019;3(95):10.3389.

28. Labes-Carrier C, Dumas J, Mendiboure B, Lachaise J. DSC as a Tool to Predict Emulsion Stability. Journal of Dispersion Science Technology. 1995;16(7):607–31.

29. Thanasukarn P, Pongsawatmanit R, McClements DJ. Influence of Emulsifier Type on Freeze-Thaw Stability of Hydrogenated Palm Oil-in-Water Emulsions. Food Hydrocolloids. 2004;18(6):1033–43.

30. Usón N, Garcia MJ, Solans C. Formation of Water-in-Oil (W/O) Nano-Emulsions in a Water/Mixed Non-Ionic Surfactant/Oil Systems Prepared by a Low-Energy Emulsification Method. Colloids and Surfaces A: Physicochemical and Engineering Aspects. 2004;250(1):415–21.

31. Li D, Li L, Xiao N, Li M, Xie X. Physical Properties of Oil-in-Water Nanoemulsions Stabilized by OSA-Modified Starch for the Encapsulation of Lycopene. Colloids and Surfaces A: Physicochemical and Engineering Aspects. 2018;552:59–66.

32. Maycotte P, Aryal S, Cummings CT, Thorburn J, Morgan MJ, Thorburn A. Chloroquine Sensitizes Breast Cancer Cells to Chemotherapy Independent of Autophagy. Autophagy. 2012;8(2):200–12.

33. Dinache A, Tozar T, Smarandache A, Andrei IR, Nistorescu S, Nastasa V, et al. Spectroscopic Characterization of Emulsions Generated with a New Laser-Assisted Device. Molecules. 2020;25(7).

34. Araújo FA, Kelmann RG, Araújo BV, Finatto RB, Teixeira HF, Koester LS. Development and Characterization of Parenteral Nanoemulsions Containing Thalidomide. European Journal of Pharmaceutical Sciences: Official Journal of the European Federation for Pharmaceutical Sciences. 2011;42(3):238–45.

35. Harwansh RK, Mukherjee PK, Biswas S. Nanoemulsion as a Novel Carrier System for Improvement of Betulinic Acid Oral Bioavailability and Hepatoprotective Activity. Journal of Molecular Liquids. 2017;237:361–71.

36. Ahmad J, Mir SR, Kohli K, Chuttani K, Mishra AK, Panda AK, et al. Solid-Nanoemulsion Preconcentrate for Oral Delivery of Paclitaxel: Formulation Design, Biodistribution, and γ Scintigraphy Imaging. BioMed Research International. 2014;2014:984756.

37. Deepika, Singh A, Chaudhari AK, Das S, Dubey NK. Zingiber zerumbet L. Essential Oil-Based Chitosan Nanoemulsion as an Efficient Green Preservative Against Fungi and Aflatoxin B(1) Contamination. Journal of Food Science. 2021;86(1):149–60.

38. Azároff LV, Kaplow R, Kato N, Weiss RJ, Wilson AJC, Young RA. X-Ray Diffraction. New York, NY: McGraw-Hill New York; 1974.

39. Connolly JR. Introduction to X-ray Powder Diffraction 2005 [Available from: http://www.xray.cz/xray/csca/kol2011/kurs/dalsi-cteni/connolly-2005/01-xrd-intro.pdf.

40. Kohli R, Mittal KL, editors. Chapter 3—Methods for Assessing Surface Cleanliness. In:. Developments in Surface Contamination and Cleaning, Volume 12. New York, NY: Elsevier; 2019. p. 23–105.

41. Mulik RS, Mönkkönen J, Juvonen RO, Mahadik KR, Paradkar AR. Transferrin Mediated Solid Lipid Nanoparticles Containing Curcumin: Enhanced In Vitro Anticancer Activity by Induction of Apoptosis. International Journal of Pharmaceutics. 2010;398(1):190–203.

42. McClements DJ. Critical Review of Techniques and Methodologies for Characterization of Emulsion Stability. Critical Reviews in Food Science and Nutrition. 2007;47(7):611–49.

43. Hathout RM, Woodman TJ. Applications of NMR in the Characterization of Pharmaceutical Microemulsions. Journal of Controlled Release. 2012;161(1):62–72.

44. Dickinson E, McClements DJ. Advances in Food Colloids. London, UK: Springer Science & Business Media; 1995.

45. Rouessac F, Rouessac A. Chemical Analysis: Modern Instrumentation Methods and Techniques. Toronto, ON: John Wiley & Sons; 2013.

46. Jenning V, Mäder K, Gohla SH. Solid Lipid Nanoparticles (SLN) Based on Binary Mixtures of Liquid and Solid Lipids: A (1)H-NMR Study. International Journal of Pharmacy. 2000;205(1-2):15–21.

47. Jenning V, Thünemann AF, Gohla SH. Characterisation of a Novel Solid Lipid Nanoparticle Carrier System Based on Binary Mixtures of Liquid and Solid Lipids. International Journal of Pharmacy. 2000;199(2):167–77.

48. Lv FF, Li N, Zheng LQ, Tung CH. Studies on the Stability of the Chloramphenicol in the Microemulsion Free of Alcohols. European Journal of Pharmaceutics and Biopharmaceutics: Official Journal of Arbeitsgemeinschaft fur Pharmazeutische Verfahrenstechnik eV. 2006;62(3):288–94.

49. Casadei MA, Cerreto F, Cesa S, Giannuzzo M, Feeney M, Marianecci C, et al. Solid Lipid Nanoparticles Incorporated in Dextran Hydrogels: A New Drug Delivery System for Oral Formulations. International Journal of Pharmacy. 2006;325(1-2):140–6.

50. Söderman O, Lönnqvist I, Balinov B. NMR Self-Diffusion Studies of Emulsion Systems. Droplet Sizes and Microstructure of the Continuous Phase. In: Sjöblom J, editor. Emulsions — A Fundamental and Practical Approach. Dordrecht. The Netherlands: Springer Netherlands; 1992. p. 239–58.

51. Packer KJ, Rees C. Pulsed NMR Studies of Restricted Diffusion. I. Droplet Size Distributions in Emulsions. Journal of Colloid and interface Science. 1972;40(2):206–18.

52. Hong Z, Xiao N, Li L, Xie X. Investigation of Nanoemulsion Interfacial Properties: A Mesoscopic Simulation. Journal of Food Engineering. 2020;276:109877.

53. Li T, Senesi AJ, Lee B. Small Angle X-ray Scattering for Nanoparticle Research. Chemical Reviews. 2016;116(18):11128–80.

54. Gradzielski M. Recent Developments in the Characterisation of Microemulsions. Current Opinion in Colloid & Interface Science. 2008;13(4):263–9.

55. Borthakur P, Boruah PK, Sharma B, Das MR. 5 - Nanoemulsion: Preparation and Its Application in Food Industry. In: Grumezescu AM, editor. Emulsions. London, UK: Academic Press; 2016. p. 153–91.

56. Zhang J, Han B, Zhao Y, Li W, Liu Y. Emulsion Inversion Induced by CO 2. Physical Chemistry Chemical Physics. 2011;13(13):6065–70.

57. Cruz L, Schaffazick SR, Costa TD, Soares LU, Mezzalira G, da Silveira NP, et al. Physico-Chemical Characterization and In Vivo Evaluation of Indomethacin Ethyl Ester-Loaded Nanocapsules By PCS, TEM, SAXS, Interfacial Alkaline Hydrolysis and Antiedematogenic Activity. Journal of Nanoscience and Nanotechnology. 2006;6(9-10):3154–62.

58. Linke C, Drusch S. Turbidity in Oil-in-Water-Emulsions — Key Factors and Visual Perception. Food Research International. 2016;89:202–10.

59. Amiri E, Aminzare M, Azar HH, Mehrasbi MR. Combined Antioxidant and Sensory Effects of Corn Starch Films with Nanoemulsion of Zataria Multiflora Essential Oil Fortified With Cinnamaldehyde On Fresh Ground Beef Patties. Meat Science. 2019;153:66–74.

60. Granato D, Masson ML. Instrumental Color and Sensory Acceptance of Soy-Based Emulsions: A Response Surface Approach. Food Science and Technology. 2010;30(4):1090–6.

61. Ozogul Y, Yuvka I, Ucar Y, Durmus M, Kösker AR, Öz M, et al. Evaluation of Effects of Nanoemulsion Based on Herb Essential Oils (Rosemary, Laurel, Thyme and Sage) on Sensory, Chemical and Microbiological Quality of Rainbow Trout (Oncorhynchus Mykiss) Fillets During Ice Storage. LWT. 2017;75:677–84.

62. Bento R, Pagán E, Berdejo D, de Carvalho RJ, García-Embid S, Maggi F, et al. Chitosan Nanoemulsions of Cold-Pressed Orange Essential Oil to Preserve Fruit Juices. International Journal of Food Microbiology. 2020;331:108786.

CHAPTER 8

Imaging Techniques for Characterization of Nanoemulsions

Anuj Garg,[1] Mohammad Zaki Ahmad,[2]
Md. Abul Barkat,[3] Pawan Kaushik,[4]
Javed Ahmad,[2] Leo M.L. Nollet[5]

[1]Institute of Pharmaceutical Research, GLA University,
Mathura, India
[2]College of Pharmacy, Najran University,
Najran, Kingdom of Saudi Arabia
[3]Department of Pharmaceutics, College of Pharmacy,
University of Hafr Al Batin, Al Jamiah, Hafr Al Batin,
Kingdom of Saudi Arabia
[4]Institute of Pharmaceutical Sciences, Kurukshetra University,
Kurukshetra, India
[5]University College, Ghent, Belgium

CONTENTS

8.1 INTRODUCTION

Emulsions are biphasic liquid preparations, comprising circular globules creating a dispersed phase, while the liquid covering the dispersed phase constitutes a continuous phase [1–3]. Nanoemulsions are nanosized liquid preparations (10–1000 nm) used extensively to encapsulate bioactive molecules and have wide applications in food and pharmaceutical industry for delivering bioactive molecules and providing them stability. There are

DOI: 10.1201/9781003121121-8

several factors which influence the stability, functionality, and the formation properties of nanoemulsions [3–5].

Nanoemulsions are widely used in functional processed foods to preserve, and supply nutraceuticals, hence gaining significant attention in last decade from the research community. There are numerous studies which elucidate the predominance of nanoemulsions in food related products and delivery of biologically active ingredients. A nanoemulsion owing to its inherent property prevents degradation of biologically active ingredients and protects them from various factors such as pH, oxidation, temperature variations, and enzymatic reactions. A large number of phytochemicals like polyphenols, flavonoids, etc. are used as a functional food due to their numerous potential health benefits. Unfortunately, most of them are lipophilic in nature and possess poor water solubility. Thus, these physicochemical properties of nutraceuticals cause hindrance in their absorption *in vivo* [5].

The notable peculiarities in functional, physicochemical, and biological characteristics of biologically active compounds have boosted research interest in the development of nano-dimensional structures for enhancing their bioavailability [6–8].

In the process of nanoemulsification, various ingredients are loaded into nanocarriers (carotenoids, phenolics, vitamins, etc.) to improve their physicochemical properties and targeted drug delivery [9, 10]. In order to study these physicochemical properties which, include particle size, state of crystallization, degree of polymerization and surface chemistry (zeta potential and type of functional group), different evaluation techniques methods are employed, and this process is called characterization of nanoemulsions [11]. Nanoemulsion characterization can provide valuable information regarding the characteristics of encapsulated food items. Nano-foods have to be characterized using multiple methods involving various processes, including molecular study through physicochemical investigations, evaluation of carriers used, and kinetics of active ingredients of bio-active foods [12]. Particle size is characterized by focusing the investigation on particle size and morphological properties. Dynamic light scattering (DLS) is a widely used technique to determine the size distribution of nanoemulsions. However, it provide hydrodynamic diameter which is calculated by using the stokes equation. The hydrodynamic particle size depends on the translation diffusion coefficient and additional surface structures adhere to the particles and ions in the medium. Furthermore, it cannot be used to study the morphology of a nanoemulsion. To address these problems, microscopic techniques are best to determine particle size distribution as well as morphology of nanoemulsions. Microscopic techniques are both effective in characterization as well as in stability studies of nanoemulsions. Microscopy techniques also provide the exact visualization and higher resolution imaging needed for the particle size determination of the nanoemulsions.

8.2 IMAGING TECHNIQUES FOR CHARACTERIZATION OF NANOEMULSIONS

Shape and surface properties (morphology) of nanoemlulsions play a significant role in delivery of bioactives/nutraceuticals. Shape of the particles and its local geometry and orientation plays an important role in determining whether or not the particle is phagocytosed. Microscopy can be used to study the details of the nanoemulsions' size, shape and aggregation status. Nowadays, transmission electron microscopy (TEM), scanning electron microscopy (SEM), and atomic force microscopy (AFM), are widely employed in studying the size and morphological characteristics of nanoemulsions [13–15]. Confocal laser scanning microscopy (CLSM) is an advanced imaging technique in which a nanoemulsion is tagged with a fluorescent dye to capture the image. This technique can be utilized to study the pathways of absorption for nanoemulsion across biological membranes. Hence, the fundamentals of imaging techniques like TEM, SEM, AFM, and CLSM are discussed in this chapter. The comparison of these different imaging techniques is also summarized in Table 8.1.

8.2.1 Transmission Electron Microscopy

Methods of electron microscopy are extensively utilized for the characterization of nanostructures using an accelerated beam focused via magnetic lenses as a lighting source [16]. A schematic representation of TEM is shown in Figure 8.1. TEM is an imaging technique with a resolution of 0.2 nm [17, 18]. It has wide application in analyzing samples in material science/metallurgy and biological sciences. Characterization of surface properties and structure via electron microscopy is immensely helpful in determining physicochemical characteristics of nano-carriers which will in turn help in developing stable nano-formulations. TEM requires samples to be thinned and can bear the high vacuum pressure inside the instrument. As with most of the techniques, this also has some drawbacks [17, 18]. As the samples are required to be extremely thin to make

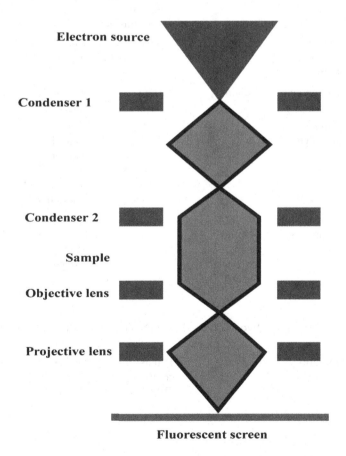

FIGURE 8.1   Schematic illustration of a transmission electron microscopy.

TABLE 8.1   Comparison of Different Imaging Techniques for Characterization of Nanoemulsions

S. No.	Parameters/ Condition	SEM	TEM	AFM	CSLM
1.	Principle	Emission of electrons	Scattering of electrons	Physical interaction with sample	Reflecting the light of the specimen or by stimulating fluorophores applied to specimen
2.	Sample preparation	Invasive, time consuming, requires coating with conducting material like gold, silver	Invasive, requires no coating, positive or negative staining can be done Copper grids are used for sample of nanoemulsion	Invasive, simple drying forming the film, ease of sample preparation	Non-invasive, easy, no sample fixation or drying
3.	Resolution/ detection limit	1 nm	0.1 nm	1 nm (XY), 0.1 nm (Z)	200 nm
4.	Environment	Vacuum	Vacuum	Vacuum/air/ liquid	Air/liquid
5.	Material sensitivity			Equal sensitivity for all material like organic and inorganic matter	Fluorophore is required
6.	Parameters studied	Size and shape, particularly surface	Size and shape	Size, shape, and roughness of surface	Structure particularly loading of drug tagged with fluorophore
7.	Dimensions	Two (X,Y)	Two (X,Y)	Three (X,Y,Z)	Two dimension
8.	S/N ratio	Low	High as compared to SEM	Highest as compared to SEM and TEM	—

them electron transparent in TEM, this may lead to loss of their structure, and there is a possibility of damage to samples by an electron beam. On the other hand, it is time-consuming also. Chuacharoen et al utilized TEM for studying the morphological parameters of curcumin loaded nanoemulsions. The TEM images revealed curcumin loaded nanoemulsion with spherical shapes and smooth surfaces having a mean particle size of approximately 193.93 nm [19]. Recently, the size and morphology of berberine loaded nanoemulsions were evaluated by using TEM which has shown their spherical shape with uniform size distribution [20]. In another study, TEM analysis showed spherical droplets of 80–150 nm in 50 min sonicated thymol loaded nanoemulsion. The results of physical size and surface architecture of thymol loaded nanoemulsion was further confirmed by performing Cryo-field emission scanning electron microscopy (Cryo-FESEM) analysis. Bright and smooth surfaced spherical nano-droplets of 90–180 nm were also noticed in Cryo-FESEM analysis [21].

Ghazy and associates evaluated the morphology and particle size of anise extract nanoemulsions by using TEM. The TEM image showed that the average size of 400 nm, is relatively smaller than that obtained from the DLS measurement. This difference in the particle size could be explained based on the principle utilized in these techniques. TEM measures the average particle size while DLS measures the hydrodynamic radius of the solvated nanodroplets surrounded by solvent molecules [22]. On the other hand, TEM observes the sample in the dry state giving a real size and shows the morphology of the dried sample. The study indicated no aggregation of the droplets, meaning that the emulsion droplets kept their identity intact during drying.

The coating of nanoemulsion with polymers is also evaluated by using TEM images. The TEM micrographs of a beta carotene loaded nanoemulsions indicated spherical shape with smooth surfaces and a mean droplet diameter around 64 nm while that of a water-soluble chitosan coated beta carotene loaded nanoemulsion showed a mean droplet diameter of around 218 nm [23]. The water-soluble chitosan coated beta carotene loaded nanoemulsions exhibited a rough surface and two distinct layers, which demonstrates that the polymer was successfully coated onto the surface of the droplets. The formation of this coating also explains the larger particle size of the coated nanoemulsions [23].

8.2.2 Scanning Electron Microscopy

SEM is a visualization technique of the surface involving the detection of scattered electrons produced from the surface of the nanobiomaterials like nanoemulsions. In SEM, electrons are scanned all over the surface of the particles and it produces high-resolution images of the particle. The nanoemulsion droplets less than 10 nm in size can also be revealed [17, 24]. Although, images produced by TEM are of higher resolution than that of SEM, but images produced by SEM provide a good view of the surface of the sample with inclusive depth of field. It can be applied to bulky samples which are impossible to analyze in TEM. This enables SEM to focus on a large number of samples at one time with a significant large depth of field available. SEM can be utilized to produce high-resolution images with higher magnification. All the above-mentioned properties of SEM such as ease of sample preparation, high magnification with better resolution, and large depth of field make SEM one of the favored techniques in scientific studies [17]. A schematic representation of SEM is given in Figure 8.2.

Now coming to drawbacks, it also requires high sample conductivity and high vacuum making it an expensive technique. As most of the nanoemulsions prepared

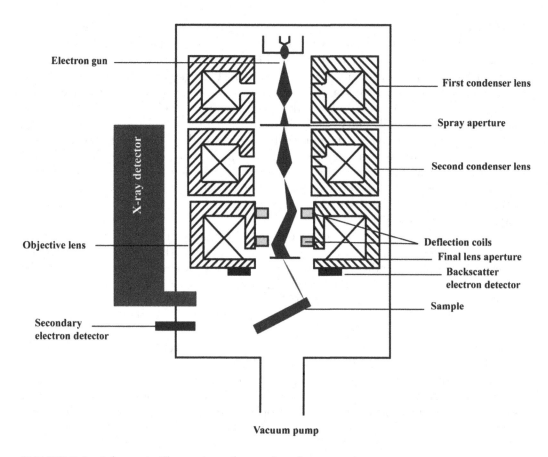

Electron gun

First condenser lens

Spray aperture

Second condenser lens

X-ray detector

Objective lens

Deflection coils
Final lens aperture
Backscatter electron detector

Sample

Secondary electron detector

Vacuum pump

FIGURE 8.2 Schematic illustration of scanning electron microscopy.

using surfactants, the presence of surfactant can hinder the characterization of the nanoemulsion as they bind to form a coating on the particle surface making it opaque for detection [17].

A nanoemulsion containing mint and parsley was studied using SEM after encapsulation in chitosan matrix. SEM images revealed the presence of chitosan resulting in suitable nucleation and the smaller size of the nanoemulsion [25].

Cryo-scanning electron microscopy (CSEM) is an advanced technique for visualization of a colloidal system like nanoemulsions in their natural forms. The oil in water pepper nanoemulsion was observed by using CSEM techniques and this technique indicated spherical droplets with sufficient uniformity. The CSEM study of pepper nanoemulsion also revealed that oil droplets are surrounded by the aqueous phase hence, the prepared nanoemulsion is an oil in water (o/w) type [26].

8.2.3 Atomic Force Microscopy

The AFM technique has been widely applied to obtain morphological information of nanobiomaterials [17, 27, 28]. It is capable of resolving surface details down to 0.01 nm and producing a contrasting three dimensional image of the sample based on the force that acts between the surface of the particles and probing tip. Additional striking advantage of using AFM as compared to electron microscopy is that vacuum is not required during operation and sample does not need to be conductive. The major advantage of AFM is that it can be used to evaluate minute details like soft surfactant layer covering the particles that can be detected while other microscopic techniques such as SEM, TEM, and DLS are not capable to determine such interactions. A schematic representation of AFM is depicted in Figure 8.3.

AFM can be employed in the study of bio-molecules such as proteins, liposomes, and polysaccharides [17, 29]. There are three common modes of AFM: contact mode, non-contact mode and tapping mode. In the contact mode, the probing tip scans the surface of the sample at a very low force. The tip exerts a mechanical load on the sample, hence this is not suitable for surface characterization of soft biomaterials. In non-contact mode, the tip of the probe does not make direct contact with the surface of the sample and this mode is preferable over the contact mode as it does not lead to sample degradation effects. However, the non-contact mode is also not commonly used for nanobiomaterials because

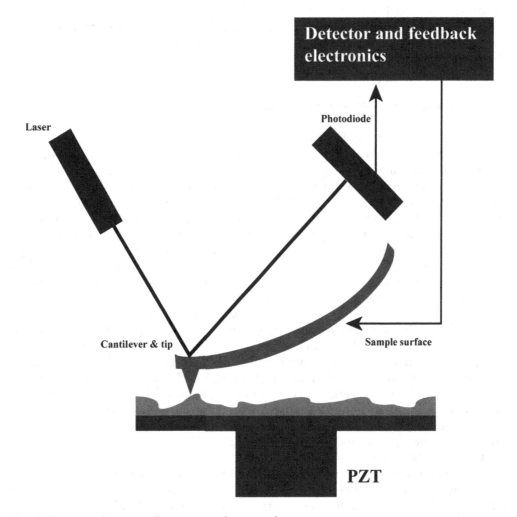

FIGURE 8.3   Schematic illustration of atomic force microscopy.

most of them are hydrophilic and because they develop a liquid meniscus layer under ambient conditions. The tapping mode, third mode of AFM, also known as intermittent contact mode is widely used to characterize nanobiomaterials. Tapping-mode AFM is even gentle enough for the visualization of supported lipid monolayers under liquid medium as confirmed by Ebeling and co-workers in 2006 [30]. In one of the studies, AFM revealed the difference between nanoemulsions and nanocapsules [31]. Galvao and associates [17] developed an o/w pepper nanoemulsion by high pressure homogenization and evaluated its surface morphology by employing AFM. They have reported two- and three-dimensional images of undefined shapes of nanoemulsions in AFM micrographs [26]. However, Cryo-SEM revealed the spherical shape of pepper nanoemulsions in the same study.

AFM images of the cinnamaldehyde (CA) loaded nanoemulsions prepared under optimal conditions also showed an irregular spherical structure. The average particle size of the CA loaded nanoemulsions in the AFM image was found to be larger than that measured by the DLS [32]. The reason of differences in the particle size might be that nanoemulsions were dried at room temperature before AFM observation, resulting in flattening of the droplets.

The shape of nanoemulsion was found to be deformed under AFM which could be due to weak adsorption of droplet on the surface of the substrate while CSEM images showed their spherical shape [26].

8.2.4 Confocal Laser Scanning Microscopy

Since the eighties, CLSM has been used in food science. A number of reviews are available discussing the application of CLSM in studying food characteristics [33–35] as compared to conventional techniques [36]. CLSM is very useful in studying the high-fat foods owing to its optical sectioning capability without the destruction of fat globules [37]. Minimum sample preparation is required for CLSM. As it is a non-invasive technique it can be engaged for studying very thin specimens and provide high-resolution three-dimensional reconstructions making it an appreciated tool for examination and characterization of functional food nanoemulsions [38, 39]. As discussed earlier CLSM is a nondestructive and comfortable technique and can differentiate between various constituents of micro and nano-capsules using multicolor fluorescence staining and labeling [40]. A schematic representation of CLSM is given in Figure 8.4.

However, this method also has some drawbacks like the assortment of appropriate lasers with well-organized fluorophore excitations. CLSM is a safe technique but it can be destructive to some viable tissues and fluorophores, due to high power laser illumination. Moreover, many fluorophores are subtle to the light intensity of laser illumination which can be a limitation in analyzing samples [41]. Kim J et al. studied the permeation profiles of a capsaicin nanoemulsion. CLSM images revealed that a capsaicin nanoemulsion could well permeate until 32 μm depth with the strongest fluorescent near the hair follicles and in all skin layers from the stratum corneum to the dermis without any physical sectioning owing to its noninvasive nature [42].

CLSM study of nanoemulsions was carried out to confirm the results of particle size by using the light scattering technique. A CLSM image of coarse emulsions loaded with mangostin extracts showed that particle size of these nanoemulsions emulsified by different oil phases and 5% mixed surfactants with varying HLB values were less than 100 nm [43].

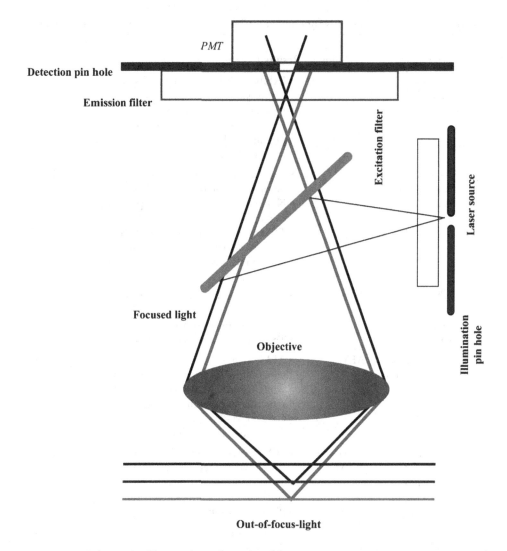

FIGURE 8.4 Schematic illustration of confocal laser scanning microscopy. PMT = photomultiplier tube.

In recent years, many research studies utilized these imaging techniques for characterization of nanoemulsions containing bioactive/nutraceuticals. Some of the research studies are summarized in Table 8.2.

8.3 CONCLUSION

Nanoemulsions can enhance stability and bioavailability of food additives and nutraceuticals. The particle size, shape, and surface characteristics of a nanoemulsion can influence its biological interaction like its cellular uptake mechanism, phagocytosis, targeting to tumor cells, etc. Several advanced and novel techniques are available for the study of these parameters of nanoemulsions, but microscopic techniques are of the utmost importance. Particle size distribution, shape, and surface characteristics can be determined by using SEM, TEM, AFM, and CLSM.

TABLE 8.2 Some Examples of Nanoemulsion Characterized by Different Imaging Techniques

S. No	Formulation	Imaging Technique	Results/Outcome	References
1.	Curcumin-loaded nanoemulsion	TEM	• Exhibits spherical shape and smooth surfaces. • Mean particle size of 194 nm. • Exhibits smaller mean particle size (87 nm) with high surfactant concentration.	[19]
2.	Quercetin-loaded nanoemulsion	TEM	• Double layered nanoemulsions are observed in the range of pH of 4–9. • Nanoemulsions prepared at pH 4 tend to clump and exhibit in TEM. • TEM image exhibits spherical and oval shape with droplet structure of nanoemulsions prepared at pH of 5,6,7, 8, and 9.	[44]
3.	(O/W) nanoemulsion of mint and parsley	SEM	• Particle size of nanoemulsions without and with chitosan is found to be 75 and 50 nm, respectively. • The nanoemulsions are spherical in shape.	[25]
4.	Pepper-loaded nanoemulsion	AFM	• AFM images reveal the droplets of nanoemulsions with undefined shapes while Cryo-SEM shows a clear spherical structure of the nanoemulsion. • Cryo-SEM also reveals the oil phase surrounded by aqueous phase.	[26]
5.	Curcumin-loaded nanoemulsion	TEM and SEM	• SEM shows the smooth and round surface of nanoemulsion. • TEM displays the spherical particle shape of <100 nm diameter, i.e. 93.64 ± 6.48 nm globule size.	[45]
6.	Rutin-loaded nanoemulsion	TEM	• TEM study reveals that most of the nanoemulsion droplets containing Rutin are in spherical shape.	[46]
7.	Eucalyptus essential oil loaded nanoemulsion	TEM	• TEM reveals that the size of nanoemulsion is in the range of nano-size and spherical in shape.	[47]
8.	Capsaicin-loaded nanoemulsions	TEM	• The TEM images indicate the presence of a droplet-type nanoemulsion structure, which exists in the single-, double-, and triple-layer forms. • The hydrophilic portions of the droplets are stained black, while the hydrophobic components are unstained and have a size range between 25 and 70 nm.	[48]

(*Continued*)

TABLE 8.2   (Continued)

S. No	Formulation	Imaging Technique	Results/Outcome	References
9.	Apigenin-loaded nanoemulsion	TEM and CLSM	• TEM reveals the nano-size of the prepared nanoemulsions and its uni-lamellar structure. • CSLM reveals the deposition of nanoemulsions containing rhodamine in the skin layers.	[49]
10.	Capsaicin-loaded nanoemulsions	TEM and CLSM	• TEM exhibits the spherical and a uniform emulsion droplet distribution of nanoemulsions. • CLSM reveals the deposition of Nile red dye in skin layers. Nile red has the ability to emit strong fluorescence, and its lipophilicity (log P = 3.10) is similar to Capsaicin (log P = 3.81). • The stratum corneum layer (10–20 μm thick) shows the strongest fluorescence intensity. • The red fluorescence intensity gradually decreases from stratum corneum to dermis. This is observed through all skin layers up to 700 μm thick.	[42]
11.	Black cumin essential oil nanoemulsions	CLSM	• Mean droplet diameter of prepared nanoemulsion containing Nile red dye is measured using CLSM. • CLSM image shows stability of a prepared nanoemulsion with no sign of flocculation and coalescence after one month of storage.	[50]
12.	Mangostin extract-loaded nanoemulsion	CLSM	• CLSM observation to confirm the results of light scattering data. • Droplets of nanoemulsions loaded with mangostin extracts made by mixed surfactants with HLB values ranging from 10.2 to 15.1 are globular and had a uniformly distributed nanoscale size.	[43]
13.	Vitamin E loaded naringenin nanoemulsion	CLSM	• CLSM study shows a higher intensity of fluorescent in brain upon administration of fluorescent tagged nanoemulsion by intranasal route than that of IV route and solution by intranasal delivery.	[51]
14.	Thymol-loaded nanoemulsion	TEM, cryo-FESEM	• TEM reveals the spherical shape of nanoemulsion droplets in size range of 80–150 nm. • Cryo-FESEM shows bright and smooth surfaced spherical nano-droplets of 90–180 nm.	[21]
15.	Cinnamaldehyde-loaded nanoemulsion	AFM	• AFM images of the cinnamaldehyde-loaded nanoemulsion exhibits an irregular spherical structure. • The average particle size of nanoemulsions in the AFM image is larger than that measured by the laser DLS.	[32]

CONFLICT OF INTEREST

The authors declare no conflict of interest.

REFERENCES

1. Acosta, Edgar. "Bioavailability of nanoparticles in nutrient and nutraceutical delivery." *Current opinion in colloid & interface science* 14, no. 1 (2009): 3–15.
2. Mcclements, David Julian. "Critical review of techniques and methodologies for characterization of emulsion stability." *Critical reviews in food science and nutrition* 47, no. 7 (2007): 611–649.
3. Tadros, Tharwat, P. Izquierdo, J. Esquena, and C. Solans. "Formation and stability of nano-emulsions." *Advances in colloid and interface science* 9 (2004): 108–109.
4. Wooster, Tim J., Matt Golding, and Peerasak Sanguansri. "Impact of oil type on nanoemulsion formation and Ostwald ripening stability." *Langmuir* 24, no. 22 (2008): 12758–12765.
5. McClements, David Julian, and Jiajia Rao. "Food-grade nanoemulsions: formulation, fabrication, properties, performance, biological fate, and potential toxicity." *Critical reviews in food science and nutrition* 51, no. 4 (2011): 285–330.
6. Jafari, Seid Mahdi, and David Julian McClements. "Nanotechnology approaches for increasing nutrient bioavailability." In *Advances in food and nutrition research*, vol. 81, pp. 1–30. Academic Press, 2017.
7. Rostamabadi, Hadis, Seid Reza Falsafi, and Seid Mahdi Jafari. "Nanoencapsulation of carotenoids within lipid-based nanocarriers." *Journal of controlled release* 298 (2019): 38–67.
8. Saravanan, P., R. Gopalan, and V. Chandrasekaran. "Synthesis and Characterisation of Nanomaterials." *Defence science journal* 58, no. 4 (2008).
9. Arpagaus, Cordin, Andreas Collenberg, David Rütti, Elham Assadpour, and Seid Mahdi Jafari. "Nano spray drying for encapsulation of pharmaceuticals." *International journal of pharmaceutics* 546, no. 1–2 (2018): 194–214.
10. Human, Chantelle, Dalene De Beer, Marieta Van Der Rijst, Marique Aucamp, and Elizabeth Joubert. "Electrospraying as a suitable method for nanoencapsulation of the hydrophilic bioactive dihydrochalcone, aspalathin." *Food chemistry* 276 (2019): 467–474.
11. Ganesan, Poovi, and Damodharan Narayanasamy. "Lipid nanoparticles: Different preparation techniques, characterization, hurdles, and strategies for the production of solid lipid nanoparticles and nanostructured lipid carriers for oral drug delivery." *Sustainable chemistry and pharmacy* 6 (2017): 37–56.
12. Pancholi, Ketan. "A review of imaging methods for measuring drug release at nanometre scale: a case for drug delivery systems." *Expert opinion on drug delivery* 9, no. 2 (2012): 203–218.
13. Lal, Ratnesh, Srinivasan Ramachandran, and Morton F. Arnsdorf. "Multidimensional atomic force microscopy: a versatile novel technology for nanopharmacology research." *The AAPS journal* 12, no. 4 (2010): 716–728.
14. Sitterberg, Johannes, Aybike Özcetin, Carsten Ehrhardt, and Udo Bakowsky. "Utilising atomic force microscopy for the characterisation of nanoscale drug delivery systems." *European journal of pharmaceutics and biopharmaceutics* 74, no. 1 (2010): 2–13.

15. Turner, Ya Tsz A., Clive J. Roberts, and Martyn C. Davies. "Scanning probe microscopy in the field of drug delivery." *Advanced drug delivery reviews* 59, no. 14 (2007): 1453–1473.

16. Ahmad, Mudasir, Priti Mudgil, Adil Gani, Fathalla Hamed, Farooq A. Masoodi, and Sajid Maqsood. "Nano-encapsulation of catechin in starch nanoparticles: Characterization, release behavior and bioactivity retention during simulated in-vitro digestion." *Food chemistry* 270 (2019): 95–104.

17. Luykx, Dion MAM, Ruud JB Peters, Saskia M. van Ruth, and Hans Bouwmeester. "A review of analytical methods for the identification and characterization of nano delivery systems in food." *Journal of agricultural and food chemistry* 56, no. 18 (2008): 8231–8247.

18. Wang, Z. L. "Transmission electron microscopy of shape-controlled nanocrystals and their assemblies." *Journal of physical chemistry B* 104 (2000): 1153–1175.

19. Chuacharoen, Thanida, Sehanat Prasongsuk, and Cristina M. Sabliov. "Effect of surfactant concentrations on physicochemical properties and functionality of curcumin nanoemulsions under conditions relevant to commercial utilization." *Molecules* 24, no. 15 (2019): 2744.

20. Sharifi-Rad, Atena, Jamshid Mehrzad, Majid Darroudi, Mohammad RezaSaberi, and Jamshidkhan Chamani (2020): Oil-in-water nanoemulsions comprising Berberinein olive oil: Biological activities, binding mechanisms to human serum albumin or holotransferrin, and QMMD simulations, *Journal of biomolecular structure and dynamics*. doi:10.1080/07391102.2020.1724568

21. Sarita Kumari, R. V. Kumaraswamy, Ram Chandra Choudhary, S. S. Sharma, Ajay Pal, Ramesh Raliya, Pratim Biswas, and Vinod Saharan. Thymol nanoemulsion exhibits potential antibacterial activity against bacterial pustule disease and growth promotory effect on soybean *Scientific reports* 2018; 8: 6650. Published online 2018 Apr 27. doi: 10.1038/s41598-018-24871-5

22. Ghazy, O.A, M.T. Fouad, H.H. Saleha, A.E. Kholif, and T.A. Morsy. Ultrasound-assisted preparation of anise extract nanoemulsion and its bioactivity against different pathogenic bacteria, *Food chemistry* 341 (2021) 128259.

23. Eun Joo Baek, Coralia V. Garcia, Gye Hwa Shin, and Jun Tae Kim. Improvement of thermal and UV-light stability of β-carotene-loaded nanoemulsions by water-soluble chitosan coating. *International journal of biological macromolecules* 165 (2020): 1156–1163.

24. Reimer, Ludwig. "Scanning electron microscopy: physics of image formation and microanalysis." *Measurement science and technology* 11, no. 12 (2000): 1826.

25. Najmeh Feizi Langaroudi, Najmeh, and Negar Motakef Kazemi. "Preparation and characterization of O/W nanoemulsion with Mint essential oil and Parsley aqueous extract and the presence effect of chitosan." *Nanomedicine research journal* 4, no. 1 (2019): 48–55.

26. Galvão, K. C. S., A. A. Vicente, and P. J. A. Sobral. "Development, characterization, and stability of O/W pepper nanoemulsions produced by high-pressure homogenization." *Food and bioprocess technology* 11, no. 2 (2018): 355–367.

27. Edwards, Katie A., and Antje J. Baeumner. "Liposomes in analyses." *Talanta* 68, no. 5 (2006): 1421–1431.

28. Ruozi, Barbara, Giovanni Tosi, Flavio Forni, Massimo Fresta, and Maria Angela Vandelli. "Atomic force microscopy and photon correlation spectroscopy: two techniques for rapid characterization of liposomes." *European journal of pharmaceutical sciences* 25, no. 1 (2005): 81–89.

29. Moraru, Carmen I., Chithra P. Panchapakesan, Qingrong Huang, Paul Takhistov, Sean Liu, and Jozef L. Kokini. "Nanotechnology: a new frontier in food science understanding the special properties of materials of nanometer size will allow food scientists to design new, healthier, tastier, and safer foods." *Nanotechnology* 57, no. 12 (2003).

30. Ebeling D, Holscher H, Fuchs H, Anczykowski B, and Schwarz UD "Imaging of biomaterials in liquids: a comparison between conventional and Q-controlled amplitude modulation ('tapping mode') atomic force microscopy" *Nanotechnology* 17 (2006): S221–226.

31. Preetz, Claudia, Anton Hauser, Gerd Hause, Armin Kramer, and Karsten Mäder. "Application of atomic force microscopy and ultrasonic resonator technology on nanoscale: Distinction of nanoemulsions from nanocapsules." *European journal of pharmaceutical sciences* 39, no. 1–3 (2010): 141–151.

32. Mingyu Ji Jiulin, Wu Xinyu Sun, Xiaoban Guo, Wenjin Zhu, Qingxiang Li Xiaodan Shi, Yongqi Tian, and Shaoyun Wang. Physical properties and bioactivities of fish gelatin films incorporated with cinnamaldehyde-loaded nanoemulsions and vitamin C. *LWT* 135 (2021): 110103.

33. Marangoni, Alejandro G., and R. W. Hartel. "Visualization and fat structural analysis of fat crystal networks." *Food technology (Chicago)* 52, no. 9 (1998): 46–51.

34. Heertje, I., P. Vlist, J. C. G. Blonk, H. A. C. M. Hendrickx, and G. J. Brakenhoff. "Confocal scanning laser microscopy in food research: some observations." *Food Structure* 6, no. 2 (1987): 2.

35. Blonk, J. C. G., and H. Van Aalst. "Confocal scanning light microscopy in food research." *Food research international* 26, no. 4 (1993): 297–311.

36. Herrera, M. L., and R. W. Hartel. "Unit D 3.2. 1-6 lipid crystalline characterization, basic protocole." In *Current protocols in food analytical chemistry (CPFA)*. New York: John Wiley & Sons, Inc., 2001.

37. Ong, Lydia, Raymond R. Dagastine, Sandra E. Kentish, and Sally L. Gras. "Microstructure of milk gel and cheese curd observed using cryo scanning electron microscopy and confocal microscopy." *LWT-food science and technology* 44, no. 5 (2011): 1291–1302.

38. Lamprecht, A., U. F. Schäfer, and C-M. Lehr. "Characterization of microcapsules by confocal laser scanning microscopy: structure, capsule wall composition and encapsulation rate." *European journal of pharmaceutics and biopharmaceutics* 49, no. 1 (2000): 1–9.

39. Mandal, Subhra, You Zhou, Annemarie Shibata, and Christopher J. Destache. "Confocal fluorescence microscopy: An ultra-sensitive tool used to evaluate intracellular antiretroviral nano-drug delivery in HeLa cells." *AIP advances* 5, no. 8 (2015): 084803.

40. Vandenbossche, Geert MR, Patric Van Oostveldt, and Jean Paul Remon. "A fluorescence method for the determination of the molecular weight cut-off of alginate-polylysine microcapsules." *Journal of pharmacy and pharmacology* 43, no. 4 (1991): 275–277.

41. Dürrenberger, Markus B., Stephan Handschin, Béatrice Conde-Petit, and Felix Escher. "Visualization of food structure by confocal laser scanning microscopy (CLSM)." *LWT-Food science and technology* 34, no. 1 (2001): 11–17.

42. Kim, Jee Hye, Jung A. Ko, Jun Tae Kim, Dong Su Cha, Jin Hun Cho, Hyun Jin Park, and Gye Hwa Shin. "Preparation of a capsaicin-loaded nanoemulsion for improving skin penetration." *Journal of agricultural and food chemistry* 62, no. 3 (2014): 725–732.

43. Sungpud, Chatchai, Worawan Panpipat, Manat Chaijan, and Attawadee Sae Yoon. "Techno-biofunctionality of mangostin extract-loaded virgin coconut oil nano-emulsion and nanoemulgel." *Plos one* 15, no. 1 (2020): e0227979.

44. Son, Hye-Yeon, Mak-Soon Lee, Eugene Chang, Seog-Young Kim, Bori Kang, Hyunmi Ko, In-Hwan Kim et al. "Formulation and characterization of quercetin-loaded oil in water nanoemulsion and evaluation of hypocholesterolemic activity in rats." *Nutrients* 11, no. 2 (2019): 244.

45. Ahmad, Niyaz, Rizwan Ahmad, Ali Al-Qudaihi, Salman Edrees Alaseel, Ibrahim Zuhair Fita, Mohammed Saifuddin Khalid, and Faheem Hyder Pottoo. "Preparation of a novel curcumin nanoemulsion by ultrasonication and its comparative effects in wound healing and the treatment of inflammation." *RSC advances* 9, no. 35 (2019): 20192–20206.

46. Ahmad, Mohammad, Juber Akhtar Sahabjada, Arshad Hussain, Md Arshad Badaruddeen, and Anuradha Mishra. "Development of a new rutin nanoemulsion and its application on prostate carcinoma PC3 cell line." *Excli journal* 16 (2017): 810.

47. Alam, Prawez, Faiyaz Shakeel, Md Khalid Anwer, Ahmed I. Foudah, and Mohammed H. Alqarni. "Wound healing study of eucalyptus essential oil containing nanoemulsion in rat model." *Journal of oleo science* (2018): ess18005.

48. Choi, Ae-Jin, Chul-Jin Kim, Yong-Jin Cho, Jae-Kwan Hwang, and Chong-Tai Kim. "Characterization of capsaicin-loaded nanoemulsions stabilized with alginate and chitosan by self-assembly." *Food and bioprocess technology* 4, no. 6 (2011): 1119–1126.

49. Jangdey, Manmohan S., Anshita Gupta, and Swarnlata Saraf. "Fabrication, in-vitro characterization, and enhanced in-vivo evaluation of carbopol-based nanoemulsion gel of apigenin for UV-induced skin carcinoma." *Drug delivery* 24, no. 1 (2017): 1026–1036.

50. Sharif, Hafiz Rizwan, Shabbar Abbas, Hamid Majeed, Waseem Safdar, Muhammad Shamoon, Muhammad Aslam Khan, Muhammad Shoaib, Husnain Raza, and Junaid Haider. "Formulation, characterization and antimicrobial properties of black cumin essential oil nanoemulsions stabilized by OSA starch." *Journal of food science and technology* 54, no. 10 (2017): 3358–3365.

51. Gaba, Bharti, Tahira Khan, Md Faheem Haider, Tausif Alam, Sanjula Baboota, Suhel Parvez, and Javed Ali. "Vitamin E loaded naringenin nanoemulsion via intra-nasal delivery for the management of oxidative stress in a 6-OHDA Parkinson's disease model." *BioMed research international* 2019 (2019).

CHAPTER 9

Separation Techniques for Characterization of Nanoemulsions

Javed Ahamad,[1] Abdul Samad,[2]
Showkat R. Mir,[3] Jamia Firdous,[4]
Javed Ahmad,[5] Leo M.L. Nollet[6]

[1]Department of Pharmacognosy, Faculty of Pharmacy,
Tishk International University, Kurdistan Region, Iraq
[2]Department of Pharmaceutical Chemistry, Faculty of Pharmacy,
Tishk International University, Kurdistan Region, Iraq
[3]Department of Pharmacognosy, School of Pharmaceutical
Education & Research, Jamia Hamdard, New Delhi, India
[4]DS College of Pharmacy, Aligarh, India
[5]Department of Pharmaceutics, College of Pharmacy, Najran
University, Najran, Kingdom of Saudi Arabia
[6]University College, Ghent, Belgium

CONTENTS

DOI: 10.1201/9781003121121-9

9.1 INTRODUCTION

Nanoemulsions are kinetically stable liquid-in-liquid dispersions with droplet sizes of the order of 100 nm (Gupta et al. 2016). In recent years, unprecedented attention has been paid to nanotechnology in large fields of science (Bhattacharyya and Singh 2009). This nano driven technology has created new possibilities and opened several doors to personalized and safe medication delivery and medical therapy. In cancer research, HIV/AIDS therapies, non-invasive imaging and nutraceutical transmission, nanotechnology has advanced. Ultimately, researchers can deliver medication with less frequent doses and more precision and penetration into tissue over a longer time, through manipulating the molecular size and surface properties.

Solid nanoparticles are the colloidal particles between 10 and 500 nm, but the preferential size is lower than 200 nm for nanomedical applications (Biswas et al. 2014). The shape and size of nanoparticles influence the way body cells recognize them, determining their distribution, toxicity and receptor capability. The fact that nanoparticles can cross the blood-brain barrier (BBB) is another particularly important aspect, which offers sustainable provision for previously hard-to-treat diseases (McMillan et al. 2011). This technique can not only be tampered with to control bioactive molecule distribution but also to achieve new objectives. It has been reported that 100 nm nanoparticles exhibited a 2.5-fold greater uptake compared to 1 µm diameter particles and a 6-fold great uptake than 10 µm particles (Desai et al. 1997).

The size of drug formulations using nano-products can influence their efficiency, but the handling of surface properties is another challenge to develop the ideal bioactive molecule delivery system (Bantz et al. 2014). Appropriate ligands that are targeting optimum physico-chemical properties and reactivity must be included to address agglomeration, stability and receptor binding prevention, and the corresponding pharmacological effects (Khanbabaie and Jahanshahi 2012) to create an optimal bioactive molecule delivery system for nanoparticles. As it is observed that the size and pharmacological role of nanoparticles are crucial for delivery, it's important to separate specific size nanoparticles. For the identification and characterization of nanoemulsions, three basic steps are involved: separation, characterization and imaging techniques. For separation field-flow fractionation (FFF) and chromatography (size-exclusion chromatography and ion-exchange chromatography) are applied, and for the characterization nuclear magnetic resonance (NMR), matrix-assisted laser desorption/ionization (MALDI), electrospray ionization mass spectrometry (ESI-MS), and X-ray diffraction (XRD) crystallography are used; and imaging techniques such as transmission electron microscopy (TEM), scanning electron microscopy (SEM) and atomic force microscopy (AFM) are applied for nanoemulsions. In this book chapter, we thoroughly review separation techniques used for the characterization of nanoemulsions that include: FFF; size-exclusion chromatography (SEC) and ion exchange chromatography (IEC).

9.2 SEPARATION TECHNIQUES FOR CHARACTERIZATION OF NANOEMULSIONS

The field of industrial bulk manufacturing is commonly referred to as a process for any mass transfer phenomenon that transforms a mixture of substances into two or more separate product mixtures typically known as fractions. Rarely will the mixture be separated into pure components through separation techniques. In colloidal food science separation technology, the active ingredients of the various compounds are primarily classified and quantified. In the nano-colloidal food chemistry, nanoemulsions, or nano encapsulated materials have been identified by several researchers. *In situ*, however, it is often not possible to detect them in food matrices, the key problem for nanoemulsion in foodstuffs is the characterization of them. Separation methods can also be used before the characterization of nanoemulsions from the food. While some research methods can be used to detect a new delivery system (NDS) *in situ*, this is not feasible in most situations because the nutritional content interferes. The presence and recognition of protein-based NDSs can interact for example with proteins in protein-containing foods. Separation methods can also be used before classification to separate the NDS from fruit. The most important techniques for isolating NDS are listed in this section. Nanoemulsions can be somewhat categorized by the different detectors can be coupled with the mentioned separation techniques (Robertson et al. 2016). This section discusses the key nanoemulsion separation techniques such as field flow fractionation and liquid chromatography.

9.2.1 Field Flow Fractionation

Field flow fractionation (FFF) is liquid chromatography, a process based on elution, where separation is made to happen in a single phase. FFF is described by an external, perpendicular channel application through a small, vacuum-like channel to the path of material flow, as shown in Figure 9.1. The laminar parabolic flow profile evolves because of the high aspect ratio of the FFF channel, with flow speeds varying from almost zero in the channel walls and reaching the most in the middle of the channel. The force applied

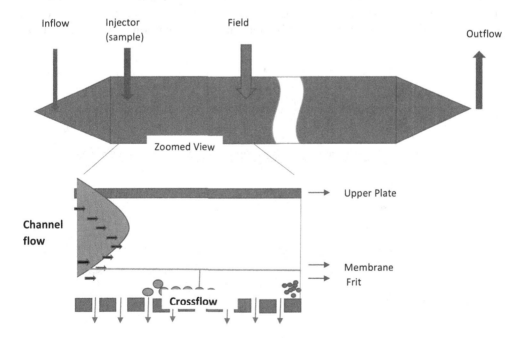

FIGURE 9.1   Schematic diagram representing FFF technique.

perpendicularly pushes the sample against the wall of accumulation. The wall becomes intensified and the analyte becomes pushed back into the middle of the channel by counteracting diffuse energy. If a balance of forces is carried out, the equilibrium is stable, and the concentration profile of the exponential analyte is established. Retention happens when analytes are found in flow rate areas which are lower than their normal flow velocity. Separation happens due to the presence of analytes in various flow-speed regions (Litzen, 1993).

The most often used mechanism, where diffusion plays a significant role in regulating channel-wide delivery of components, is the normal separation mode (Giddings 1993). Figure 9.1 explains the fundamental concept of the operation of FFF in standard mode. A fractogram is indeed the representation of an elution time curve. Separation of analytes depends on the applied method (operation modes) in FFF by various opposite forces in addition to other separation characteristics such as selectivity and resolution. The mode of operation sets the elution of analytes.

Three common, implied modes are: The fractogram is the expression of the detector against the elution time curve. The analytes may be separated by defined processes in FFF (operative modes) emerging out of different counterfeiting forces. The mode shall determine the sample component's elution sequence and other attributes of separation also including selection criteria with resolution capacity. The *regular, steric, and hyper layer* modes that can be applied in any FFF technique are the three most frequently used ones as depicted by Figure 9.2. The standard common method is utilized for analytical samples having sizes smaller than <1 µm, by following the principle of Brownian motion (Myers 1997; Chmelik 1999).

The populational species with smaller components aggregates faster than a larger constituent in places of higher parabola speed sources and elute populations. The steric mode refers to materials greater than 1 nm, where there is insignificant diffusion, although retention is confined to the nearest wall of aggregation. Tiny particles will come closer than big particles to the aggregation surface, and the centre of mass of tiny particle's is in the slowest streamlines. In steric mode, the elution order is from larger

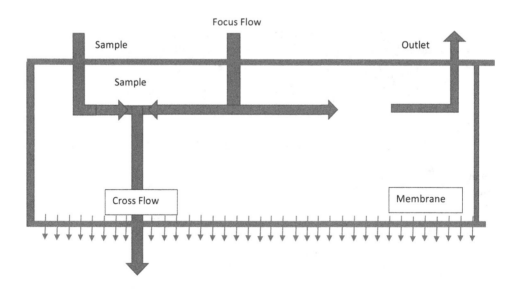

FIGURE 9.2   Schematic diagram indicating flow direction of AsFl-FFF.

to smaller. Finally, lifting or hyper-layer mode contributes to higher-speed streams of sample components positioned in more than one particle radius. The hydrodynamic lifting forces are exerted when high flow rates are used. The elution sequence is similar to that of sterics. The standard mode process is used to isolate most nanoparticles (Giddings 1993; Schimpf et al. 2000).

9.2.1.1 Types of FFF Techniques (Based on Fields and Geometries)

In FFF several fields are used in accordance with the essence of the substance to be studied. For efficient separation and convenience of operation, the optimal field strength is considered as the prime criteria. Each field type interacts with the analytes various physical and chemical properties (Giddings 1993; Schimpf et al. 2000). Typical fields of FFF technique include crossflow current, differential temperature, electric potency, centrifuge, gravity power, dielectrophorous waves and electromagnetism. This is the reason of the origin of a range of FFF techniques such as:

- Fluid FFF (Fl-FFF)
- Thermal FFF (Th-FFF)
- Electrical FFF (El-FFF)
- Sediment FFF (Sd-FFF)
- Gravitational FFF (Gr-FFF)
- Dielectrophoric FFF (DEP-FFF)
- Acoustic (Ac-FFF)
- Magnetic FFF (Mg-FFF)

In these various techniques of FFF, the analyte retention and separation are accomplished according to the various analyte properties such as thickness, thermal diffusion, load, density, mass and magnetic sensitivity (Chmelik 1999; Beckett et al. 2007; Semyonov and Maslow 1988).

The field intensity is the most relevant experimental state in FFF because it plays an instrumental role in attaining optimum time bound separation along with finest resolution possible. The intensity of the field is controlled by automated programmes in accordance with the decay function during the separation. Controlling the field strength is beneficial while analyzing large molar mass or particle distributions (Moon et al. 2002; Williams and Giddings 1994). It optimizes resolution and processing time for fractionated analyte and improves their detectability. Various forms of field programming, such as linear, parabolic, exponential and step-by-step decay functions have been used (Moon et al. 2002; Williams and Giddings 1994; Leeman et al. 2006). Crossflow and temperature gradient parameters respectively are added to field programming for Fl-FFF and Th-FFF. In general, for low mass and lighter particles greater field intensity is needed. The channel is formed with the geometrical cut-out of two flat surfaces blocks by bracing a thin spacer (usually of Mylar or polyimide). The substance of the block must be aligned with the liquid carrier and express the field applied. Since diffusion is a slow business, the channel thickness (w) must be small enough for the sample to be maintained within a suitable time period. The size of the channel should be such, that it enables optimum retention time variations amongst the analytes. This type of rectangular channel can be seen in two endless parallel plates between which the parabolic laminar flow speed can be conveniently evolved and the edge fluid disturbance reduced due to the strong aspect ratio (breadth to thickness) of 100.

In 1997, Williams et al. (1997) established a diminishing channel width, exponentially, to have consistent volumetric flow rates over the duration of the asymmetric canal. FFF also utilizes an alternative channel system in cylindrical form for the use of hollow fibres (HF). A radical cross flow would be applied to transfer the components of the fibre surface (Park et al. 2005; Se-Jong et al. 2003).

The asymmetrical field-flow fraction (As-Fl-FFF) prototype was first adopted in 1987 (Janca 2002). In this method a membrane is utilized as a mean (wall) of aggregating analytes. As-Fl-FFF is differentiated with Fl-FFF only by presence of a single permeable wall in the channel. A solid wall impermeable to the career liquid replaces the upper porous wall.

As-Fl-FFF differs from Fl-FFF as its channel only has a single, permeable wall. A solid wall, impermeable to the carrier liquid, replaces the upper porous wall. The flow is divided from a single channel into the crossflow and the standard flow of the channel. The amount of resistance felt during the inline flow shall decide the ratio of the cross flow and channel flow. As-Fl-FFF gains from Fl-FFF by a simpler design and simulation of a prototype through a transparent top wall (Wahlund and Giddings 1987). Figures 9.1 and 9.2 displays a graphical representation contrasting Fl-FFF and As-Fl-FFF. As-Fl-FFF uses sample relaxation and concentration, although only the former is used in Fl-FFF. These methods increase resolution and lower peak widths. It was noticed that As-Fl-FFF was having lesser peak heights, smaller than Fl-FFF. Fl-FFF and As-Fl-FFF suit and frit outlet have also been used to lessen the touch point of the sample with the membrane and concentrate it on-line (Li et al. 1997).

9.2.1.1.1 Analyte Retention   The velocity profile of the parabolic flow across the channel as depicted by Figures 9.1 and 9.2, can be denoted as

$$v(x) = 6v\left\{x/w - (x/w)^2\right\}, \tag{9.1}$$

where the average velocity of the liquid carrier is the width from the deposition wall (x=0, on an accumulation wall) across a thickness pipe of w and x. The analyte concentration profile (c(x)) is determined by transfer of the analyte to the wall caused by the field and diffusion of the wall in the regions with lower concentration. If these two contrary distribution processes are in equilibrium, there is no net material stream and the exponential distribution of the concentrations as defined by Eq. (9.2).

$$c(x) = c_0 e^{(-U/D)x} \tag{9.2}$$

Analyte concentration at the aggregation wall (x = 0) is denoted by C_0, D (diffusion coefficient), while U is particle velocity caused by field force F. The propagation of a brown particle is based on the frictions (f) and can be expressed by the equation of Nernst–Einstein:

d = kT/f, where k is Boltzmann, and T is temperature (absolute).

The mean layer thickness (l) is determined by the distance from the sample region to the middle region.

F = fU, so the diffusion coefficient is: D = [kT (U/F)]. The mean layer thickness (l) is determined by the distance from the middle sample region to the aggregation area. The thermal energy ratio is expressed with the applied strength: l = kT/F = D/U. The thickness (also known as the retention parameter) of the dimensionless zone is defined as the proportion (l/w). In comparison to the sample layer thickness and sample layer volume fraction, it reveals the level of compacted in areas. The concentration profile for general FFF can be seen in the expression as:

$$\lambda = kT/wF = D/Uw \qquad (9.3)$$

Eq. (9.2) can also be written as:

$$c(x) = c_o e^{\,(-x/l)}$$

or the new form becomes Eq. (9.4) as

$$c(x) = c_o e^{(-x/\lambda w)} \qquad (9.4)$$

As the distance to the accumulation wall increases, the concentration of the analyte decreases exponentially. Thus, with enough interaction with the applied external surface, the analyte's bulk should be within a few micrometres from the aggregation wall. The term represented in Eq. (9.3) follows different types of field-specific structures and describes the field-specific associations with the analyte 's various physicochemical properties. Retention is a delay in analysis areas by limiting them to streamlines that expedite at speeds less than the average speed of the liquid carrier. The retention ratio (R_r) can be defined as the ratio of the analyte area (area mean)'s velocity to the liquid carrier's average speed. The speed of R and fluid flow profiles can be calculated by:

$$R_r = V_{zone}/(v) = c(x)V(x)/c(x)v(x) \qquad (9.5)$$

Putting the values of Eqs. (9.1) and (9.2) in Eq. (9.5), an equation referring to R_r may be derived as

$$R_r = 6\lambda\{Coth(1/2\lambda\text{-}2\lambda)\} \qquad (9.6)$$

R may also be empirically determined by taking the approximate void time ratio t_0 to retention time tr. The retention times calculated at maximum are excellent approximations under non-overloading conditions when elution profiles are Gaussian distributions (Schimpf et al. 2000). For values of less than 0.02, the value R can be determined by using 6 approximations with an error of 5%:

$$R = t_0/t_r = V_0/V_r = 6\lambda, \qquad (9.7)$$

where V_0 represents the volume of the channel whereas V_r is retention volume

To measure different physiochemical parameters of analyte, the values gathered from study retention time can be used. The empirical progressive preservation of FFF was intended to be built on a variety of hypotheses, including an inter-infinite parallel flow profile. Many nanoparticles are broken in the normal mode in which smaller molecules first get eluted. The reverse is SEC, where larger molecules first elute.

9.2.1.1.2 Efficiency and Resolution   FFF retention theory uses the assumption of an exponential concentration profile of the analyte and of the parabolic flow profile. Errors in retention parameters may arise under certain conditions. FFF measurements also involve deviations between the experimental results and the expected theoretical behaviour due to various effects. Some effects, e.g., zone broadening, are unavoidable and can only be corrected empirically. Certain results, such as zone extension, are inevitable and can be self-controlled by trial and error only. The careful monitoring and calibration of the various parameters such as flow rates, concentration and temperature will reduce several effects. These deviations can be defined using online detectors, including light dispersion that provides an idea about separate size and/or molecular weight measurements. The empirical plate height (H) is a dispersion calculation expressed in the sample peak. Non-equilibrium (Hn), axial diffusion (Hd), polydispersion sample (Hp) and Certain results, such as zone extension, are inevitable and can be only empirically changed. The careful monitoring and calibration of the various parameters such as flow speeds, concentration and temperature will reduce a number of effects. Instrumentation operating effects (Hi) are key factors controlling zone expansion. The height of the plates is additive (H = Hn + Hp + Hd + Hi) and determined by these factors. Hn is the leading contributor to the calculated peak width of the non-equilibrium expression. This influence is due to the axial differentials of the zone elements, since they lie around the thickness of the channel in different speed lines. The flux rate and the measurements of the channel are based mainly on Hn. The retention parameter Hn has been found to be a complex function (Schimpf et al. 2000):

$$Hn = \mu(\lambda)\,w^2 v/D = 24\lambda^3 w^2\,(v)/D \tag{9.8}$$

In a case $\mu(\lambda)$ turn out to be equal to 24 λ3 (because λ has turned zero). Axial diffusion (Hd) describes the effect of longitudinal diffusion on the axial concentration gradients. Hd is only important when the flow rate is very low. With high molecular weights, the diffusion rate is slow and even Hd makes no or minimal contribution to the plate's height. For highly polydispersed samples, Hp's polydissipation contribution to plate height may be significant. The Hi impact can be decreased in a well-designed framework to a negligible number. Provided the peaks with a Gaussian form, plate height can be estimated from the possible number of N plate as well as from average standard deviation of "n" as seen in

$$n = L/h = (L/\delta)^2 \tag{9.9}$$

It is easy to align various instruments with selected lower h or higher n. The resolution or split power refers to the device's separation capability. It is the difference between the two pinnacles and can be expressed mathematically by the resolution index (Rs) as shown in

$$Rs = \Delta tr/4\sigma r = \Delta Vr/4\sigma v \tag{9.10}$$

σr is the standard deviation in time function and σv is standard deviation in Volume function, while ΔVr and Δtr are the volume and retention time variations, respectively. Peaks are overlapping when Rs are near zero and start separating as Rs increases. The resolution index is based on the length, height of the channel, relative weight and specificity of molecular differences. FFF's advantage is its ability to strongly differentiate materials. The calculated resolution was higher for ThFFF than SEC for three distinct binary polymer mixtures. Calculated resolution values showed that ThFFF had a substantial benefit over the SEC, with polydispersity corrections (Janca 2003).

9.2.1.1.3 Flow and Thermal Field-Flow Fractionation   This study is concerned with two forms of flow FFF: symmetric field-flow fractionation (FlFF) and asymmetrical field-flow fraction (As-FlFF). In 1976, Giddings classified symmetric Fl-FFF. The channel spacer, in this first type of flow FFF, is covered in each wall by two parallel, ceramic-fried, parallel blocks (2–5 m pores) (Schimpf et al. 2000). The "field," which is opposite to the canal path, is used as a cross-flow. On the upper wall, the crossflow runs through the pores of the pipe and escapes using an ultrafiltration membrane, replacing the bottom wall with a second permeable membrane, on the aggregation wall. The molecular weight cut off values of ultrafiltration membranes are classified: low nominal membranes prohibit smaller membrane permeating analytes. Ultrafiltration membranes of multiple types have been used in FlFF (Schimpf et al. 2000); cellulose regeneration, polyimide/poly (ethylene terephthalate), polystyrene sulfonate, polypropylene, polyether sulfone and polyacrylamide. The choice of the membrane involves close attention to membrane thickness, fluidity, mechanical and chemical uniformity, and solvent/molecular weight, and solvent/membrane relation. An asymmetrical variant of AsFlFF was first released in 1987 (Janca 2002). Here, too, a membrane is used as a storage panel. AsFl-FFF varies from Fl-FFF because the channel is only permeable to a single wall (the wall of accumulation). A sturdy wall that is resistant to the carrier liquid substitutes the upper porous wall. The current flow and crossflow are originated from a single inlet channel flow. The ratio is based on inline flow regulated by the operator. AsFl-FFF has the following benefits over Fl-FFF: streamlined design and the opportunity to display the sample through a transparent, elevated wall (Wahlund and Giddings 1987). As previously stated, it is special for any form of FFF field and interacts in various dimensions with the physicochemical properties of different analytes. At speed U (equal to V/c/Aac), and cross-fluxes of the two Fl-FFF and As-FlFF, the external field and the analyte components are pushed against the aggregation wall. Aac is proportional to channel volume (v^0) determined by the channel width on the surface region of the built-up wall. The accumulation walls surface area, i.e., Aac is equal to the volume of the channel (V0) divided by the thickness of the channel. Eq. (9.3) can be explained by:

$$= DV^0/V_c W^2, \tag{9.11}$$

where V_c is the volumetric flow rate of the cross-flow and D represents analyte's diffusion coefficient. If we put the values from Eq. (9.7) into Eq. (9.11), it generates another fresh equation as shown in

$$tr = t^0/6\lambda = T^0 W^2 V^c/6DV^0, \tag{9.12}$$

where V is volumetric flow rate along the channel: $V = V^0/t^0$. According to the Stokes-Einstein equation: $D = kT/3\pi\eta d_h$ where dynamic viscosity of the carrier fluid is presented by η, the hydrodynamic diameter (dh) can be directly calculated from the retention data.

Stokes-Einstein's replacement for D again would give a dh in t_r: $t_r = W^2/6D$ ln $\{1 + (Vc/V_{out})\}$ oscillates and concentrates in As-Fl-FFF while only the former is used for Fl-FFF. Specimen relaxation and concentrating are used in As-Fl-FFF. These approaches increase resolution and minimize peak widths. Of course, As-Fl-FFFF created smaller heights of plates, i.e., narrower strips than Fl-FFF. In addition to decreased relaxation and loading times, Fl-FFF and As-Fl-FFF have been used to minimize sample contacts with the membrane and to concentrate the isolated sample online. (Li et al. 1997).

The latter is a key feature of operating with a detector, for example MALS (Multi Angle Light Scattering), which involves higher concentrations of samples.

9.2.1.1.4 High Temperature Asymmetric Field-Flow Fractionation   Temperature elevations will increase the separation and test time (Schimpf et al. 2000). The high temperature increases the solubility of many polymers, contributes to fast diffusion and produces reduction in band growth. "Postnova Analytics (Landsberg am Lech, Germany)" developed and distributed medium-temperature asymmetric field-flow (MT AsF1FFF) and (HT AsF1FF) system of fractionation. For separation and characterization of the high molecular mass polyolefin resins, HT AsF1FFF was specifically developed. Several detectors such as IR, RI, MALS and viscometry were used. A steel channel and a flexible ceramic accumulation wall diaphragm are used in HT AsFl-FFF. This enables measuring of 1,2,4-trichlorobenzene (TCB) with chlorinated organic solvents up to 220°C. The trapezoidal channel is formed from a 250- to 350-μm Mylar spacer. With a maximum width of 2 cm, the length of the channel is 27.8 cm. The membrane was composed of a 10 nm porous ceramic foil. A special focusing flow was introduced to improve the efficiency of the polymer separation. This flow is the second flow of input that enters the channel near the centre and divides into two sub-streams. The injection flow on the channel begins with one component of the flow. Both streams create a solid barrier and cross the membrane to leave the canal. In the area where both flows come into contact, the injection flow sample will concentrate laterally and stay in the same place until the concentrate flow ceases. The second flow sensor exits the channel through the outlet and ensures a steady flow of the sensor in the focusing step. The polymer molecules can be preserved at the start of the channel after injection by this new technique. As a consequence, with minimum length diffusion, the molecule can be isolated, resulting in lower range enlargement. HT AsFl-FFF is typically an alternative technique to "high-temperature size-exclusion chromatography" (HT SEC). Sensitivity and effectiveness of separation have been demonstrated to be superior for high molecular weight species. SEC does however work better when the molecular weight content is less segregated. The low molecular cuts of the ceramic membrane prevent the breaking of molar polymers to HT AsFlFF. The present pores in the membrane cause low molar polyethylene cuts of approximately 5×10^4 g/mol (Stegeman et al. 1994), which results in low recovery of large quantities of small molecular polydisperse samples. This is because of the ceramic membrane's low molecular weight limit. To solve this issue, it is important to develop new membrane architectures and materials. In the near future, HT AsFl-FFF should have the slot outlet technology. The sample is placed close to the membrane during elution. The sample layer thickness is normally between 1 and 10 μm. There is only pure solvent in the remaining pipe. During passing through the detector, the sample layer and the solvent layer are combined and the level is lower. With the slot outlet technology, a solvent layer may be isolated by a single pump. The residual solvent layer would then be crossed and the peak height amplified. AsFl-FFF has boosted the signal to noise ratio 5–8 times (Leeman et al. 2007). In the future the HT AsFl-FFF should be a universal and sufficient solution in combination with HT SEC. The problem of the lack of small molecules for linear polyolefins is now partly removed by the combination of HT SEC and HT AsFl-FFF measurements (Stegeman et al. 1994).

9.2.1.1.5 Thermal FFF (ThFFF)   A temperature gradient is used to conduct samples to an accumulation wall using the initially experimentally designed FFF technique. The ribbon-like channel is sandwiched between two independently controlled, the one with heating elements and the other with flowing chilled water or coolant. The heat/cold wall variations can reach up to ~100 K, which translates into gradients of ~104 km–¹, when extended over a channel thickness of ~76 µm. ThFFF was initially used in organic solvents for plastic polymers, but has been enlarged into nanoparticles in aqueous and non-aqueous solvents (Ratanathanawongs et al. 1995).

$$F = kT \frac{DT}{D} \frac{dT}{dx}$$

The applied temperature gradient is the coefficient of thermal diffusion where D_T and dT/dx. Combining Eqs. (9.1) and (9.2) it is shown that t_r is proportionate to the power of the field (dT/dx) and the DT/D Soret coefficient. If DT is known, D can be measured, which then produces molecular weight (MW), due to dependency D = A(MW)-b where A is a constant of proportionality, which is experimentally defined and where b is a sum of 0,6 for thermodynamically advantageous dissolvent. If specified, DT can also be determined. Alternatively, DT can be used for analytical problems in relation to polymer and nanoparticles interfacial matrix and polymer microstructure and architecture (Messaud et al. 2009). However, retention times are challenging to measure as thermal diffusion in fluids are not well known and DT values are unique to the individual polyurethane. DFFF is a very fascinating strategy for the FFF family. This is overcome by integrating ThFFF with detectors to separately calculate MW and by continuing attempts to test the hypotheses of current DT prediction (Mes et al. 2003).

9.2.2 High Performance Liquid Chromatography

High-Pressure Liquid Chromatography (HPLC) method is a sophisticated analytical tool used for detection, isolation, and quantification of analytes in complex mixtures. HPLC instrumentation consists of a storage tank, high-pressure pump/compressor, injector, analytical column and wide range of detectors (Ahamad et al. 2014). The injection of the sample blend (ported in mobile phases) in the column separates the compounds. Because of variations in the separation of the moving liquid phase and the stationary phase (column material), the individual components in the mixture move through the column at various speeds. The separation of components/analytes may be dependent on load: weak/strong cation/anion (IEC), molecular mass (SEC), polarity (reverse phase and normal phase HPLC), basic characteristics (affinity chromatography) and depending on the form of stationary phase (Vezocnik et al. 2015). An ultraviolet-visible (UV-screen) detector, a fluorescence detector, and photo diode array (PDA) are the most common detectors in HPLC. HPLC is regularly utilized for analysis of pharmaceutical and food products such as carbohydrates, vitamins, additional additive materials, mycotoxins, amino-acids, proteins, fats and oils, chiral compounds, and pigments, etc. HPLC is a simple, durable, economic, and replaceable technology. Based on the compound to be examined, sensitive selective detectors may be used in HPLC. Another value of HPLC is that it can pick both the stationary phases and the mobile phases to ensure that the sample components are isolated.

As size and/or charging are the standard features of nanoemulsions, the most appropriate forms of fluid chromatography to distinguish nanoemulsions from the food matrix are the SEC and IEC. SEC is size-dependent, larger compounds are elucidated faster than the smaller compounds, for example, nanoemulsions, nano-drug systems. The molecular weight of a compound can be measured by the use of reference mass specifications.

Caution must be taken, however, because the elution can be influenced by the compound's nature. Furthermore, if the compound appears to interfere with the matrix of the column, the elusion location will shift (Youssof et al. 2016). For the separation of nanoemulsions or nano-drug systems size-exclusion and ion-exchange chromatography are commonly used.

9.2.2.1 Size Exclusion Chromatography

Size exclusion chromatography is a chromatographical approach in which solution molecules are differentiated by their size and molecular weight (in certain instances). It is typically used for small molecules and macromolecular complexes, including protein materials and industrial polymerization products. Size is the key attribute of nanoemulsions, and it is therefore the most suitable form for the separation of nanoemulsions from the food matrix (Saifullah et al. 2016; Wei et al. 1999). Size exclusion chromatography was used for the study of bioactive ingredient distribution based on nanoemulsion. The creation of a nanoemulsion should therefore concentrate on droplets of a particular size, i.e., a limited distribution. If not, the accuracy and volume of for example a fluorocarbon mark can be compromised by contact with large-scale, droplets or with a higher perfluorocarbon load. The perfluorocarbon nanoemulsion developed using phospholipids generally shows a wide-ranging distribution – a problem which the standard production process cannot solve. The production of specified perfluorocarbon nanoemulsion can be accomplished, however, by combining centrifugation with chromatography for size exclusion. Similarly, this approach was used to describe bioactive ingredient distribution by ketene-based polyester synthesized with high carbon/silica electron-rich composite surface (Swarnalatha et al. 2008). Exclusion of the dimension nanoemulsion characterization methods by chromatography are rare and are often used in the active ingredient setting (Bae and Chung 2014; Vezocnik et al. 2015).

9.2.2.2 Ion-Exchange Chromatography

Ion-exchange chromatography is based on its association with an ion exchanger and distinguishes ions and polar molecules. It is used for several different molecules – including large proteins, short nucleotides and amino acids. As the charging scale is also the most effective form of liquid chromatography for the isolation of nanoemulsions from a food matrix, it is also a key characteristic of nanoemulsion (Saifullah et al. 2016). Few researchers have used IEC with special engineered materials over the past decade.

9.3 APPLICATIONS OF SEPARATION TECHNIQUES IN CHARACTERIZATION OF NANOEMULSIONS

9.3.1 Application of FFF in the Characterization of Nanoemulsions

Size and shape are important parameters for a nanoemulsion-based bioactive compound delivery system. These parameters control the kinetics of bio-distribution, and cellular membrane deformability. It has been shown that smaller nanoparticles escape natural body clearance mechanisms more efficiently and, hence, circulate longer in the blood.

In addition to size, the shape plays an important role. Studies demonstrate that cylindrical nanoparticles interact with cells very differently to spherical ones, resulting in dramatic changes in bioavailability (Robertson et al. 2016). FFF is successfully applied for the separation and characterization of nano-drug delivery systems and nanoemulsions (Robert et al. 2012). Esposito et al. (2015) applied the sedimentation field flow fractionation (SdFFF) technique for the separation of cannabinoid antagonists in lipid nanocarrier (LNC) system. The study finding suggested that nanoparticles ranged from 82 nm to 400 nm sizes. The sizing of high density, medium density, and very low-density lipoproteins and the isolation of 50 and 100 base single-strand DNA are two special applications of mAs-Fl-FFF. DNA bundling with a single strand and unbounding to replicate a protein, suggested that mAs-Fl-FFF may be used to test DNA protein binding properties (Yohannes et al. 2006). Most applications reported have been plodded using an aqueous carrier fluid to isolate particles from sizes varying from 60 to multiple micrometres. The spectrum of MW and specific size which the micro-thermal FFF channel can calculate was shown by the results (Janca 2008).

9.3.2 Application of Different Types of Chromatography Techniques in the Characterization of Nanoemulsions

SEC is a well-established method for the separation of nanoemulsions and macromolecules in solution according to their sizes. SEC reveals that well-defined fractions of a perfluorodecalin emulsion were successfully isolated through a column of Toyopearl HW-75S (Grapentin et al. 2015). The nanoemulsions were defined in this analysis as a partition coefficient in an outer emulsion aqueous phase is a function of the time between emulsion phases, active ingredient preparation, free substance, micellar solubilization, encapsulation quality and vitamin E. Zhang et al. (2007) also studied encapsulation efficiency of plasmid nanoemulsions using anion-exchange chromatographic method. The ion-exchange chromatography has been used to cleanse the fatty ester sucrose in the nanoemulsion (Bromley 2011). Nandita et al. (2016) have employed the HPLC technique to measure vitamin E that is not encapsulated and have measured the encapsulation efficacy of mustard oil nanoemulsion. Other researchers have also used HPLC to classify various nanoemulsions, and to determine the stability of a particular compound in nanoemulsions other than encapsulation efficiency (Li et al. 2015; Panatieri et al. 2016; Youssof et al. 2016).

9.4 CONCLUSION

Nanoemulsions have gained popularity over the past decade because of their exceptional properties such as small size, stability, appearance and tunable rheology, and by applications in nutraceutical fields with respect to targeted bioactive molecule delivery. Characterization techniques for nanoemulsions are advanced in recent decades due to advancement in separation and imaging technologies. In this book chapter, we focused on separation techniques for nanoemulsions. FFF and liquid chromatography are most commonly used for the separation of nanoemulsions and nano drug systems. FFF techniques are used for the separation of several types of nanoemulsions and have wide applications in the pharmaceutical and food industry. Liquid chromatography especially size-exclusion and ion-exchange chromatography are applied for separation nanoemulsions.

CONFLICT OF INTEREST

The authors declared no conflict of interest.

LIST OF ABBREVIATIONS

AFM	Atomic Force Microscopy
BBB	Blood Brain Barrier
ESI-MS	Electrospray Ionization Mass Spectrometry
FFF	Field Flow Fractionation
HPLC	High Pressure Liquid Chromatography
IEC	Ion Exchange Chromatography
MALDI	Matrix-Assisted Laser Desorption/Ionization
NDDS	New Drug Delivery System
NMR	Nuclear Magnetic Resonance
NLCs	Nanostructured Lipid Carriers
SEC	Size-Exclusion Chromatography
SEM	Scanning Electron Microscope
SdFFF	Sedimentation Field Flow Fractionation
TEM	Transmission electron microscopy
XRD	X-Ray Diffraction

REFERENCES

Ahamad, J., Amin, S., Mir S.R. 2014. Development and validation of HPLC-UV method for estimation of swertiamarin in *Enicostemma littorale*. *J Pharm BioSci* 1: 9–16.

Bae, P., Chung, B. 2014. Multiplexed detection of various breast cancer cells by perfluoro-carbon/quantum dot nanoemulsions conjugated with antibodies. *Nano Converg* 1: 23.

Bantz, C., Koshkina, O., et al, 2014. The surface properties of nanoparticles determine the agglomeration state and the size of the particles under physiological conditions. *Beilstein J Nanotechnol* 5: 1774–86.

Beckett, R., Sharma, R., Andric, G., Chantiwas, R., Jakmunee, J., Grudpan, K. 2007. Illustrating some principles of separation science through gravitational field-flow fractionation. *J Chem Educ* 84: 1955–62.

Bhattacharyya, D., Singh, S. 2009. Nanotechnology, big things from a tiny world: a review. *Int J u- and e- Serv, Sci Technol* 2(3): 29–38.

Biswas, A.K., Islam, M.R., et al, 2014. Nanotechnology based approaches in cancer therapeutics. *Adv Nat Sci: Nanosci Nanotechnol* 5: 043001.

Bromley, P.J. 2011. Nanoemulsion including sucrose fatty acid ester. EP2563164A1, European Patent Office.

Chmelik, J. 1999. Different elution modes and field programming in gravitational field-flow fractionation: A theoretical approach. *J Chromatogr A* 845: 285–91.

Desai, M.P., Labhasetwar, V., et al, 1997. The mechanism of uptake of biodegradable microparticles in Caco-2 cells is size dependent. *Pharm Res* 14: 1568–73.

Esposito, E., Ravani, L., Drechsler, M., et al, 2015 Cannabinoid antagonist in nanostructured lipid carriers (NLCs): design, characterization and *in vivo* study. *Mater Sci Eng C Mater Biol Appl* 48: 328–36.

Giddings, J.C. 1993. Field-flow fractionation: analysis of macromolecular, colloidal, and particulate materials. *Science* 260: 1456–66.

Grapentin, C., Barnert, S., Schubert, R. 2015 Monitoring the stability of perfluorocarbon nanoemulsions by Cryo-TEM image analysis and dynamic light scattering. *PLoS One* 10:e0130674.

Gupta, A., Eral, H.B., Hattona, T.A., Doyle, P.S. 2016. Nanoemulsions: formation, properties and applications. *Soft Matter* 12: 2826.

Janca, J. 2002. Micro-channel thermal field-flow fractionation: new challenge in analysis of macromolecules and particles. *J Liq Chromatogr Rel Technol* 25: 683–704.

Janca, J. 2003. Micro-channel thermal field-flow fractionation: analysis of ultra-high molar mass polymers and colloidal particle with constant and programmed field force operation. *J Liq Chromatogr Rel Technol* 26: 2173–91.

Janca, J. 2008. Microthermal field-flow fractionation: analysis of synthetic, natural, and biological macromolecules and particles. New York: HNB Publishing.

Khanbabaie, R., Jahanshahi, M. 2012. Revolutionary impact of nanodrug delivery on neuroscience. *Curr Neuropharmacol* 10(4): 370–92.

Leeman, M., Islam, M.T., Haseltine, W.G. 2007. Asymmetric flow field flow fractionation coupled with multiangle light scattering and refractive index detectors for characterization of ultra-high molar mass poly(acrylamide) flocculants. *J Chromatogr A* 1172: 194–203.

Leeman, M., Wahlund, K.G., Wittgren, B. 2006. Programmed cross flow asymmetrical flow field-flow fractionation for the size separation of pullulans and hydroxypropyl cellulose. *J Chromatogr A* 1134: 236–45.

Li, P., Hanson, M., Giddings, J.C. 1997. Advances in frit-inlet and frit-outlet flow field-flow fractionation. *J Microcolumn Sep* 10: 7–18.

Li, Y., Teng, Z., Chen, P., et al, 2015. Enhancement of aqueous stability of allyl isothiocyanate using nanoemulsions prepared by an emulsion inversion point method. *J Colloid Interface Sci* 438: 130–7.

Litzen, A. 1993. Separation speed, retention, and dispersion in asymmetrical flow field-flow fractionation as functions of channel dimensions and flow rates. *Anal Chem* 65: 461–70.

McMillan, J., Batrakova, E., et al, 2011. Cell delivery of therapeutic nanoparticles. *Prog Mol Biol Transl Sci* 104: 563–601.

Mes, E.P.C., Kok, W.Th., Tijssen, R. 2003. Prediction of polymer thermal diffusion coefficients from polymer-solvent interaction parameters: comparison with thermal field flow fractionation and thermal diffusion forced Rayleigh scattering experiments. *Int J Polym Anal Charact* 8: 133–53.

Messaud, F.A., Sanderson, R.D., Runyon, J.R., Otte, T., Pasch, H., Williams, S.K.R. 2009. An overview on field-flow fractionation techniques and their applications in the separation and characterization of polymers. *Prog Polym Sci* 34: 351–68.

Moon, M.H., Williams, P.S., Kang, D., Hwang, I. 2002. Field and flow programming in frit inlet asymmetrical flow field-flow fractionation. *J Chromatogr A* 955: 263–72.

Myers, M.N. 1997. Overview of field-flow fractionation. *J Microcolumn Sep* 9: 151–62.

Nandita, D., Ranjan, S., Mundra, S., et al, 2016. Fabrication of food grade vitamin E nanoemulsion by low energy approach, characterization and its application. *Int J Food Prop* 19: 700–8.

Panatieri, L.F., Brazil, N.T., Faber, K., et al, 2016. Nanoemulsions containing a coumarin-rich extract from *Pterocaulon balansae* (Asteraceae) for the treatment of ocular acanthamoeba keratitis. *AAPS Pharm Sci Tech* 18(3): 1–8.

Park, I., Paeng, K.J., Kang, D., Moon, M.H. 2005. Performance of hollowfiber flow field-flow fractionation in protein separation. *J Sep Sci* 28: 2043–49.

Ratanathanawongs, S.K., Shiundu, P.M., Giddings, J.C. 1995. Size and compositional studies of core-shell latexes using flow and thermal field-flow fractionation. *Colloids Surf A* 105: 243–50.

Robert, S., Lacroix, R., Poncelet, P., Harhouri, K., Bouriche, T., Judicone, C., Wischhusen, J. 2012. High-sensitivity flow cytometry provides access to standardized measurement of small-size microparticles-brief report. *Arterioscler Thromb Vasc Biol* 32: 1054–58.

Robertson, J.D., Rizzello, L., Avila-Olias, M., Gaitzsch, J., Contini, C., Magon, M.S. 2016. Purification of nanoparticles by size and shape. *Sci Rep* 6: 27494. doi: 10.1038/srep27494.

Saifullah, M., Ahsan, A., Shishir, M.R.I. 2016. Production stability and applications of micro- and nano-emulsion in food processing industry. In: Grumezescu A (ed), Emulsions, 1st edn. Academic Press, London, pp. 405–33.

Schimpf ME, Caldwell KD, Giddings JC, editors. 2000. Field-flow fractionation handbook. New York: Wiley.

Se-Jong, S., Hyun-Hee, N., Byoung-Ryul, M., Jin-Won, P., Ik-Sung, A., Kangtaek, L. 2003. Separation of proteins mixtures in hollow fiber flow field-flow fractionation. *Bull Korean Chem Soc* 24: 1339–44.

Semyonov, S.N., Maslow, K.I. 1988. Acoustic field-flow fractionation. *J Chromatogr* 446: 151–56.

Stegeman, G., van Asten, A.C., Kraak, J.C., Poppe, H., Tijssen, R. 1994. Comparison of resolving power and separation time in thermal field-flow fractionation, hydrodynamic chromatography, and size-exclusion chromatography. *Anal Chem* 66: 1147–60.

Swarnalatha, S., Selvi, P.K., Ganesh, K.A., Sekaran, G. 2008. Nanoemulsion drug delivery by ketene based polyester synthesized using electron rich carbon/silica composite surface. *Colloids Surf B Biointerfaces* 65: 292–9.

Vezocnik, V., Rebolj, K., Sitar, S., et al, 2015. Size fractionation and size characterization of nanoemulsions of lipid droplets and large unilamellar lipid vesicles by asymmetric-flow field flow fractionation/multi-angle light scattering and dynamic light scattering. *J Chromatogr A* 1418: 185–91.

Wahlund, K.G., Giddings, J.C. 1987. Properties of an asymmetrical flow field flow fractionation channel having one permeable wall. *Anal Chem* 59: 1332–39.

Wei, G.T., Liu, F.K., Wang, C.R.C. 1999. Shape Separation of Nanometer Gold Particles by Size-Exclusion Chromatography. *Analytical Chemistry* 71: 2085–91.

Williams, P.S. 1997. Design of asymmetric flow field-flow fractionation channel for uniform channel flow velocity. *J Microcolumn Sep* 9: 459–67.

Williams, P.S., Giddings, J.C. 1994. Theory of field programming field flow fractionation with corrections for steric effects. *Anal Chem* 66: 4215–28.

Yohannes, G., Sneck, M., Varjo, S.J.O., Jussila, M., Wiedmer, S.K., Kovanen, P.T., et al, 2006. Miniaturization of asymmetric flow field-flow fractionation and application to studies on lipoprotein aggregation and fusion. *Anal Biochem* 354: 255–65.

Youssof, A.M.E., Salem-Bekhit, M.M., Shakeel, F. et al, 2016. Analysis of anti-neoplastic drug in bacterial ghost matrix, w/o/w double nanoemulsion and w/o nanoemulsion by a validated "green" liquid chromatographic method. *Talanta* 154: 292–8.

Zhang, A., Wu, D., Liu, F., Zhang, F. 2007. Preparation of plasmid nanoemulsion and its character determination by anion exchange chromatography method. *Sheng Wu Gong Cheng Xue Bao* 23: 1135–39.

Applications

Nanoemulsions as Delivery Vehicle for Nutraceuticals and Improving Food Nutrition Properties

Syed Sarim Imam,[1]
Mohammed Asadullah Jahangir,[2]
Sadaf Jamal Gilani,[3] Ameeduzzafar Zafar,[4]
Sultan Alshehri[1,5]

[1]Department of Pharmaceutics, College of Pharmacy,
King Saud University, Riyadh, Kingdom of Saudi Arabia
[2]Department of Pharmaceutics, Nibha Institute of
Pharmaceutical Sciences, Rajgir, India
[3]Department of Basic Health Sciences, Princess Nourah bint Abdulrahman
University, Riyadh, Kingdom of Saudi Arabia
[4]Department of Pharmaceutics, College of Pharmacy,
Jouf University, Sakaka, Al-Jouf, Kingdom of Saudi Arabia
[5]Department of Pharmaceutical Sciences, College of Pharmacy,
Almaarefa University, Riyadh, Kingdom of Saudi Arabia

CONTENTS

DOI: 10.1201/9781003121121-10

10.1 INTRODUCTION

Emulsions have a long history of being used in formulation as a metastable dispersion system of two immiscible fluids [Windhab et al., 2005]. There are wide applications reported in pharmaceutical, cosmetics, agrochemical, nutraceutical and food industries [Tadros, 2009; Donsi, 2018]. Different food products are in the form of emulsion like spreads and butter are W/O emulsion while desserts, beverages, mayonnaise are O/W emulsions [McClements, 2015]. Multiple emulsion of w/o/w or o/w/o systems are also used in the food industries. The w/o/w emulsions are made up of large oil droplets which has water droplets dispersed in an aqueous phase, vice-versa large water droplets having oil droplets dispersed in oil phase makes up the o/w/o system [Weiss et al., 2006]. The emulsion system has been categorized into different types like coarse, micro and nano-emulsions on the basis of their size. Coarse emulsions are usually a thermodynamically metastable formulation and the particle size found in the range of more than 200 nm. Various destabilizing factors break them down over time. They are optically turbid because the dimension of the droplets falls similar to the wavelength of light and hence appears opaque by scattering the incident light. Microemulsions have droplets in the diameter range of less than 100 nm and are found in thermodynamically stable state. But even slight variation in the environmental factors like temperature inadequately affects their stability. Unlike macroemulsions, microemulsions are optically transparent as their particle size is less than the wavelength of light [Anton and Vandamme, 2011]. In case of nanoemulsions the diameter of droplets of less than 200 nm and in some cases less than 100 nm are also found. They are also thermodynamically metastable which shows phase separation in course of time. But they are kinetically stable and do not separate due to gravity and also does not show aggregation of droplet due to low attractive force between droplets of small size [McClements and Rao, 2011; Nakabayashi et al., 2011]. They are not affected by variations in chemical and physical factors. The small droplet size of nanoemulsions influences the release behavior and rheological properties. Thus, supporting the applications of nanoemulsions in a more diverse way.

The application of nanotechnology has significantly increased which eventually has improved the product development process. However, nanotechnology has limited application in food industries [Weiss et al., 2006]. Recently, the bottom up technology has immensely impressed food technology and nutraceutical industry by improving different aspects of food processing techniques to ensure safety and molecular blending of products [Chen et al., 2006]. It has improved texture, taste, durability and appearance of food products. Development of nanoemulsions-based food products have improved the stability, solubility and bioavailability of nutraceuticals [Huang et al., 2010]. Nutraceuticals or functional foods act as antioxidants, antimicrobials and health promoting agents. They are also used as preservative, coloring and flavoring agents. Incorporation of functional ingredients in foods depends upon their physiochemical compatibility with the food matrix. However, most of the food stuffs show poor solubility, volatility, low bioavailability and shows chemical and biological degradation making it difficult to directly incorporate them into the food matrix. Development of nanoemulsions overcomes all these shortcomings. They are capable of stabilizing functional foods. Apart from transporting, bioactive lipids nanoemulsions efficiently protect the nutraceuticals from biological and chemical degradation. Overall, nanoemulsions significantly enhance the health factor of nutraceuticals and functional foods. Nanoemulsions have been extensively applied as delivery agents for phytosterols, vitamins of lipophilic nature, carotenoids, polyunsaturated fatty acids, etc. Nanoemulsions provide enhanced drug loading capacity, improved solubility, bioavailability, controlled release and provide protection from degradation by enzymes. This chapter provides an overview of nanoemulsions, components, formulation techniques, and application in food and nutraceutical industry.

10.2 NUTRACEUTICALS AND FUNCTIONAL FOODS – AN OVERVIEW

The word "nutraceutical" is derived from two words which are "nutrition" and "pharmaceutical". Such products are primarily food or part of food whose function is to modify normal physiological functions and to maintain the health of human beings [Das et al. 2012]. It is a broad terminology which includes the use of minerals, vitamins, amino-acids, botanicals or herbs. 2500 years ago, the father of medicine Hippocrates (460–377 BC) defines the relation of food and its essence for the treatment of different kinds of disorders in a very classical way optimizing various benefits [Dickinson, 2011]. Historical civilization like Roman, Egyptian, Greek and others had practiced the use of herbal medicines for the treatment and prevention of the disease [Ruchi, 2017; Adusei-Mensah et al., 2019]. The precise and accurate definition of a nutraceutical is yet in controversies and there is no established definition which has a worldwide acceptance. In China, ginseng had been used for the prevention and treatment of several diseases. Thyme, garlic, cumin, turmeric, juniper, cinnamon were extensively used by ancient Egyptians and Roman civilizations [Helal et al., 2019]. Honey stands as one of the most extensively used natural products in many ancient civilizations; its significance is very well described in the holy Quran and Bible. Nutraceutical of animal and plant origin can provide an exciting platform for the food industries to come up with novel products having health and medicinal benefits. Food industry is now focusing on the protective and therapeutic potentials of natural products rather than imputing concentrations of inorganic pollutants, adulterants, micro-organisms and fatty acids [Kaur and Kapoor, 2001]. Nutraceuticals are brought to the market by different industries like herbal and dietary supplement industry, food industry, and newly merged nutrition/agribusiness/pharmaceutical conglomerate [Jahangir et al., 2020]. Natural products from these industries range from isolated nutrients, dietary supplements, herbal products, to novel genetically engineered "designer" and processed food products like beverages, cereals, soups, etc. [Dureja et al., 2003; Malik, 2008]. Various reports have been published related to the application of nutraceuticals as a therapeutic agent in diseases like atherosclerosis, cancer, cardiovascular disease, inflammation, diabetes, hypertension, obesity and others [Effo et al., 2018]. Nutraceuticals like curcumin, silymarin, docosahexaenoic acid, vitamin E, choline and phosphatidylcholine have been extensively used in the prevention and treatment of steatosis, similarly caffeine, gallic acid, curcumin have been exploited as effective anti-aging and antioxidant agents [Bourbon et al., 2018; Stellavato et al., 2018; Zhang et al., 2019]. The purposed mechanism is neuroprotection such as free radical scavenging and ROS, chelation of iron, modulation of cell-signaling pathways, and inhibition of inflammation [Jahangir et al., 2020]. Nutraceuticals like beta-carotene, vitamin E (or tocopherol), vitamin C are found to have antioxidative effect. Vitamin D has been reported to attenuate MPP+ and 6-OHDA-induced neurotoxicity. Ubiquinone or Coenzyme Q10 (CoQ10) has been reported to inhibit the loss of dopaminergic neurons and thus, it is a readily available dietary supplement. Creatine has also been investigated for its possible role in the treatment and prevention of dopaminergic neuronal loss. Omega-3 poly unsaturated fatty acid has also been investigated for its neuroprotective effect against dopamine loss. (–)-epigallocatechin-3-gallate (EGCG) is highly abundant in green tea and has weight loss effect. Baicalein is a flavonoid obtained from *Scutellaria baicalensis*, a traditional Chinese herb of which root preparations are known as Huang-Qin which is reported to have extensive application in insomnia, hypertension, dysentery, diarrhea, respiratory infections hemorrhagic conditions and inflammations. Stilbenes, like resveratrol commonly found in peanuts, grapes, pines, and berries falls in the class of antioxidants and have been reported with neuroprotective effects [Chao et al., 2012]. Curcumin based nanosuspension formulated using d-α-Tocopherol Polyethylene Glycol 1000 Succinate (TPGS) have been demonstrated with potential in the treatment of Alzheimer's disease [Shin et al., 2016]. The ω-3 Polyunsaturated Fatty Acids (PUFA) are abundant in fish of which oil-in-water emulsions formulated by dual-channel microfluidization technique were found to enhance brain functions and reduce the risk of hypertension and coronary heart disease [Liu et al., 2016]. In a similar study, omega-3 o/w nanoemulsion prepared using the aforementioned technique was found to have potential in treating brain disorders [Uluata et al., 2015]. Casein-caffeine based controlled released tablet formulation has been used as a natural psychoactive substance, brain stimulant, in the treatment and prevention of neuro-degenerative disease and neurological disorders and in improving the cognition rejuvenation of the brain [Kay et al., 2018; Tan et al., 2019].

10.3 APPLICATION OF DIFFERENT FORMULATION TECHNIQUES IN PREPARATION OF FOOD-GRADE NANOEMULSION

The radius of the nanoemulsion droplet is almost equivalent to the thickness of the emulsifier layer [McClements and Rao, 2011]. For nanoemulsions in food and nutraceuticals, the thickness of shell falls in the range of 1–2 nm for small surfactant molecules, 2–10 nm for protein monolayers and 10–50 nm for multiple layers of biopolymer. Thus, there is a significant difference in the structure and compositions of the droplets of conventional emulsions and nanoemulsions. Nanoemulsions are prepared by either opting high energy or low energy methods (Figure 10.1) [Anton et al., 2008; Date et al., 2010]. The process involves breaking up of larger droplets into smaller droplets, adsorption of surfactants and collision of droplets [Silva et al., 2015]. Food industries usually employ high energy methods to develop oil-in-water nanoemulsions [Gutierrez et al., 2008]. For preparing nanoemulsions by low energy methods, the composition and temperature of oil-in-water system is altered. The energy input is accomplished by chemical potential of its constituents [Bouchemal et al., 2004]. Another important factor which influences the formulation of nanoemulsion is the fluidity at the interface oil/water [Bhosale et al., 2014]

10.3.1 High-Energy Methods

The instrument like microfluidizers, ultrasonicators and high-pressure valve homogenizers are used in the preparation of nanoemulsions by high energy methods. These devices

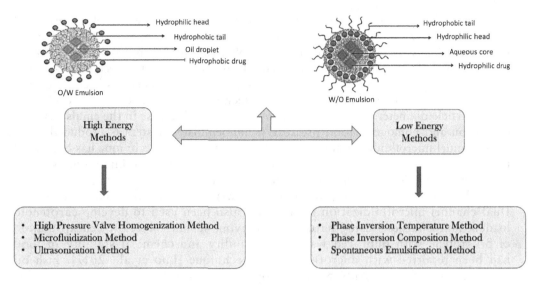

FIGURE 10.1   High-energy and low-energy techniques for developing nanoemulsion system.

generate high disruptive force to break the dispersed phase into small droplets [Maali and Hamed Mosavian, 2013]. The different techniques exploited for incorporating functional foods, nutraceuticals, preservatives, coloring and flavoring agents into nanoemulsion systems are discussed in detail as below.

10.3.1.1 High-Pressure Homogenization

The homogenization chamber can be of different types like radial diffuser, colloidal jet or simple orifice plate type [Donsì et al., 2009]. Very high pressure of about 300 MPa is generated inside the chamber which causes the coarser emulsion to be pushed out through a minute orifice of micrometer range by the homogenizer valve. During this stage the coarser emulsion is broken into fine droplets by shear stress, turbulence, and cavitation [Schultz et al., 2004]. In the food and nutraceutical industries the high pressure valve homogenization method is extensively used for producing nanoemulsions owing to their ease of production, scalability and reproducibility, efficiency and production of better quality of product [Liu et al., 2019]. This technique has been exploited to produce nanoemulsions of lemon, mandarin, carvacrol and bergamot. It has been reported that carvacrol nanoemulsion had significant inhibitory effect against *E. coli* and *S. typhimurium* [Severino et al., 2015]. Nanodispersions of lycopene having narrow polydispersity index and enhanced stability have been developed for its application in beverages by homogenization technique [Shariffa et al., 2017]. This technique has been initially used in the dairy industries for fat globule disruption. For now, they are now an important technique for application of nanoemulsions in food processing.

10.3.1.2 Microfluidization Technique

Formation of emulsion by microfluidization technique is done by following similar principle to that of high-pressure valve homogenizer. It differs only in the use of microchannels with dimensions of 50–300 μm to form droplets by this technique. This process is repeated in order to increase the emulsification time and pressure [Mao et al., 2010; Jafari et al., 2007]. NE of lemon grass, thyme, sage oil dispersed in sodium alginate solution are developed by exploiting microfluidization technique. Nanoemulsion of sage essential oil has been reported to possess properties like better transparency, resistance to water vapor, and flexibility. NE based edible films of thyme were reported to have strong antimicrobial effect against *E. coli*. Essential oils of ginger-based NE was also prepared by microfluidization method. NE from natural emulsifiers has been efficiently produced by dual channel microfluidization technique. It has been reported that the mean particle diameter was found to decrease with the increase in the emulsifier concentration and homogenization pressure if nanoemulsion system is produced using dual channel microfluidization technique. Microfluidization technique has been used to develop NE of whey protein isolate and quillaja saponin with fine droplets better than the one developed using soy lecithin and gum Arabic as it required less amount of emulsifying agent and the droplets formed were also very fine [Bai et al., 2016]. Dual channel microfluidization method has also been used to develop carotenoid loaded nanoemulsions using natural emulsifying agents. Successful encapsulation of β-carotene with improved water dispersibility and chemical stability in food had been reported with microfluidization technique [Luo et al., 2017]. Fish oil nanoemulsions have already been developed by microfluidization technique. The droplet produced were very fine which were found to provide better absorption in the digestive tract [García-Márquez et al., 2017]. However, as like other methods microfluidization technique comes with the drawback of increase in the droplet size due to coalescence as the result of longer emulsification time and high pressure involved in the production. But the droplets formed by microfluidization technique are very fine and are of uniform size.

10.3.1.3 Ultrasonication Method

In the ultrasonication method high frequency sound waves of about 20 kHz agitate and break the coarser emulsion droplets into nanoemulsions. Sonotrode produces high mechanical vibration and acoustic cavitation. As the cavitation gets collapsed strong waves are generated which break the coarser droplets into fine droplets [Behrend et al., 2000]. In a study by Kentish et al., flax seed oil and Tween 40-based nanoemulsion was prepared by high intensity ultrasonication method with a droplet diameter of 140 nm [Kentish et al., 2008]. In a similar attempt by Leong and his colleagues nanoemulsions with a droplet radius of 20 nm were produced by high intensity ultrasound technique using emulsifying agents like Span 80, sunflower oil and Tween 80 [Leong et al., 2009]. Nanoemulsions of essential oils of *Zataria multiflora* were prepared using high-intensity ultrasound technique. It was reported that the bioactivity of the essential oil was found to increase with the preparation of nanoemulsion [Gahruie et al., 2017]. Resveratrol cyclodextrin inclusion complex as well as only resveratrol-based nanoemulsion was prepared by high intensity ultrasonic method. The researchers reported average droplet size of 20 and 24 nm, respectively, for resveratrol and inclusion complex-based nanoemulsion. It was also concluded that nanoemulsions prepared by ultrasonication method had a good loading and release efficiency [Kumar et al., 2017].

10.3.2 Low-Energy Methods

Energy from the chemical potential of emulsifying components is used to form nanoemulsion in the low energy methods. By gentle mixing nanoemulsions are formed spontaneously at the interface of oil and water. The spontaneity of the process can be controlled by either changing the temperature without altering the composition or by changing the composition and interfacial properties without altering the temperature [Anton and Vandamme, 2009].

10.3.2.1 Phase Inversion Composition

The composition of NE is varied which changes the hydrophilic-lipophilic behavior of the emulsifying agent [Maestro et al., 2008]. However, for highly hydrophobic compounds the phase inversion is quite difficult [Witthayapanyanon et al., 2006]. Food grade NE loaded with Vit-E acetate has been successfully developed by the phase inversion method. The authors concluded that the developed nanoemulsion droplets had mean particle diameter of about 40 nm. This method is more effective than microfluidization technique in case of high surfactant concentration.

10.3.2.2 Phase Inversion Temperature

In this technique the inversion of phase occurs as a function of temperature without change in the composition [Salager et al., 2004]. The emulsifying agent changes its lipophilicity or hydrophilicity with variation in temperature. Cinnamon oil based nanoemulsions have been developed using phase inversion temperature method. Non-ionic surfactant, cinnamon oil and water are heated above the phase inversion temperature followed by rapid cooling and continuous stirring which forms small oil droplets with a diameter of 101 nm [Chuesiang et al., 2018]. It was also reported in the study that the antibacterial activity of the cinnamon oil nanoemulsion was influenced by the concentration of the surfactant. With 15–20% of weight increase in surfactant concentration enhances the antimicrobial activity of nanoemulsion is enhanced. [Chuesiang et al., 2019].

10.3.2.3 Spontaneous Emulsification Method

Emulsions are formed spontaneously by mixing water and oil together with an emulsifying agent at a particular temperature and with continuous stirring. Vit-E acetate nanoemulsions have been produced using the spontaneous emulsification technique. The researchers reported low polydispersity index with the diameter of droplets being in the range of less than 50 nm. Surfactant to emulsion ratio along with the composition of oily phase is needed to be optimized for developing finer and smaller droplets. By increasing the temperature and the speed of stirring the particle size could be made to reduce [Saberi et al., 2013]. In a similar way Vit-D nanoemulsions have been prepared by utilizing the spontaneous emulsification technique. By this method nanoemulsion of the size range less than 200 nm were obtained which were stable to droplet growth at ambient temperature but were found to be unstable with the increase in temperature. Use of co-surfactant could enhance the thermal stability of nanoemulsions [Guttoff et al., 2015]. Fish oil nanoemulsions were also produced using spontaneous emulsification method. The developed formulation was reported to be transparent with physical stability at 37°C and stability against oxidation at 55°C for 14 days [Walker et al., 2015]. Cinnamaldehyde nanoemulsions were prepared by self-emulsifying emulsion system. Medium chain triglyceride was added to form stable nanoemulsions of cinnamaldehyde. It was reported that the encapsulation efficiency of nanoemulsion was 80% for one week and it was expected that the release will be slow. Phase separation was found after 12 days of storage at 37°C [Tian et al., 2016]. Table 10.1 enlists the different techniques exploited for incorporating nutraceuticals, preservatives, flavoring and coloring agent into nanoemulsion system.

TABLE 10.1 Different Techniques for Incorporating Nutraceuticals, Preservatives, Flavoring and Coloring Agents into Nanoemulsion System

Method Classification	Techniques	Bioactive Agent	Function	References
High energy methods	High pressure homogenization	Citral	Flavoring agent	Tian et al., 2017
		β-carotene	Coloring agent	Wei and Gao, 2016
		Vit-D	Nutraceutical	Golfomitsou et al., 2018
		Vit-E, phytosterol	Nutraceutical	Cheong et al., 2018
		Curcumin	Nutraceutical	Silva et al., 2018
	Microfluidization	Ginger essential oil	Preservative	Acevedo Fani et al., 2015
		Curcumin	Nutraceutical	Artiga-Artigas et al., 2018
		β-carotene	Coloring agent	Chu et al., 2007
	Ultrasonication	Oregano essential oil	Preservative	Bhargava et al., 2015
		Orange essential oil	Preservative	Sugumar et al., 2015
		Ginger essential oil	Preservative	Noori et al., 2018
		Capsaicin	Preservative	Akbas et al., 2018
Low energy methods	Spontaneous emulsification	Resveratrol	Nutraceutical	Davidov-Pardo and McClements 2015
		Astaxanthin	Nutraceutical	Alarcon-Alarcon et al., 2018
		Cinnamaldehyde	Preservative	Otoni et al., 2014a
		Curcumin	Preservative	Abdou et al., 2018
	Solvent evaporation	β-carotene	Coloring agent	Tan and Nakajima, 2005a
	Phase inversion temperature	Cinnamon oil	Preservative	Chuesiang et al., 2018
	Phase Inversion composition	Vit-E acetate	Nutraceutical	Witthayapanyanon et al., 2006

10.4 ENCAPSULATION OF NUTRACEUTICALS, FLAVORING AND COLORING AGENTS IN NANOEMULSION SYSTEM

Resveratrol is a naturally obtained polyphenol from the raspberries, blueberries and grapes' skin which is found to have anticancer, antioxidant and anti-obesity activity. Resveratrol is encapsulated into a nanoemulsion system. It is achieved by exploiting spontaneous emulsification technique. It is developed by using 10% grape seed and orange oil as the oily phase, 10% Tween

80 as the surfactant and remaining 80% as the aqueous phase. The nanoemulsion thus formed was found to have a droplet size of 100 nm. The developed formulation was reported to be stable against UV-light degradation [Davidov-Pardo and McClements, 2015]. Cholecalciferol or Vit-D based oil in water edible nanoemulsion has been used for fortification of diary emulsions. Polysorbate 20 was used as emulsifying agent, soybean lecithin was mixed with cocoa butter to develop nanoemulsions having mean droplet diameter of less than 200 nm with high pressure homogenization technique. Nanoemulsions loaded with Vit-D3 in its oily core were developed. Fortification of the whole fat milk was done with nanoemulsions of Vit-D3. These were found to be stable for 10 days against gravitational separation and particle growth [Golfomitsou et al., 2018]. Kenaf seed oil-based oil in water nanoemulsion was developed using sodium caseinate as the emulsifying agent, Tween 20 and β-cyclodextrin and it has shown to improve physicochemical and *in vitro* stability of the antioxidant and bioactive ingredients. The researchers reported that the developed product was found to be stable for 8 weeks at 4°C with enhanced retention of phytosterols and Vit-E [Cheong et al., 2018]. Photostabilization of Astaxanthin nanoemulsion was achieved to make it able to be used in foods. The carrageenan coating and chitosan-coating was found to protect astaxanthin from UV-light and prevents photodegradation of the same [Alarcon-Alarcon et al., 2018]. The bioaccessibility of hydrophobic bioactive compounds have been reported to be increased by the use of nanoemulsions. Nanoemulsions of curcumin prepared using alginate and chitosan were found to improve the antioxidant property of curcumin and were also found to control the lipid digestibility of the curcumin. Thus, these bioactive nanoemulsion systems can be used to target obesity [Silva et al., 2018]. Curcumin nanoemulsions stabilized with lecithin have been formulated and were reported with 75% encapsulation efficiency and were found to be stable for 86 days [Bhosale et al., 2014].

Ketones, aldehydes and esters are present as coloring and flavoring agents in food but are susceptible to photolytic and oxidative degradation. Encapsulating these materials into nanoemulsion systems may prevent the degradative effect and enhance the shelf life [Goindi et al., 2016]. Citral is used as flavoring agent in cosmetics and food industry which easily gets degraded forming bad flavor compounds. Citral developed as oil in water nanoemulsion system with additionally added β-carotene or black tea extract as antioxidant has shown enhanced stability [Yang et al., 2011]. Ubiquinol-10 and citral nanoemulsion system has shown enhanced chemical stability. Ubiquinol-10 was found to protect citral from chemical and oxidative degradation. It reduces the production of off flavored compounds [Zhao et al., 2013]. In a similar study Tween 20 and gelatin were successfully used as emulsifying agent and were found to stabilize citrate from acidic degradation in the nutraceutical and food industry [Tian et al., 2017]. β-carotene is a naturally occurring coloring agent and widely used antioxidant in the food and nutraceutical industry. However, it is susceptible to degradation by light, oxygen and heat. A nanodispersion system of β-carotene was prepared by exploiting emulsification-evaporation method which was reported with enhanced stability. Emulsification of organic β-carotene solution was done in aqueous phase having emulsifying agent. The mean diameter of droplets for developed nanodispersion system of β-carotene was found to be 60–140 nm [Tan and Nakajima, 2005a]. Oil in water β-carotene nanodispersion system was developed using polyglycerol esters of fatty acids as non-ionic emulsifying agent. The researchers concluded that the mean diameter of the droplets of developed nanodispersion system was found to be 85–132 nm with physical and physicochemical stability [Tan and Nakajima, 2005b]. β-carotene nanodispersion stabilized with protein was developed by emulsification-evaporation method was found to have the mean diameter of nanodispersion droplets of 17 nm. Sodium caseinate used as emulsifying agent was found to decrease the particle size at higher concentration. It also helps in improving the polydispersity index of the nanodispersion system [Chu et al., 2007]. Tween 20 as emulsifying agent was used to develop oil in water β-carotene nanodispersion system exploiting high pressure homogenization technique. The researchers concluded that the developed nanodispersion system was having droplets of mean diameter in the range of 130–185 nm. The formulation was found to be stable and showed only 25% degradation after 4 weeks at 4 and 25°C [Yuan et al., 2008]. Tween 20, sucrose fatty acid ester, decaglycerol monolaurate, sodium caseinate are used as emulsifying agents to prepare β-carotene nanodispersion system. The developed formulation showed 30–210 nm of mean diameter. Nanodispersion systems stabilized by starch casein were reported to be more stable against oxidation. It works by creating a physical barrier and protects by antioxidant property of caseins [Yin et al., 2009]. Spray dried powders and modified starch are added to β-carotene nanoemulsion after emulsification method are reported to have enhanced storage stability. Starch shows low oxygen permeability and thus assists in β-carotene retention during storage [Liang et al., 2013].

Nanoemulsion systems have been extensively used in coatings and films for possible application in nutraceutical and food packaging. The continuous phase of films and coatings is made up of biopolymer matrix which provides them with monodispersity index and stability. The coalescence of droplets tends to decrease with the increase in the viscosity of the continuous phase [Artiga-Artigas et al., 2017]. Bioedible nanoemulsion films include the dispersion of bioactive agents into continuous phase which tend to form the matrix of the packaging film. High or low energy methods are used to add food grade emulsion. It makes the formulation homogenized which is then casted into films having controlled thickness and is further dried. Characterization of the developed films for their morphological, structural, mechanical, thermal and functional properties are done [Otoni et al., 2014a]. Edible nanoemulsion films of clove and cinnamaldehyde using biopolymer pectin were produced with enhanced antimicrobial activity. Pectin films have low hydrophilic and hydrophobic ratio which allows decreased vapor and water permeability [Otoni et al., 2014a; Sasaki et al., 2016]. Oregano and clove essential oil based nanoemulsions have been developed using cellulose and its derivatives [Otoni et al., 2014b]. Essential oils of mandarin, lemon, bergamot, carvacrol have been developed into nanoemulsion coatings and films using chitosan and were shown to have enhanced antimicrobial activity [Severino et al., 2015]. Essential oils of lemongrass, corn sage and thyme were developed in nanoemulsion films using sodium alginate [Acevedo Fani et al., 2015; Artiga-Artigas et al., 2017]. In a similar study canola oil based nanoemulsion films were developed using porcine gelatin [Alexandre et al., 2016]. Food coatings with nanoemulsions have been done and tested for their effective ability to protect food material from degradation. Chitosan coating containing nanoemulsions of essential oils of mandarin on green bean was found to inhibit the growth of *Listeria innocua* [Severino et al., 2014]. Thymol nanoemulsion coated with chitosan and Quinoa protein edible coating containing thymol nanoemulsion were found to significantly inhibit the growth of fungi on cherry tomatoes after 7 days at a storage temperature of 5°C [Robledo et al., 2018].

10.5 APPLICATION OF NANOEMULSION IN FOOD AND NUTRACEUTICAL INDUSTRY

Functional foods of the bioactive compounds are a bit difficult to be designed due to the low solubility and bioavailability and stability issues. These components are easily susceptible to degradation and deterioration due to oxidation [Shahidi and Zhong, 2010].

Some of the bioactive compounds show poor solubility and get metabolized frequently eventually reducing its bioavailability [Jin et al., 2016]. Nanoemulsion system provides a novel platform for encapsulating bioactive compounds into an oily phase using emulsifying agent which ensures the solubility, stability, release ability and bioavailability of the bioactive compound [McClements et al., 2007]. For successful development of nanoemulsions in functional foods, there should be compatibility between the components and the organoleptic properties must not be hampered. Encapsulation protects the bioactive materials from different processing conditions and inhibits the degradation of the material from oxidation, temperature variations, light, pH and other manufacturing and storage conditions. However, economic feasibility for industrial scale up production must also be considered [Pathak, 2017].

Numerous functional foods have been developed as nanoemulsion system and various in vitro and in vivo studies have also been reported. But still their application in commercial functional food products is extremely limited. Sustainable food processing can be done by developing nanoemulsion systems. Different nutraceutical industries like Nestle and Unilever are extensively exploiting nanoemulsion systems in their nutraceutical products [Salvia-Trujillo et al., 2017]. Water in oil nanoemulsions are being developed by Nestle. They have also patented micelle forming and polysorbate emulsions for uniform and rapid thawing of frozen foods in the microwave [Möller et al., 2009]. Unilever used nanoemulsion systems to reduce fat content in ice creams [Unilever, 2011]. NutraLease has developed nanoencapsulation technique which is known as nano-sized self-assembled structured lipids. Minute micelles known as nanodrops are developed which serve as carriers for lipophilic bioactive agents. It protects the bioactive agents from degradation in the digestive tract and assists in absorption. These are used in developing beverages having functional compounds. With the nano-sized self-assembled structured lipids technology omega-3, phytosterols, β-carotene, vitamins and isoflavones have been incorporated in the food products and were found to have enhanced bioavailability and improved shelf life [NutraLease, 2011]. NovaSol beverages having nanoemulsions of curcumin, sweet pepper, apocarotenal, chlorophyll, lutein are developed by Aquanova with enhanced stability [AquaNova, 2011]. Similarly, bottled water enriched with electrolytes and flavors have also been developed using the nanoemulsion technology [Piorkowski and McClements, 2014]. Natural antioxidants extracted from fruits, flowers, cereals and vegetables have been developed into a nanoemulsion system to be added as food preservative. This technique has been patented. These systems were freeze dried and was later used for the preservation of foods. Nanoemulsions are applied as a thin layer over the food to inhibit fluid and gas exchange with the outer environment of the food product and thus keeps it stable for a longer duration of time. It was also suggested that upon thawing such frozen foods show better organoleptic properties [Malnati et al., 2019].

10.6 CONCLUSION AND FUTURE PERSPECTIVE

In the last decade numerous studies have been conducted to understand the advantages of nanoemulsion for encapsulating bioactive compounds or functional foods. Nanoemulsions were found to increase the bioavailability of nutraceutical compounds and were confirmed through different in vitro studies. However, still very limited results are available for reporting the actual health benefits of nanoemulsions in food. Different studies for understanding and evaluating the application of low and high energy approach to successfully formulating nanoemulsion. Fewer reports are available on reducing the cost of nanoemulsion production. Similarly, the risk associated with the application of nanoemulsion technique in the food and nutraceutical industry. No elucidation is provided for potential biological and toxicological fate of nanoemulsions after digestion.

Nanoemulsion has shown great potential for encapsulating foods and bioactive compounds and thus opened numerous applications in the nutraceutical industry. Coating of foods and bioactive compounds with edible nanoemulsion coating have been found to enhance the stability, quality, and shelf-life. However, their potential application in the food industry is dependent on the cost effectiveness in its development and their safety profile which needs to be considered before commercialization of the product. Thus, it becomes utmost important to optimize the encapsulated bioactive component for scaled up industrial production. The future of this innovative technique depends upon the risk assessment, safety profile and cost effectiveness in producing and commercializing nanoemulsion-based food, and nutraceutical products.

REFERENCES

Abdou, E. S., Galhoum, G. F., and Mohamed, E. N. (2018). Curcumin loaded nanoemulsions/pectin coatings for refrigerated chicken fillets. Food Hydrocoll. 83, 445–453.

Acevedo Fani, A., Salvia Trujillo, L., Rojas Graü, M. A., and Martín Belloso, O. (2015). Edible films from essential oil loaded nanoemulsions: Physicochemical characterization and antimicrobial properties. Food Hydrocoll. 47, 168–177.

Adusei-Mensah, F., Haaranen, A., Kauhanen, J., et al. (2019). Post-market safety and efficacy surveillance of herbal medicinal products from users' perspective: a qualitative semi-structured interview study in Kumasi, Ghana. Int. J. Pharm. Pharmacol. 3, 136. doi: 10.31531/2581-3080.1000136

Akbas, E., Soyler, B., and Oztop, M. H. (2018). Formation of capsaicin loaded nanoemulsions with high pressure homogenization and ultrasonication. LWT. 96, 266–273.

Alarcon-Alarcon, C., Inostroza-Riquelme, M., Torres-Gallegos, C., Araya, C., Miguel, M., Sanchez-Caamano, C., et al. (2018). Protection of astaxanthin from photodegradation by its inclusion in hierarchically assembled nano and microstructures with potential as food. Food Hydrocoll. 83, 36–44.

Alexandre, E. M. C., Lourenço, R. V., Bittante, A. M. Q. B., Moraes, I. C. F., and Sobral, P. J. A. (2016). Gelatine based films reinforced with montmorillonite and activated with nanoemulsion of ginger essential oil for food packaging applications. Food Pack. Shelf Life. 10, 87–96.

Anton, N., and Vandamme, T. F. (2009). The universality of low-energy nanoemulsification. Int. J. Pharm. 377, 142–147.

Anton, N., and Vandamme, T. F. (2011). Nano-emulsions and microemulsions: clarifications of the critical differences. Pharm. Res. 28, 978–985.

Anton, N., Benoit, J.-P., and Saulnier, P. (2008). Design and production of nanoparticles formulated from nano-emulsion templates—a review. J. Control. Release 128, 185–199.

AquaNova (2011). Available online at: http://www.aquanova.de/media/public/pdf_produkte~unkosher/NovaSOL_beverage

Artiga-Artigas, M., Acevedo-Fani, A., and Martín-Belloso, O. (2017). Effect of sodium alginate incorporation procedure on the physicochemical properties of nanoemulsions. Food Hydrocoll. 70, 191–200.

Artiga-Artigas, M., Lanjari-Pérez, Y., and Martín-Belloso, O. (2018). Curcumin loaded nanoemulsions stability as affected by the nature and concentration of surfactant. Food Chem. 266, 466–474.

Bai, L., Huan, S., Gu, J., and McClements, D. J. (2016). Fabrication of oil-in-water nano-emulsions by dual-channel microfluidization using natural emulsifiers: Saponins, phospholipids, proteins, and polysaccharides. Food Hydrocoll. 61, 703–711.

Behrend, O., Ax, K., and Schubert, H. (2000). Influence of continuous phase viscosity on emulsification by ultrasound. Ultrason Sonochem. 7, 77–85.

Bhargava, K., Conti, D. S., da Rocha, S. R. P., and Zhang, Y. (2015). Application of an oregano oil nanoemulsion to the control of foodborne bacteria on fresh lettuce. Food Microbiol. 47, 69–73

Bhosale, R. R., Osmani, R. A., Ghodake, P. P., Shaikh, S. M., and Chavan, S. R. (2014). Nanoemulsion: a review on novel profusion in advanced drug delivery. Indian J. Pharm. Biol. Res. 2, 122–127.

Bouchemal, K., Briançon, S., Perrier, E., and Fessi, H. (2004). Nano-emulsion formulation using spontaneous emulsification: solvent, oil and surfactant optimisation. Int. J. Pharm. 280, 241–251.

Bourbon, A.I., Pinheiro, A.C., Cerqueira, M.A., and Vicente, A.A. (2018). In vitro digestion of lactoferrin-glycomacropeptide nanohydrogels incorporating bioactive compounds: Effect of a chitosan coating. Food Hydrocoll. 84, 267–75.

Chao, J., Leung, Y., Wang, M., and Chang, R.C. (2012). Nutraceuticals and their preventive or potential therapeutic value in Parkinson's disease. Nutr Rev. Jul 1;70(7):373–86.

Chen, H., Weiss, J., and Shahidi, F. (2006). Nanotechnology in nutraceuticals and functional foods. Food Tech. 60, 30e36.

Cheong, A. M., Tan, C. P., and Nyam, K. L. (2018). Stability of bioactive compounds and antioxidant activities of kenaf seed oil-in-water nanoemulsions under different storage temperatures. J. Food Sci. 83, 2457–2465.

Chu, B. S., Ichikawa, S., Kanafusa, S., and Nakajima, M. (2007). Preparation of proteinstabilized β-carotene nanodispersions by emulsification–evaporation method. J. Am. Oil Chem. Soc. 84, 1053–1062.

Chuesiang, P., Siripatrawan, U., Sanguandeekul, R., McClements, D. J., and McLandsborough, L. (2019). Antimicrobial activity of PIT-fabricated cinnamon oil nanoemulsions: effect of surfactant concentration on morphology of foodborne pathogens. Food Control. 98, 405–411.

Chuesiang, P., Siripatrawan, U., Sanguandeekul, R., McLandsborough, L., and McClements, D.J. (2018). Optimization of cinnamon oil nanoemulsions using phase inversion temperature method: Impact of oil phase composition and surfactant concentration. J. Colloid Interface Sci. 514, 208–216.

Das, L., Bhaumik, E., Raychaudhuri, U., and Chakraborty, R. (2012). Role of nutraceuticals in human health. J Food Sci. Technol. 49(2), 173–183.

Date, A. A., Desai, N., Dixit, R., and Nagarsenker, M. (2010). Self-nanoemulsifying drug delivery systems: formulation insights, applications and advances. Nanomedicine 5, 1595–1616.

Davidov-Pardo, G., and McClements, D. J. (2015). Nutraceutical delivery systems: resveratrol encapsulation in grape seed oil nanoemulsions formed by spontaneous emulsification. Food Chem. 167, 205–212.

Dickinson, A. (2011). History and overview of DSHEA. Fitoterapia. 82(1), 5–10.

Donsi, F. (2018). Applications of nanoemulsions in foods. In S. M. Jafari and D. J. McClements (eds.), Nanoemulsions: Formulation, Applications, and Characterization (pp. 349–377). Cambridge, MA: Academic Press.

Donsì, F., Ferrari, G., and Maresca, P. (2009). High-pressure homogenization for food sanitization. In: Barbosa-Canovas, G.V., Mortimer, A., Lineback, D., Spiess, W., Buckle, K. (eds.), Global Issues in Food Science and Technology (pp. 309–352). Burlington, MA: Academic Press.

Dureja, H., Kaushik, D., and Kumar, V. (2003). Developments in nutraceuticals. Indian J. Pharmacol. 35, 363–372.

Effo, K.E., Djadji, A.T.L., N'Guessan, B.N., et al. (2018). Evaluation of the anti-inflammatory activity and ulcerogenic risk of "sarenta", an ivorian herbal preparation. Int. J. Pharm. Pharmacol. 2, 131.doi: 10.31531/2581-3080.1000131

Gahruie, H. H., Ziaee, E., Eskandari, M. H., and Hosseini, S. M. H. (2017). Characterization of basil seed gum-based edible films incorporated with Zataria multiflora essential oil nanoemulsion. Carbohydr. Polym. 166, 93–103.

García-Márquez, E., Higuera-Ciapara, I., and Espinosa-Andrews, H. (2017). Design of fish oil-in-water nanoemulsion by microfluidization. Innov. Food Sci. Emerging. Technol. 40, 87–91.

Goindi, S., Kaur, A., Kaur, R., Kalra, A., and Chauhan, P. (2016). Nanoemulsions: an emerging technology in the food industry. Emulsions. 3, 651–688.

Golfomitsou, I., Mitsou, E., Xenakis, A., and Papadimitriou, V. (2018). Development of food grade O/W nanoemulsions as carriers of vitamin D for the fortification of emulsion based food matrices: a structural and activity study. J. Mol. Liquids. 268, 734–742.

Gutiérrez, J. M., González, C., Maestro, A., Solè, I., Pey, C. M., and Nolla, J. (2008). Nano-emulsions: new applications and optimization of their preparation. Curr. Opin. Colloid Interface Sci. 13, 245–251.

Guttoff, M., Saberi, A. H., and McClements, D. J. (2015). Formation of vitamin D nano-emulsion-based delivery systems by spontaneous emulsification: factors affecting particle size and stability. Food Chem. 171, 117–122.

Helal, N.A., Eassa, H.A., Amer, A.M., Eltokhy, M.A., Edafiogho, I., and Nounou, M.I. (2019). Nutraceuticals' novel formulations: the good, the bad, the unknown and patents involved. Recent Pat Drug Deliv. Formul. 13(2), 105–156. doi:10.2174/1872211313666190503112040.

Huang, Q., Yu, H., and Ru, Q. (2010). Bioavailability and delivery of nutraceuticals using nanotechnology. J. Food Sci. 75, R50eR57.

Jafari, S., He, Y., and Bhandari, B. (2007). Optimization of nano-emulsions production by microfluidization. Eur. Food Res. Technol. 225, 733–741.

Jahangir, M.A., Anand, C., Muheem, A., Gilani, S.J., Taleuzzaman, M., Zafar, A., Jaffer, M., Verma, S., and Barkat, M.A. (2020). Nano phytomedicine based delivery system for CNS disease. Curr. Drug Metab. 21(9), 661–673.

Jahangir, M.A., Muheem, A., Anand, C., and Imam, S.S. (2020). Recent advancements in transdermal drug delivery system. In: Pharmaceutical Drug Product Development and Process Optimization: Effective Use of Quality by Design. San Diego, CA: Apple Academic Press. 191–216.

Jin, W., Xu, W., Liang, H., Li, Y., Liu, S., and Li, B. (2016). Nanoemulsions for food: properties, production, characterization, and applications. Emulsions 3, 1–36.

Kaur, C., and Kapoor, H.C. (2001) Antioxidants in fruits and vegetables-the millenium's health. Int J Food Sci. Technol. 36, 703–725.

Kay, D.G., and Maclellan, A. (2018). Composition and method for improving cognitive function and brain bioavailability of ginseng and ginsenosides and treating neuro-degenerative disease and neurological disorders. WO2018148821A1.

Kentish, S., Wooster, T., Ashokkumar, M., Balachandran, S., Mawson, R., and Simons, L. (2008). The use of ultrasonics for nanoemulsion preparation. Innov. Food Sci. Emerg. Technol. 9, 170–175.

Kumar, R., Kaur, K., Uppal, S., and Mehta, S. K. (2017). Ultrasound processed nano-emulsion: a comparative approach between resveratrol and resveratrol cyclodextrin inclusion complex to study its binding interactions, antioxidant activity and UV light stability. Ultrason. Sonochem. 37, 478–489.

Leong, T. S. H., Wooster, T. J., Kentish, S. E., and Ashokkumar, M. (2009). Minimising oil droplet size using ultrasonic emulsification. Ultrason. Sonochem. 16, 721–727.

Liang, R., Huang, Q., Ma, J., Shoemaker, C. F., and Zhong, F. (2013). Effect of relative humidity on the store stability of spraydried beta-carotene nanoemulsions. Food Hydrocoll. 33, 225–233.

Liu, F., Zhu, Z., Ma, C., et al. (2016). Fabrication of concentrated fish oil emulsions using dual-channel microfluidization: Impact of droplet concentration on physical properties and lipid oxidation. J. Agric. Food Chem. 64(50), 9532–9541.

Liu, J., Bi, J., Liu, X., Zhang, B., Wu, X., Wellala, C. K. D., et al. (2019). Effects of high pressure homogenization and addition of oil on the carotenoid bioaccessibility of carrot juice. Food Funct. 10, 458–468.

Luo, X., Zhou, Y., Bai, L., Liu, F., and McClements, D. J. (2017). Fabrication of β-carotene nanoemulsion-based delivery systems using dual-channel microfluidization: physical and chemical stability. J. Colloid Interface Sci. 490, 328–335.

Maali, A., and Hamed Mosavian, M. T. (2013), Preparation and application of nano-emulsions in the last decade (2000–2010). J. Dispers. Sci Tech. 34, 92–105.

Maestro, I., Sole, C., Gonzalez, C., Solans, C., and Gutierrez, J. M. (2008). Influence of the phase behavior on the properties of ionic nanoemulsions prepared by the phase inversion composition method. Colloid Interface Sci. 327, 433–439.

Malik, A. (2008). The potentials of nutraceuticals. In Pharmainfo.net 6 Metchinkoff E (1907): The Prolongation of Life (pp. 151–183). New York: Putmans Sons.

Malnati, R. M. E. J., Du-Pont, M. X. A., and Morales, D. A. O. (2019). Method for producing a nanoemulsion with encapsulated natural antioxidants for preserving fresh and minimally processed foods, and the nanoemulsion thus produced. WIPO WO2019039947A1. Available online at: https://patentscope.wipo.int/search/en/detail.jsf?docId=WO2019039947

Mao, L., Yang, J., Xu, D., Yuan, F., and Gao, Y. (2010). Effects of homogenization models and emulsifiers on the physicochemical properties of b-carotene nanoemulsions. J. Dispers. Sci. Technol. 31, 986–993.

McClements, D.J. (2015). Food Emulsions: Principles, Practices, and Techniques. Boca Raton, FL: CRC Press.

McClements, D. J., Decker, E. A., and Weiss, J. (2007). Emulsion-based delivery systems for lipophilic bioactive components. J. Food Sci. 72, 109–124.

McClements, D. J., and Rao, J. (2011). Food-grade nanoemulsions: formulation, fabrication, properties, performance, biological fate, and potential toxicity. Crit. Rev. Food Sci. Nutr. 51, 285–330.

Möller, M., Eberle, U., Hermann, A., Moch, K., and Stratmann, B. (2009). Nanotechnology in the Food Sector. Zürich: TA-SWISS.

Nakabayashi, K., Amemiya, F., Fuchigami, T., Machida, K., Takeda, S., Tamamitsub, K., et al. (2011). Highly clear and transparent nanoemulsion preparation under surfactant-free conditions using tandem acoustic emulsification. Chem. Commun. 47, 5765–5767.

Noori, S., Zeynali, F., and Almasi, H. (2018). Antimicrobial and antioxidant efficiency of nanoemulsion-based edible coating containing ginger (Zingiber officinale) essential oil and its effect on safety and quality attributes of chicken breast fillets. Food Control. 84, 312–320.

NutraLease (2011). Available online at: http://www.nutralease.com/Nutra/Templates/showpage.asp?DBID=1&LNGID=1&TMID=84&FID=768

Otoni, C. G., de Moura, M. R., Aouada, F. A., Camilloto, G. P., Cruz, R. S., Lorevice, M. V., et al. (2014a). Antimicrobial and physicalmechanical properties of pectin/papaya puree/cinnamaldehyde nanoemulsion edible composite films. Food Hydrocoll. 41, 188–194

Otoni, C. G., Pontes, S. F., Medeiros, E. A., and Soares, N. de F. (2014b). Edible films from methylcellulose and nanoemulsions of clove bud (Syzygium aromaticum) and oregano (Origanum vulgare) essential oils as shelf life extenders for sliced bread. J. Agric. Food Chem. 62, 5214–5219.

Pathak, M. (2017). Nanoemulsions and their stability for enhancing functional properties of food ingredients. Nanotechnol. Appl. Food. 2017, 87–106.

Piorkowski, D. T., and McClements, D. J. (2014). Beverage emulsions: recent developments in formulation, production, and applications. Food Hydrocoll. 42, 5–41.

Robledo, N., Vera, P., López, L., Yazdani-Pedram, M., Tapia, C., and Abugoch, L. (2018). Thymol nanoemulsions incorporated in quinoa protein/chitosan edible films; antifungal effect in cherry tomatoes. Food Chem. 246, 211–219.

Ruchi, S. (2017). Role of nutraceuticals in health care: A review. Int. J. Green Pharm. 11(3). doi: http://dx.doi.org/10.22377/ijgp.v11i03.1146.

Saberi, A. H., Fang, Y., and McClements, D. J. (2013). Fabrication of vitamin E-enriched nanoemulsions: factors affecting particle size using spontaneous emulsification. J. Colloid Interface Sci. 391, 95–102.

Salager, J. L., Forgiarini, A., Marquez, L., Pena, A., Pizzino, A., Rodriguez, M. P., et al. (2004). Using emulsion inversion in industrial processes. Adv. Colloid Interface Sci. 108, 259–272.

Salvia-Trujillo, L., Soliva-Fortuny, R., Rojas-Graü, M. A., McClements, D. J., and Martín-Belloso, O. (2017). Edible nanoemulsions as carriers of active ingredients: A review. Annu. Rev. Food Sci. Technol. 8, 439–466.

Sasaki, R. S., Mattoso, L. H. C., and de Moura, M. R. (2016). New edible bionanocomposite prepared by pectin and clove essential oil nanoemulsions. J. Nanosci. Nanotechnol. 16, 6540–6544.

Schultz, S., Wagner, G., Urban, K., and Ulrich, J. (2004). High-pressure homogenization as a process for emulsion formation. Chem. Eng. Technol. 27, 361–368.

Severino, R., Ferrari, G., Vu, K. D., Donsi, F., Salmieri, S., and Lacroix, M. (2015). Antimicrobial effects of modified chitosan based coating containing nanoemulsion of essential oils, modified atmosphere packaging and gamma irradiation against Escherichia coli O157:H7 and Salmonella Typhimurium on green beans. Food Control. 50, 215–222.

Severino, R., Vu, K. D., Donsì, F., Salmieri, S., Ferrari, G., and Lacroix, M. (2014). Antibacterial and physical effects of modified chitosan based-coating containing nanoemulsion of mandarin essential oil and three non-thermal treatments against Listeria innocua in green beans. Int. J. Food Microbiol. 191, 82–88.

Shahidi, F., and Zhong, Y. (2010). Lipid oxidation and improving the oxidative stability. Chem. Soc. Rev. 39, 4067–4079

Shariffa, Y. N., Tan, T. B., Uthumporn, U., Abas, F., Mirhosseini, H., Nehdi, I. A., et al. (2017). Producing a lycopene nanodispersion: formulation development and the effects of high pressure homogenization. Food Res. Int. 101, 165–172.

Shin, G.H., Li, J., Cho, J.H., Kim, J.T., and Park, H.J. (2016). Enhancement of curcumin solubility by phase change from crystalline to amorphous in Cur-TPGS nanosuspension. J. Food Sci. 81(2), N494–N501.

Silva, H. D., Cerqueira, M. A., and Vicente, A. A. (2015). Influence of surfactant and processing conditions in the stability of oil-in-water nanoemulsions. J. Food Eng. 167, 89–98.

Silva, H. D., Poejo, J., Pinheiro, A. C., Donsi, F., and Vicente, A. A. (2018). Evaluating the behaviour of curcumin nanoemulsions and multilayer nanoemulsions during dynamic in vitro digestion. J. Functional Foods. 48, 605–613

Stellavato, A., Pirozzi, A.V.A., de Novellis, F., et al. (2018). In vitro assessment of nutraceutical compounds and novel nutraceutical formulations in a liver-steatosis-based model. Lipids Health Dis.17, 24.

Sugumar, S., Singh, S., Mukherjee, A., and Chandrasekaran, N. (2015). Nanoemulsion of orange oil with non ionic surfactant produced emulsion using ultrasonication technique: evaluating against food spoilage yeast. Appl. Nanosci. 6, 113–120.

Tadros, T.F., editor (2009). Emulsion science and technology: a general introduction. In Emulsion Science and Technology (pp. 1–56). Weinheim: Wiley-VCH Verlag GmbH & Co. KGaA.

Tan, S., Ebrahimi, A., and Langrish, T. (2019). Controlled release of caffeine from tablets of spray-dried casein gels. Food Hydrocoll. 88, 13–20.

Tan, C. P., and Nakajima, M. (2005a). β-Carotene nanodispersions: preparation, characterization and stability evaluation. Food Chem. 92, 661–671.

Tan, C. P., and Nakajima, M. (2005b). Effect of polyglycerol esters of fatty acids on physicochemical properties and stability of β-carotene nanodispersions prepared by emulsification/evaporation method. J. Sci. Food Agric. 85, 121–126.

Tian, H., Li, D., Xu, T., Hu, J., Rong, Y., and Zhao, B. (2017). Citral stabilization and characterization of nanoemulsions stabilized by a mixture of gelatin and Tween 20 in an acidic system. J. Sci. Food Agric. 97, 2991–2998.

Tian, W. L., Lei, L. L., Zhang, Q., and Li, Y. (2016). Physical stability and antimicrobial activity of encapsulated cinnamaldehyde by self-emulsifying nanoemulsion. J. Food Process Eng. 39, 462–471.

Uluata, S., McClements, D.J., and Decker, E.A. (2015). Physical stability, autoxidation, and photosensitized oxidation of omega-3 oils in nanoemulsions prepared with natural and synthetic surfactants. J. Agric. Food Chem. 63(42), 9333–9340.

Unilever (2011). Available online at: http://www.unilever.com/innovation/productinnovations/coolicecreaminnovations/?WT.LHNAV=Cool_ice_cream_innovations

Walker, R. M., Decker, E. A., and McClements, D. J. (2015). Physical and oxidative stability of fish oil nanoemulsions produced by spontaneous emulsification: effect of surfactant concentration and particle size. J. Food Eng. 164, 10–20.

Wei, Z., and Gao, Y. (2016). Physicochemical properties of β-carotene bilayer emulsions coated by milk proteins and chitosan–EGCG conjugates. Food Hydrocoll. 52, 590–599.

Weiss, J., Takhistov, P., and McClements, D. J. (2006). Functional materials in food nanotechnology. J. Food Sci. 71, 107–116.

Windhab, E. J., Dressler, M., Feigl, K., Fischer, P., and Megias-Alguacil, D. (2005). Emulsion processing—from single-drop deformation to design of complex processes and products. Chem. Eng. Sci. 60, 2101–2113.

Witthayapanyanon, A., Acosta, E. J., Harwell, J. H., and Sabatini, D. A. (2006). Formulation of ultralow interfacial tension systems using extended surfactants. J. Surfactants Deterg. 9, 331–339.

Yang, X., Tian, H., Ho, C. T., and Huang, Q. (2011). Inhibition of citral degradation by oil-in-water nanoemulsions combined with antioxidants. J. Agric. Food Chem. 59, 6113–6119.

Yin, L. J., Chu, B. S., Kobayashi, I., and Nakajima, M. (2009). Performance of selected emulsifiers and their combinations in the preparation of β-carotene nanodispersions. Food Hydrocoll. 23, 1617–1622.

Yuan, Y., Gao, Y., Zhao, J., and Mao, L. (2008). Characterization and stability evaluation of β-carotene nanoemulsions prepared by high pressure homogenization under various emulsifying conditions. Food Res. Int. 41, 61–68.

Zhang, X., Liu, J., Qian, C., Kan, J., and Jin, C.H. (2019). Effect of grafting method on the physical property and antioxidant potential of chitosan film functionalized with gallic acid. Food Hydrocoll. 89, 1–10.

Zhao, Q., Ho, C.T., Huang, Q. (2013). Effect of ubiquinol-10 on citral stability and off-flavor formation in oil-in-water (O/W) nanoemulsions. J. Agric. Food Chem. 61(31), 7462–7469.

CHAPTER 11

Application of Nanoemulsions to Improve Food System Color, Flavor, Texture, and Preservation

Sabna Kotta,[1] Hibah Mubarak Aldawsari,[1]
Shaimaa M. Badr-Eldin,[1] Nabil Abdulhafiz Alhakamy,[1]
Rasheed Ahemad Shaik[2]

[1]Department of Pharmaceutics, Faculty of Pharmacy,
King Abdulaziz University, Jeddah, Kingdom of Saudi Arabia
[2]Department of Pharmacology & Toxicology,
Faculty of Pharmacy, King Abdulaziz University, Jeddah,
Kingdom of Saudi Arabia

CONTENTS

DOI: 10.1201/9781003121121-11

11.1 INTRODUCTION

Food industry is in search of new technologies for enhancing flavor, taste, color, stability as well as nutritional value of foodstuffs for better consumer demand. Nowadays there is a great demand on food products with more health benefits. Consumer demand is high for food products with natural ingredients rather than synthetic or artificial ingredients. Also food fortified with natural bioactive ingredients and micro nutrients are in great demand. Most of the bioactive substances are physically and chemically not stable in all stages of food processing and this leads to unfavorable effects like change in the appearance or decrease in the bioavailability of added bioactive agents (Martins et al., 2015).

Nanotechnology can address most of these issues and improve the overall properties of food products by various approaches. It offers protection of bioactive agents, controlling the release pattern, increasing the bioavailability, modification of texture, improving the stability, etc. (Vinogradov, Bronich and Kabanov, 2002). It should be considered that material manipulation in nanometer size range may have potential toxicity issues like modification in the cellular redox balance or inflammation in tissues. Since present researches are not enough to predict the toxicity profile of nanostructures, further researches and studies are mandatory to evaluate the toxicity of these nanomaterials (Arora, Rajwade and Paknikar, 2012).

Nanoemulsions technology is a comparatively easier approach to answer the issues like stability, bioavailability, rheological property, etc., in food industry. Researches in the field of development of food grade nanoemulsions are extensively increased in last few years to develop effective delivery systems for encapsulating and thereby protecting and modifying the release of bio active ingredients. These bioactive ingredients in food include vitamins, nutraceuticals, fatty acids, etc. The global food market for food encapsulation estimated a 33.9 billion US dollar in 2020 and projecting to achieve a size of 50.8 billion US dollar in 2027 (Global Industry Analysts, Inc., San Jose, CA).

For specific applications in food and beverage industry nanoemulsions possesses many prospective advantages as compared to conventional emulsions. They exhibit a better stability profile against particle aggregation as well as gravitational separation (Tadros et al., 2004).

Since nanoemulsions are mostly transparent in nature it can be added to food products which need to be transparent or translucent in nature like soft drinks, juices, soups, etc. (Velikov and Pelan, 2008; McClements and Rao, 2011). Formation of gel like systems at very less droplet concentration is also possible with nanoemulsions technology which can be utilized in some food products (McClements and Rao, 2011). Many studies have been done on nanoemulsions in food systems to increase bioavailabilty of the incorporated bio active ingredients. In this chapter, we are detailing various applications of nanoemulsions in food products like delivery of lipophilic ingredients, increasing the stability and bioaccessibility, texture modification, etc. Also we are describing various formulation methods of edible nanoemulsion briefly. Main challenges of nanoemulsion technology in food industry and regulatory issues are also discussed. Nanoemulsions based food products in market are also discussed.

11.2 FORMULATION OF FOOD-GRADE NANOEMULSIONS

Food-grade nanoemulsions can be manufactured by different approaches and they can be generally classified broadly into two, as low energy methods and high energy methods based on the underlying mechanism. The details of production methods, mechanism of formation,

TABLE 11.1 Various Production Methods of Food Grade Nanoemulsions

Method	Advantages	Examples of Food Grade Nanoemulsions Formulated
High-Energy Methods		
High-pressure homogenization	Comparatively low polydispersibility and high-volume throughput of nanoscale droplets	Pan et al., 2014b; Jo, Chun, et al., 2015; Flores et al., 2016; Majeed et al., 2016; Ricaurte et al., 2016; Tastan et al., 2016
Microfluidization	High emulsification efficiency and produce stable nanoemulsions	Jo and Kwon, 2014; Gomes et al., 2017
Ultrasonication	Very efficient in reducing the size of droplets and results in the formation of stable nanoemulsions	Mahdi Jafari, He and Bhandari, 2006; Liu et al., 2013; Mustafa et al., 2014; Borthakur et al., 2016
High-shear stirring	Fast and cost-effective method	Scholz and Keck, 2015; Negahdaripour, 2020
Low-Energy Methods		
Phase inversion composition	Kinetically stable nanoemulsions formation. Low energy required, cost and ease of operation	Pan et al., 2014a; Elgammal, Schneider and Gradzielski, 2015; Kwon et al., 2015
Phase inversion temperature	Economical and no need of any sophisticated equipments. Physically stable nanoemulsions formation. Excellent in protecting encapsulated sensitive bioactive ingredients	Dario et al., 2016; Gao and McClements, 2016; Su and Zhong, 2016; Gomes et al., 2017; Chuesiang et al., 2018
Spontaneous emulsification	Simple and inexpensive method, no need of any sophisticated equipments. Less operating cost	Saberi, Fang and McClements, 2013; Davidov-Pardo and McClements, 2015; Guttoff, Saberi and McClements, 2015

properties and instabilities of nanoemulsion are extensively reviewed by many researchers (Mason et al., 2006; McClements and Rao, 2011; Kotta et al., 2012). Table 11.1 summarizes production methods of food grade nanoemulsions and advantages of each along with some works done using each technique.

Another way of classification of nanoemulsions fabrication is top-down approach and bottom-up approach. The top-down approach deals with disruption of oil phase to homogeneous nanosized droplets while the bottom-up approach intended at assembling of building blocks from molecular level to a structured system. Figure 11.1 shows the schematic diagram of different techniques of nanoemulsions fabrication in terms of surfactant to oil ratio and energy required with the approximate droplet size expected.

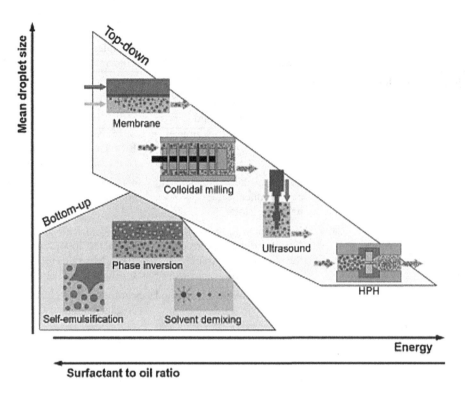

FIGURE 11.1   Schematics diagram of different fabrication methods of nanoemulsions.
Reprinted from (Donsì and Ferrari, 2016). Copyright (2019), with permission from Elsevier.

Top-down methods generally deal with mechanical size reduction processes of oil droplets in the process fluid to generate mechanical stress. So here the emulsification happens by two procedures; firstly breaking up of coarse droplets to finer ones and then coating of the new interface created by absorption of emulsifier in order to prevent re-coalescence and to increase kinetic stability. Here the type and amount of emulsifier play an important role in emulsification efficiency along with processing conditions (Donsì, Sessa and Ferrari, 2012; Silva, Cerqueira and Vicente, 2015). High pressure homogenization, colloidal milling and ultrasonication are mainly used methods for the production of food grade essential oil nanoemulsions (Donsì, Annunziata and Ferrari, 2013).

Spontaneous association of surfactants around the oil globules for achieving a thermodynamic equilibrium is utilized in the emulsification process by bottom-up fabrication approaches. The self-assembly oil droplets mainly depends on the properties like solubility, surface activity, geometry, etc. of the molecules involved (McClements and Rao, 2011).

11.3 APPLICATIONS OF NANOEMULSIONS IN FOOD INDUSTRY

The technological limitation in the formulation of functional foods is the low stability solubility as well as low bioavailability of the bioactives. The bioactives are more vulnerable to degradation due to processing and oxidative damage upon storage. Also some bioactive agents have very less solubility and rapid metabolism and this leads to reduced bioavailability. Some bioactives are volatile in nature and sensitive to processing conditions. All these challenges can be addressed by the use of nanoemulsions for the encapsulation of bioactive agents for their efficient use in food products (Xianquan et al., 2005; Aswathanarayan and Vittal, 2019).

11.3.1 Nanoemulsions as Antimicrobial Agents

Nowadays essential oils, such as eugenol, thymol, and *trans*-cinnamaldehyde, as natural antimicrobial agents in food products have been on high demand due to consumers' requirements of food safety and quality (Inouye, Takizawa and Yamaguchi, 2001). To increase safety, nanoemulsions formulated with essential oils as natural antimicrobials in food substances are of great interest (Acevedo-Fani, Soliva-Fortuny and Martín-Belloso, 2017). The excellent antimicrobial activity of essential oils can be utilized in food preservation by emulsification techniques to overcome their water solubility issue. Nanosized emulsions not only protect these oils from degradation, but also enhance their water dispersibilty.

There are mainly four different ways by which essential oil nanoemulsions interact with microbial cell membrane. Firstly small nanodroplets with increased surface area and hydrophilic surface can pass through hydrophilic transmembrane channels of porin proteins of Gram negative bacteria (Nazzaro et al., 2013). Nanodroplets may facilitate the close contact of essential oil with microbial cell membrane to disrupt phospholipid bilayer in case of Gram positive bacteria and yeast cells (Moghimi, Ghaderi, et al., 2016). Secondly fusion of emulsifier with the cell membrane phospholipid bilayer results in the release of essential oils to desired site in a targeted manner (Li et al., 2015). Also specific direct interactions between cell membrane and emulsifier have been reported (Salvia-Trujillo et al., 2014a, b). Thirdly, the sustained release of the essential oil from the nanodroplets due to partition between the oil phase and the aqueous phase, extends their antimicrobial activity. The nanodroplets act as reservoirs, with oil molecules in dynamic equilibrium between water phase and the oil phase (Donsì et al., 2012). And the fourth proposed mechanism is the electrostatic interaction between the positively charged nanodroplets with negatively charged cell membrane of microbes lead to a higher concentration of essential oil at the site of action (Chang, McLandsborough and McClements, 2013). The four proposed mechanisms are shown as a schematic diagram in Figure 11.2 (Donsì and Ferrari, 2016).

Trans-cinnamaldehyde nanoemulsions incorporated in watermelon juice which contains abundant nutrients could efficiently inhibit the microbial growth. These trans-cinnamaldehyde nanoemulsions were produced by high-energy emulsification technique using optimum ratios of tween and trans-cinnamaldehyde (Jo, Chun, et al., 2015). In another work it was found that microbial growth in orange juice, which was incorporated with eugenol nanoemulson significantly reduced after 6 h of storage at 25°C and 4°C (Ghosh, Mukherjee and Chandrasekaran, 2014). Patrignani et al., formulated nanoemulsions of hexanal and trans-2-hexenal by high pressure homogenization method and incorporated it into apple juice which was deliberately inoculated with pathogenic species. It was found that cell loads were decreased under detection limit in juice containing nanoemulsions which under gone high pressure homogenization as compared to controls (Patrignani et al., 2020). Application of oregano oil nanoemulsions has proved to inhibit foodborne microbial growth in fresh lettuce. Artificially inoculated lettuce with

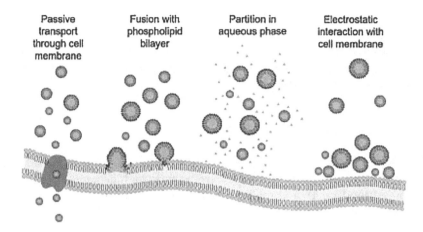

FIGURE 11.2 Schematic representation of interaction of essential oil nanoemulsions with microbial cell membrane.

Reprinted from (Donsì and Ferrari, 2016). Copyright (2019), with permission from Elsevier.

Escherichia coli, Listeria monocytogenes and *Salmonella typhimurium* was dipped into oregano oil nanoemulsions produced by ultrasonication technique to evaluate the efficacy of bacterial inactivation and found that 0.05% or 0.1% nanoemulsions are effective in bacterial inhibition (Bhargava et al., 2015). O/W emulsions from lemon myrtle oil possess excellent antibacterial activity against *E. coli, Salmonella typhimurium, Listeria monocytogenes, B. cereus* and *P. aeruginosa* (Buranasuksombat et al., 2011). Physically stable thyme oil nanoemulsions with corn oil as ripening inhibitor show significant antimicrobial activity against *Zygosaccharomyces bailii* (acid-resistant spoilage yeast). These types of nanoemulsions can be utilized as antimicrobial delivery systems in food industry (Chang, McLandsborough and McClements, 2012). It was found that replacement of pork back fat with olive oil nanoemulsion enhanced the nutrional value and turned the product to healthier in terms of quantity and quality of fat. The treatment also enhanced the appearance, texture, and juiciness of pork patties (López-López et al., 2010).

Sage oil nanoemulsion possesses anti-bacterial activity against some food-borne bacteria namely *E. coli, Salmonella typhi* and *Shigella dysentery*. The nanoemulsions treatment could produce extensive cell membrane damage (Moghimi, Aliahmadi, et al., 2016). Anise oil nanoemulsions can also act as excellent antimicrobial agents against common foodborne pathogens like *E. coli* and *Listeria monocytogenes*. This type of nanoemulsions based delivery systems for essential oils, not only act as anti-microbial agents but also slow down and sustain the release of the bioactives and increase their solubility in aqueous based food products (Topuz et al., 2016).

Carvacrol and peanut oil nanoemulsions prepared by high pressure homogenization method and stabilized using soy lecithin or Tween 20 or monoolein or sucrose palmitate or pea protein isolate was found to be effective against *E. coli* in cooked pork meat sausage and zucchini. The antibacterial activity of nanoemulsions was measured on *E. coli* inoculated pork meat sausage and zucchini cylinders. The antibacterial activity was significantly related to emulsion droplet size rather than formulation ingredients. Infusion of essential oils in vegetable or animal food products as emulsions formulation is highly dependent on the emulsion droplet size in terms of oil diffusion and antimicrobial activity. The study suggests that charge and surface composition of nanoemulsions do not have significant role in controlling the diffusion rate into food stuffs (Donsì et al., 2014). In contrast to this another study suggests that the essential oil type determine the physico-chemical properties and antimicrobial activity of nanoemulsions rather than droplet size. Nanoemulsions were formulated using essential oils from thyme, palmarosa, lemongrass, tea tree, geranium, clove, rosewood, mint, marjoram, or sage stabilized with tween 80 and sodium alginate using high shear homogenization. Out of this nanoemulsions with clove, lemongrass, palmarosa, or thyme showed higher antibacterial activity against *E. coli* (Salvia-Trujillo et al., 2015). Nanoemulsions of carvacrol with medium chain triglyceride and Tween showed good physical stability and antimicrobial activity and hence can be utilized as antimicrobial delivery systems in the food industry (Chang, McLandsborough and McClements, 2013). Another example is peppermint oil nanoemulsions with medium-chain triacylglycerol and biopolymer formulated by high pressure homogenization possess excellent inhibitory effect against *Listeria monocytogenes* and *Staphylococcus aureus*. This can be utilized to extend the shelf life of aqueous based food stuffs (Liang et al., 2012).

11.3.2 Nanoemulsions for Washing the Food Surface

Essential oil containing solutions for washing are not new and have been widely explored. But this method is limited by the low water solubility of most essential oils (Gutierrez et al., 2009). Essential oil nanoemulsions can be effectively utilized as washing solution for vegetables owing to their antimicrobial effect as well as improved wettability. An oregano oil nanoemulsion with improved antimicrobial effect was effectively used for decontamination of fresh lettuce leaves (Bhargava et al., 2015). Carvacrol and eugenol nanoemulsions have also proved efficacy in washing of spinach leaves which were inoculated with *E. coli* and *S. enterica* (Ruengvisesh et al., 2015). Nanoemulsions as washing solutions need further extensive investigations to minimize the effect on organoleptic properties of the product as well as to avoid any associated toxicity issues.

11.3.3 Nanoemulsions to Deliver Lipophilic Ingredients

Lipophilic bioactive compounds like essential oils, antimicrobials, antioxidants, vitamins, etc. can be encapsulated and effectively delivered using nanoemulsions technology. Usually the lipohphilic components are dissolved in the oil phase prior to emulsification. Based on the molecular as well as physicochemical properties, like o/w partition coefficient, hydrophobicity, solubility, surface activity, melting point, etc. the physical location of the solubilized components may vary within the nanoemulsions. More polar lipophilic components will stay inside the amphiphilic shell while more non-polar lipophilic components will be distributed within the hydrophobic core. The physical and chemical stability of the active ingredient is highly significant since it can affect the chemical as well as physical stability of the system. Lipophilic ingredients which are prone to chemical degradation in the presence of water soluble components should be trapped or enclosed inside the oil core (Choi et al., 2009). Even though it is difficult to determine the exact location of the lipophilic ingredient within the emulsion using analytical methods, some techniques like nuclear magnetic resonance (Shen et al., 2005), Raman scattering microspectroscopy (Day et al., 2010) and fluorescence spectroscopy (Tanaka et al., 2003) are proved to be useful.

Apart from preventing degradation from oxidation and light, the solubility and their bioavailability can also be improved. The rheological properties and release of nano-emulsions can be altered as per the requirements (Martins et al., 2015). Nanoemulsions increase bioavailability by different mechanisms. The larger surface area of small droplets makes the digestion quick by digestive enzymes so that the release and absorption of bio-actives will be easy. The small droplets can penetrate to the mucous membrane and this will increase residence time. Also there is chance of direct transport of the smaller par-ticles across epithelial membrane by paracellular or transcellular transport (McClements and Rao, 2011).

11.3.4 Nanoemulsions to Increase Bioaccessibility

Bioaccessibility as well as permeability of green tea catechins can be enhanced by con-verting these into nanoemulsions. Nanoemulsions were prepared by dispersing green tea catechins powder in sunflower oil and then homogenizing with soy protein dispersion with the help of a high-speed homogenizer. Apart from stability, the formulation could improve the bioaccessibility to 2.78-fold and also a significant increase in intestinal per-meability of catechins in Caco-2 cell model. Figure 11.3 shows the schematic representa-tion of the production procedure of soy protein stabilized o/w nanoemulsions of catechins (Bhushani, Karthik and Anandharamakrishnan, 2016).

Nanoemulsions of vitamin E made out of sunflower oil and saponin using high-pressure homogenization technique showed increase in the oral bioavailability to that of conventional emulsion. The chyme of the nanoemulsion was less flocculated in GIT as compared to conventional emulsion. Pharmacokinetic studies proved a 3-fold increase in the bioavailability of vitamin E and thus this type of formulation design could be utilized in food and pharmaceutical industries (Parthasarathi, Muthukumar and Anandharamakrishnan, 2016).

In another work, stable nanoemulsions of vitamin E-acetate were formulated with orange oil, whey protein isolate and biopolymers using high pressure homogenization technique. This is a promising approach for the development of food grade vitamin-enriched delivery systems (Ozturk et al., 2015a). In a different study the influence of carrier oil type on the bioaccessibility of vitamin D_3 nanoemulsions was studied. The

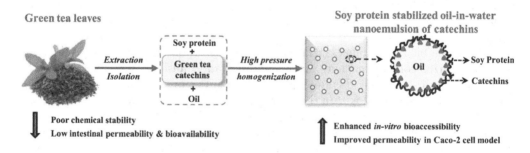

FIGURE 11.3   Schematic representation of the production of soy protein stabilized cat-echins emulsion.

Reprinted from (Bhushani, Karthik and Anandharamakrishnan, 2016). Copyright (2019), with permission from Elsevier.

nanoemulsions were prepared using high pressure homogenization technique with the help of quillaja saponin with various oils and it was found that the bioaccessibility of vitamin D_3 was increased in the order: medium chain triglycerides < mineral oil < orange oil < fish oil ≈ corn oil. The result proved that the chemistry of carrier oil have a significant impact on vitamin bioaccessibility as well as lipid digestion. It was also found that nanoemulsions formulated using long chain triglycerides like corn oil or fish oil are most effective in enhancing the bioaccessibility of vitamin D_3 (Ozturk et al., 2015b).

The bioavailability and stability of resveratrol can be increased through nanoemulsions based delivery system. Several bioactive food components in nutraceuticals and functional foods can be effectively delivered in this manner. Nanoemulsions were formulated by encapsulating resveratrol in peanut oil with hydrophilic and lipophilic emulsifier with the help of high pressure homogenizer. Lecithin-based nanoemulsion was able to deliver resveratrol in shorter times through cell monolayers in a sustained release fashion (Sessa et al., 2014).

Due to potential health benefits incorporation of carotenoids like β-carotene in beverages and food are highly recommended. The poor water solubility and as a result poor bioavailability can be overcome by using nanoemulsions formulation strategy. Nanoemulsions of β-carotene was formulated in corn oil with Tween 20 as surfactant using a high pressure microfluidizer and a significantly higher bioaccessibility was observed (Salvia-Trujillo et al., 2013). Curcumin nanoemulsions were prepared with corn oil and lactoferrin (an aqueous emulsifier) using a homogenizer followed by passage through a high pressure homogenizer and can be used for incorporation into food products (Pinheiro, Coimbra and Vicente, 2016). Various works have been reported to improve bioavailability of different lipophilic nutraceuticals (Nielsen, Petersen and Müllertz, 2008; Choi et al., 2009; Vishwanathan, Wilson and Nicolosi, 2009; Ozaki et al., 2010). MultiSal™ is a delivery system for delivering hydrophilic and lipophilic nutrients developed by Salvona Company. This system can control the release pattern and can extend the contact time in mouth for prolonging the sensation of flavors.

11.3.5 Nanoemulsions to Modify Texture

Since the droplets size in nanoemulsions are very small, the interfacial coating around it contribute to overall particle volume. As a result, gelation of a nanoemulsion with very less amount of oil may be achieved as compared to a conventional emulsion. This can be utilized for the production of highly viscous or gel type fat products. There is not much work done on nanoemulsion to modify the texture of food products. Nanoemulsions gel can be produced by electrostatic repulsion effects if the electrical double layer thickness is on the order of nanoemulsions droplet radius. Even though this type of systems have been developed in non-edible systems, this approach can be potentially explored in food industry also (Wilking and Mason, 2007; Kawada et al., 2010).

11.3.6 Nanoemulsions to Increase Shelf-Life

11.3.6.1 Edible Films

Foods with low amount of synthetic preservatives are always a priority for consumers. Edible films as packaging materials in food industry are on high demand due to their organic nature. Water soluble film forming polymers are incorporated in nanoemulsions

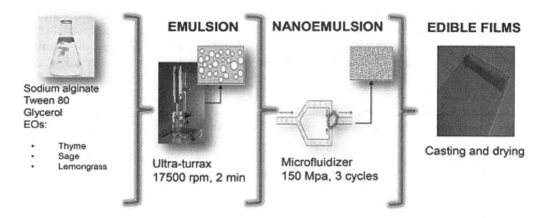

FIGURE 11.4   Schematic representation of the production of edible films from essential oil nanoemulsions.

Reprinted from (Acevedo-Fani et al., 2015). Copyright (2019), with permission from Elsevier.

formulated with antimicrobial essential oils to produce antimicrobial films. One such example is antimicrobial methyl cellulose film produced from nanoemulsions of clove and oregano essential oils to increase the shelf life of packaged bread slices. The films could inhibit the yeast and mold counts in sliced bread for 15 days (Otoni, Pontes, et al., 2014). Similarly, cinnamaldehyde nanoemulsions with pectin as film former in papaya fruit puree has shown excellent antimicrobial properties. These edible films were prepared by incorporating the film forming polysaccharide into aqueous phase of nanoemulsion (Otoni, Moura, et al., 2014). Edible films of sodium alginate from thyme, lemongrass or sage oil nanoemulsions have excellent antimicrobial activity against *E coli*. Microfluidization technique was used to formulate nanaoemulsion. The formulated edible films were with satisfactory mechanical as well as physical properties. Figure 11.4 shows the schematic representation of the procedure for the production of edible films from essential oil nanoemulsions (Acevedo-Fani et al., 2015).

11.3.6.2 Edible Coatings

Similar to edible films, edible coating can also be formulated from nanoemulsions. Nanoemulsions coating was found to be more effective in inhibition of microbial growth as compared to conventional emulsion-based coating. A film coating of lemon grass oil nanoemulsions from carnauba wax based solution was proved to inhibit *Salmonella typhimurium* and *E. coli* on plums without affecting the flavor, glossiness and fracturability of plums (Kim et al., 2013). Similarly carnauba wax coating from lemongrass oil nanoemulsions was able to inhibit microbial growth and extent the shelf life of grapes without effecting their flavor or glossiness (Kim et al., 2014). Edible coatings can increase the shelf life of fresh-cut apples up to two weeks. This edible coatings of sodium alginate formed from lemon grass oil nanoemulsions were able to inhibit *E. coli* growth for a period of two weeks on fresh cut Fuji apples (Salvia-Trujillo et al., 2015).

The shelf life of rucola leaves was increased from 3 to 7 days by the incorporation of chitosan coating containing lemon oil nanoemulsions. The anti-microbial effect was achieved without hindering the organoleptic characteristics of the product (Sessa, Ferrari and Donsi, 2015). Carvacrol nanoemulsions based chitosan coating can show excellent inhibitory effect against *E. coli* and *S. Typhimurium* on green beans. The coating along with gamma irradiation and modified atmosphere packaging was able to reduce the microbial count to undetectable levels for 13 days for *E. coli* and from day 7 to 13 for *S. typhimurium* on the storage of green beans (Severino et al., 2015). The composite edible films made out of oregano oil nanoemulsions possess excellent mechanical properties as well as efficient moisture barrier (Bilbao-Sáinz et al., 2010). Chitosan-based edible films from basil and thyme oil emulsions prepared by microfluidization technique were softer, less rigid, glossy as well as stretchable in nature (Dhama et al., 2020).

11.3.7 Nanoemulsions to Improve Color and Flavor

The aldehyde, ketone, and ester functional groups in coloring and flavoring agents used in food make it susceptible to photolytic degradation and oxidation. Encapsulation of these agents inside nanoemulsions can avoid these problems to improve its shelf life (Goindi et al., 2016). Citral nanoemulsions along with natural antioxidants like tanshinone, β carotene and extracts of black tea were able to achieve very good chemical stability during storage (Yang et al., 2011).

Similarly, citral and ubiquinol-10 ($Q_{10}H_2$) nanoemulsions also showed a good chemical stability (Zhao, Ho and Huang, 2013).

β-Carotene precursor of vitamin A is a commonly used antioxidant and colorant in various food stuffs. But, it undergoes degradation in presence of light, oxygen and heat. β-carotene nanoemulsions were prepared using emulsification-evaporation technique showed excellent stability (Tan and Nakajima, 2005). β-carotene nanoemulsions were made using high pressure homogenization with Tween also reduced β-carotene degradation up to 4 weeks of storage (Yuan et al., 2008). Spray-dried and starch stabilized β-carotene nanoemulsions showed better stability during storage (Liang et al., 2013). Starch caseinate and chitosan-epigallocatechin-3-gallate coating improved physicochemical characteristics of β-carotene nanoemulsions (Wei and Gao, 2016). High water dispersibilty and chemical stability was achieved for β-carotene freeze dried nanoemulsions produced by sonication method (Chen et al., 2017). The yogurt containing carotenoid nanoemulsions with gelatin and whey protein isolate as natural coloring agent showed good stability up to 60 days (Medeiros et al., 2019).

11.3.8 Multilayer Emulsions

Multilayer emulsions contain minimum two interfacial layers composed of surfactants and some biopolymers. The interfacial membranes are produced by assembling oppositely charged biopolymers in a layer-by-layer (LbL) fashion (Zeeb, Thongkaew and Weiss, 2014). They have more physical stability as compared to normal emulsions. In food industry the practical application of multilayer emulsions is still limited. However, multilayer emulsions can incorporate different bioactive lipids and can be utilized as a strategy to increase the fat quality of different foodstuffs. Double layer nanoemulsions containing biopolymers can enhance the stability of nanoemulsion, and can be utilized in the production of functional foods containing bioactive ingredients.

The layer-by-layer coating technology offers possible advantages due to the capability to control the exact location of multiple polymer layers with nano size precision; the capacity to specify the concentrations of bioactive ingredient simply by changing the number of polyelectrolyte coatings; and also the chance to utilize a variety of polymers, which increase its possible applications. For the coating, biopolymers like polysaccharides or proteins are of high demand. Non-toxicity, natural availability and biocompatibility are the main reason behind the acceptance of biopolymers. The strong electrostatic interaction between the surface and the charged polyelectrolytes helps to build the new layer. Further layers can be added by simple addition of oppositely charged polyelectrolytes in solution, which promote the adsorption of the polyelectrolytes over the first layer. The number of layers can be adjusted as per the application of the food system (McClements, Decker and Weiss, 2007; Pinheiro, Coimbra and Vicente, 2016).

Capsaicin loaded nanoemulsions stabilized with chitosan and alginate prepared by self-assembly emulsification technique is a promising functional ingredient delivery system. The droplet size of this double-layer nanoemulsions was 20 nm or lower with excellent stability (Choi et al., 2011). Pork patties can be enriched with multilayer nanoemulsion of fish oil made by LbL deposition method using Tween 20, chitosan, and low methoxyl pectin. Improvement in chemical stability and reduction of bacterial count can also be achieved by using this emulsion coating. The emulsion treatment did not make any change in hardness, springiness and cohesiveness during storage period (Jo, Kwon, et al., 2015).

A study was conducted to understand the behavior of multilayer nanoemulsions inside GIT and found that the biopolymer coating may be able to control lipid digestion rate as well as absorption of free fatty acid. For this study, curcumin multilayer nanoemulsions were prepared by drop-wise addition of curcumin nanoemulsion to an alginate solution with stirring. Sonication was done to disrupt the flocculated droplets (Pinheiro, Coimbra and Vicente, 2016).

In another study bilayer emulsions of β-carotene were prepared by LbL electrostatic deposition of lactoferrin and lactoferrin–polyphenol conjugates. It was found that the lactoferrin–polyphenol stabilized emulsion has enhanced physical stability and chemical stability of β-carotene (Liu et al., 2016). In a work by Salminen and Weiss (Salminen and Weiss, 2014), whey protein isolate and apple pectin complexes were adsorbed on the emulsion interface by means of electrostatic interactions and accessed the emulsion stability. This technique succeeded in improving the emulsion stability. The emulsion was stable at pH 4.5, after the deposition of the coating. Also the biopolymer coated nanoemulsions were resistant to heat treatments and salt additions up to 200 mM level of NaCl. However, the biopolymer coating could not prevent the instability after freeze-thawing cycles.

Deposition of alginate layer on Tween 20 and β-lactoglobulin nanoemulsions reduced the lipid digestion by inhibiting the access of enzyme lipase to the lipid in the nanoemulsion (Li and McClements, 2014). The influence of the anionic and cationic polysaccharides coating on carotenoid nanoemulsions was investigated in terms of bioaccessibility, *in vitro* digestion and physicochemical properties. It was found that degradation of carotenoid is slightly inhibited by coating. Also chitosan coating reduced lipids digestibility and bioaccessibility. Also it was found that stability of emulsion was improved by coating. So the use of interfacial layers in emulsions has a good potential in the development of delivery systems for nutrients and also offer an enhanced protection of emulsified bioactives against degradation while processing and storage. So this strategy can be utilized in food and pharmaceutical industry (Zhang et al., 2015).

In a study of multilayering of emulsions using LbL electrostatic deposition of oppositively charged biopolymers were carried out along with an enzymatic treatment for studying the influence of layer thickness and composition on the release behavior of lutein. Primary whey protein isolate stabilized emulsion exhibited a 5 fold increase in lutein release than non-cross linked secondary emulsions. Dodecyltrimethylammonium bromide stabilized and beet pectin coated secondary emulsion released only 0.13% of lutein as compared to uncoated primary emulsion (7.2%). The results of the study highlights that adjusting the interfacial properties with the help of biopolymers either by cross linking or deposition is a useful approach for the development of specific release-over-time profile delivery systems in food industry (Beicht et al., 2013; Salvia-Trujillo et al., 2015).

11.4 NANOEMULSIONS-BASED FOOD PRODUCTS IN MARKET AND ASSOCIATED CHALLENGES

Some nanoemulsions based food products which are available in market (McClements and Rao, 2011) and summarized in Table 11.2.

There are many challenges in utilizing nanoemulsions in the food industry. The major ones are discussed here. Firstly the limitation in the selection of food grade emulsifier restricts selection of emulsifier for diverse product design. Due to this limitation, productions of nanoemulsions with different release profile are restricted. The second limitation is chances of instability during extreme processing or storage conditions like heating, freezing-thawing, salt content, chilling, dehydration, etc. Another limitation is a chance of oxidation of the incorporated bioactive agent due a very thin interfacial layer. And finally failure may happen in offering improved protection and stability of incorporated bioactive ingredients during the gastro intestinal passage (Drusch, 2007; McClements, Decker and Weiss, 2007; Martins et al., 2015).

TABLE 11.2   Nanoemulsions-Based Food Products in Market

Name of the Food Product	Product Description	Company
Canola active oil	Phytosterols fortified canola oil by encapsulation technique.	Shemen, Haifa, Israel
Fortified fruit juice	Fruit juice fortified with theanine, vitamin, SunActive iron and lycopene by nanoencapsulation technology	High vive, USA
Color emulsion	Nanoemulsions composed of paprika, beta-carotenal or apocarotenal	Wild Flavors, Inc, USA
Nutri-Nano™CoQ-10 3.1x Softgels	Nanosized oil in water systems	Solgar, USA
"Daily boost"	Nanoencapsulation of fortified bioactive components or vitamin or beverage	Jamba Juice Hawaii, USA
Vitamin supplements, Spray for Life®	Fortified vitamin beverage by nanoencapsulation technique	Health Plus International®, Inc., USA

One approach to overcome these limitations is LbL encapsulation technique. This technique utilizes multilayer encapsulation together with nano emulsification for enhancing the characteristics of lipid bioactive agents. Jung et al. utilized this technique to enhance the uptake and utilization of fish oil omega-3 fatty acids. The skin permeability as well as intestinal absorption pattern are well improved. Therefore lecithin/chitosan/ low methoxypectin multilayer encapsulation together with emulsification can be utilized for the effective delivery of lipid bioactive agents (Jung et al., 2016). More works related to this technique are mentioned above in section 3.7.

11.5 REGULATORY PERSPECTIVES

Another important aspect to be considered is safety issue associated with consumption of foods containing nanoemulsions (Donsì and Ferrari, 2016). The regulatory aspects are associated with formulation ingredients as well as the use of nanotechnology for their synthesis. The formulation ingredients in food grade nanoemulsions are usually well consolidated and enlisted by different authorities like Food and Drug Administration (www.fda.gov), World Health Organization (www.who.int) and the European Food Safety Authority (www.efsa.europa.eu) (McClements and Rao, 2011). But there is not an exact globally recognized definition for nanoengineered substances in terms of regulatory aspects. FDA has not yet established a regulatory definition other than a nonbinding guideline (http://www.fda.gov/regulatoryinformation/guidances/ucm257698.htm) to determine whether it falls in nanoscale range or it possess peculiar biological, chemical, or physical properties because of its nanosize range (Martins et al., 2015).

However, European Union provides a definition (Regulation (EU) No 1363/2013) of engineered nanomaterials as "any intentionally manufactured material, which contains 50% or more of with one or more external dimensions in the nanoscale range (1–100 nm)". European Union regulatory frame work includes the potential uses of nanotechnologies in the food by precise approval procedures or by the principles of the general food law (Regulation (EC) No 178/2002).

Natural antimicrobial agents used in food products are commonly regulated as food additives in definite database along with their allowed concentration and use by many authorities. In Europe the European Food Safety Authority (www.efsa.europa.eu), in U.S.A the Food and Drug Administration (www.fda.gov), and in China the China Food Additives & Ingredients Association in China (www.cfaa.cn) are some of them (Malhotra, Keshwani and Kharkwal, 2015).

FDA has a list of natural extractives, oleoresins and essential oils that are generally recognized as safe for their intended use (Code of Federal Regulations, Title 21, Volume 3, 1998) while The European Food Safety Authority defines the individual food ingredients and their eventual restriction of use in various food categories (Regulation (EC) No 2232/96 and Regulation (EC) No 1334/2008).

It is actually difficult to standardize plant extracts in terms acceptable daily intake or no observed adverse effect level, due to their batch to batch variation in composition. In case of a new compound which has not yet approved, the regulatory agencies should address it in terms of its identification, usage, limits and short-term and long-term health consequences and other safety concerns (Malhotra, Keshwani and Kharkwal, 2015). Apart from this the use of very reactive substance in essential oils should be constantly monitored due to associated health risk. Because some compound like carvacrol has recently shown some concentration related toxic effects *in vitro* even though included in various authorities like FDA, FAO/WHO, Council of Europe, etc. (LLana-Ruiz-Cabello et al., 2014; Donsì and Ferrari, 2016). It is highly recommended that development of specific ISO standards to judge the legal aspect for setting out the definition, labeling requirements, rules of use, maximum allowed level, etc. in consideration with high potential use of essential oils as preservative in food industry (Del Nobile et al., 2012).

11.6 CONCLUSIONS AND FUTURE DIRECTIONS

These days, consumers demand for high-quality, safe as well as healthy food products are in great demand. Consumers are avoiding food stuffs with synthetic ingredients and looking for products with natural alternatives. For example food stuffs with essential oils as antimicrobial agent are a better choice for consumers than one with synthetic preservatives. There is a global change nowadays for the intake of food stuffs which promote health, that means this change in consumption pattern requires extensive research in food industry to meet the consumer demands (Del Nobile et al., 2012; Donsì et al., 2012).

Nanoemulsions are one of the vast growing delivery systems in food as well as pharmaceutical industry especially in products which need to be optically translucent or transparent. And also in food system incorporated with bioactive ingredients in order to improve bioavailability and release characteristics. Selections of food grade surfactant system, suitable processing operations, economic aspects, scale-up procedure are some of the major challenges to overcome. In addition, toxicity concern of nano scale material is another big challenge.

The use of essential oil nanoemulsions as antimicrobial agent have many limitations or challenges like the efficiency should be high enough to control the growth of microbial flora but it should be mild enough to avoid interference with natural organoleptic properties of the product. Another one is the encapsulation of essential oil should be sufficient to mask the taste of essential oil at the same time preserving its antimicrobial property. Costs should be carefully reviewed, since essential oils are expensive and also the nanoemulsions production technology costs should be carefully reviewed, since essential oils are expensive and also the nanoemulsions production technology should be wisely selected. To promote shelf life of foodstuffs while maintaining the freshness the combined application of emulsion technology with ozone, UV or gamma irradiation (Severino et al., 2014), pulsed light (Donsì et al., 2015), modified-atmosphere packaging (Severino et al., 2015), etc. can be used (Tajkarimi, Ibrahim and Cliver, 2010; Donsì et al., 2012).

CONFLICT OF INTEREST

The authors declare no conflict of interest.

REFERENCES

Acevedo-Fani, A. et al. (2015) 'Edible films from essential-oil-loaded nanoemulsions: Physicochemical characterization and antimicrobial properties', *Food Hydrocolloids*, 47, pp. 168–177. doi: https://doi.org/10.1016/j.foodhyd.2015.01.032.

Acevedo-Fani, A., Soliva-Fortuny, R. and Martín-Belloso, O. (2017) 'Nanostructured emulsions and nanolaminates for delivery of active ingredients: Improving food safety and functionality', *Trends in Food Science & Technology*, 60, pp. 12–22. doi: https://doi.org/10.1016/j.tifs.2016.10.027.

Arora, S., Rajwade, J. M. and Paknikar, K. M. (2012) 'Nanotoxicology and in vitro studies: the need of the hour', *Toxicology and applied pharmacology*. Elsevier, 258(2), pp. 151–165.

Aswathanarayan, J. B. and Vittal, R. R. (2019) 'Nanoemulsions and Their Potential Applications in Food Industry ', *Frontiers in Sustainable Food Systems*, p. 95. Available at: https://www.frontiersin.org/article/10.3389/fsufs.2019.00095.

Beicht, J. et al. (2013) 'Influence of layer thickness and composition of cross-linked multilayered oil-in-water emulsions on the release behavior of lutein', *Food & Function*. The Royal Society of Chemistry, 4(10), pp. 1457–1467. doi: 10.1039/C3FO60220F.

Bhargava, K. et al. (2015) 'Application of an oregano oil nanoemulsion to the control of foodborne bacteria on fresh lettuce', *Food Microbiology*, 47, pp. 69–73. doi: https://doi.org/10.1016/j.fm.2014.11.007.

Bhushani, J. A., Karthik, P. and Anandharamakrishnan, C. (2016) 'Nanoemulsion based delivery system for improved bioaccessibility and Caco-2 cell monolayer permeability of green tea catechins', *Food Hydrocolloids*, 56, pp. 372–382. doi: https://doi.org/10.1016/j.foodhyd.2015.12.035.

Bilbao-Sáinz, C. et al. (2010) 'Nanoemulsions prepared by a low-energy emulsification method applied to edible films', *Journal of Agricultural and Food Chemistry*. American Chemical Society, 58(22), pp. 11932–11938. doi: 10.1021/jf102341r.

Borthakur, P. et al. (2016) '5—Nanoemulsion: Preparation and its application in food industry', in Grumezescu, A. M. B. T.-E. (ed.) *Nanotechnology in the Agri-Food Industry*. Academic Press, pp. 153–191. doi: https://doi.org/10.1016/B978-0-12-804306-6.00005-2.

Buranasuksombat, U. et al. (2011) 'Influence of emulsion droplet size on antimicrobial properties', *Food Science and Biotechnology*. Springer, 20(3), pp. 793–800.

Chang, Y., McLandsborough, L. and McClements, D. J. (2012) 'Physical properties and antimicrobial efficacy of thyme oil nanoemulsions: Influence of ripening inhibitors', *Journal of Agricultural and Food Chemistry*. American Chemical Society, 60(48), pp. 12056–12063. doi: 10.1021/jf304045a.

Chang, Y., McLandsborough, L. and McClements, D. J. (2013) 'Physicochemical properties and antimicrobial efficacy of carvacrol nanoemulsions formed by spontaneous emulsification', *Journal of Agricultural and Food Chemistry*. ACS Publications, 61(37), pp. 8906–8913.

Chen, J. et al. (2017) 'Encapsulation of carotenoids in emulsion-based delivery systems: Enhancement of β-carotene water-dispersibility and chemical stability', *Food Hydrocolloids*. Elsevier, 69, pp. 49–55.

Choi, A.-J. et al. (2009) 'Effects of surfactants on the formation and stability of capsaicin-loaded nanoemulsions', *Food Science and Biotechnology*, 18(5), pp. 1161–1172.

Choi, A.-J. et al. (2011) 'Characterization of capsaicin-loaded nanoemulsions stabilized with alginate and chitosan by self-assembly', *Food and Bioprocess Technology*, 4(6), pp. 1119–1126. doi: 10.1007/s11947-011-0568-9.

Chuesiang, P. et al. (2018) 'Optimization of cinnamon oil nanoemulsions using phase inversion temperature method: Impact of oil phase composition and surfactant concentration', *Journal of Colloid and Interface Science*, 514, pp. 208–216. doi: https://doi.org/10.1016/j.jcis.2017.11.084.

Dario, M. F. et al. (2016) 'A high loaded cationic nanoemulsion for quercetin delivery obtained by sub-PIT method', *Colloids and Surfaces A: Physicochemical and Engineering Aspects*, 489, pp. 256–264. doi: https://doi.org/10.1016/j.colsurfa.2015.10.031.

Davidov-Pardo, G. and McClements, D. J. (2015) 'Nutraceutical delivery systems: Resveratrol encapsulation in grape seed oil nanoemulsions formed by spontaneous emulsification', *Food Chemistry*, 167, pp. 205–212. doi: https://doi.org/10.1016/j.foodchem.2014.06.082.

Day, J. P. R. et al. (2010) 'Label-free imaging of lipophilic bioactive molecules during lipid digestion by multiplex coherent anti-Stokes Raman scattering microspectroscopy', *Journal of the American Chemical Society*. ACS Publications, 132(24), pp. 8433–8439.

Del Nobile, M. A. et al. (2012) 'Food applications of natural antimicrobial compounds', *Frontiers in Microbiology*. Frontiers, 3, p. 287.

Dhama, K. et al. (2020) 'An update on SARS-CoV-2/COVID-19 with particular reference to its clinical pathology, pathogenesis, immunopathology and mitigation strategies', *Travel Medicine and Infectious Disease*, 37, p. 101755. doi: https://doi.org/10.1016/j.tmaid.2020.101755.

Donsì, F., Annunziata, M. and Ferrari, G. (2013) 'Microbial inactivation by high pressure homogenization: Effect of the disruption valve geometry', *Journal of Food Engineering*. Elsevier, 115(3), pp. 362–370.

Donsì, F. et al. (2012) 'Design of nanoemulsion-based delivery systems of natural antimicrobials: effect of the emulsifier', *Journal of Biotechnology*. Elsevier, 159(4), pp. 342–350.

Donsì, F. et al. (2014) 'Infusion of essential oils for food stabilization: Unraveling the role of nanoemulsion-based delivery systems on mass transfer and antimicrobial activity', *Innovative Food Science & Emerging Technologies*, 22, pp. 212–220. doi: https://doi.org/10.1016/j.ifset.2014.01.008.

Donsì, F. et al. (2015) 'Green beans preservation by combination of a modified chitosan based-coating containing nanoemulsion of mandarin essential oil with high pressure or pulsed light processing', *Postharvest Biology and Technology*, 106, pp. 21–32. doi: https://doi.org/10.1016/j.postharvbio.2015.02.006.

Donsì, F. and Ferrari, G. (2016) 'Essential oil nanoemulsions as antimicrobial agents in food', *Journal of Biotechnology*, 233, pp. 106–120. doi: https://doi.org/10.1016/j.jbiotec.2016.07.005.

Donsì, F., Sessa, M. and Ferrari, G. (2012) 'Effect of emulsifier type and disruption chamber geometry on the fabrication of food nanoemulsions by high pressure homogenization', *Industrial & Engineering Chemistry Research*. ACS Publications, 51(22), pp. 7606–7618.

Drusch, S. (2007) 'Sugar beet pectin: A novel emulsifying wall component for microencapsulation of lipophilic food ingredients by spray-drying', *Food Hydrocolloids*. Elsevier, 21(7), pp. 1223–1228.

Elgammal, M., Schneider, R. and Gradzielski, M. (2015) 'Preparation of latex nanoparticles using nanoemulsions obtained by the phase inversion composition (PIC) method and their application in textile printing', *Colloids and Surfaces A: Physicochemical and Engineering Aspects*, 470, pp. 70–79. doi: https://doi.org/10.1016/j.colsurfa.2015.01.064.

Flores, Z. et al. (2016) 'Physicochemical characterization of chitosan-based coating-forming emulsions: Effect of homogenization method and carvacrol content', *Food Hydrocolloids*, 61, pp. 851–857. doi: https://doi.org/10.1016/j.foodhyd.2016.07.007.

Gao, S. and McClements, D. J. (2016) 'Formation and stability of solid lipid nanoparticles fabricated using phase inversion temperature method', *Colloids and Surfaces A: Physicochemical and Engineering Aspects*, 499, pp. 79–87. doi: https://doi.org/10.1016/j.colsurfa.2016.03.065.

Ghosh, V., Mukherjee, A. and Chandrasekaran, N. (2014) 'Eugenol-loaded antimicrobial nanoemulsion preserves fruit juice against, microbial spoilage', *Colloids and Surfaces B: Biointerfaces*, 114, pp. 392–397. doi: https://doi.org/10.1016/j.colsurfb.2013.10.034.

Goindi, S. et al. (2016) 'Nanoemulsions: An emerging technology in the food industry', in *Emulsions*. Elsevier, 3, pp. 651–688.

Gomes, G. V. L. et al. (2017) 'Physico-chemical stability and in vitro digestibility of beta-carotene-loaded lipid nanoparticles of cupuacu butter (Theobroma grandiflorum) produced by the phase inversion temperature (PIT) method', *Journal of Food Engineering*, 192, pp. 93–102. doi: https://doi.org/10.1016/j.jfoodeng.2016.08.001.

Gutierrez, J. et al. (2009) 'Impact of plant essential oils on microbiological, organoleptic and quality markers of minimally processed vegetables', *Innovative Food Science & Emerging Technologies*. Elsevier, 10(2), pp. 195–202.

Guttoff, M., Saberi, A. H. and McClements, D. J. (2015) 'Formation of vitamin D nanoemulsion-based delivery systems by spontaneous emulsification: Factors affecting particle size and stability', *Food Chemistry*, 171, pp. 117–122. doi: https://doi.org/10.1016/j.foodchem.2014.08.087.

Inouye, S., Takizawa, T. and Yamaguchi, H. (2001) 'Antibacterial activity of essential oils and their major constituents against respiratory tract pathogens by gaseous contact', *Journal of Antimicrobial Chemotherapy*, 47(5), pp. 565–573. doi: 10.1093/jac/47.5.565.

Jo, Y.-J., Chun, J.-Y., et al. (2015) 'Physical and antimicrobial properties of trans-cinnamaldehyde nanoemulsions in water melon juice', *LWT—Food Science and Technology*, 60(1), pp. 444–451. doi: https://doi.org/10.1016/j.lwt.2014.09.041.

Jo, Y.-J. and Kwon, Y.-J. (2014) 'Characterization of β-carotene nanoemulsions prepared by microfluidization technique', *Food Science and Biotechnology*, 23(1), pp. 107–113. doi: 10.1007/s10068-014-0014-7.

Jo, Y.-J., Kwon, Y.-J., et al. (2015) 'Changes in quality characteristics of pork patties containing multilayered fish oil emulsion during refrigerated storage', *Korean journal for food science of animal resources*. The Korean Society for Food Science of Animal Resources, 35(1), p. 71.

Jung, E. Y. et al. (2016) 'Effect of layer-by-layer (LbL) encapsulation of nano-emulsified fish oil on their digestibility ex vivo and skin permeability in vitro', *Preventive Nutrition and Food Science*. The Korean Society of Food Science and Nutrition, 21(2), pp. 85–89. doi: 10.3746/pnf.2016.21.2.85.

Kawada, H. et al. (2010) 'Structure and rheology of a self-standing nanoemulsion', *Langmuir*. ACS Publications, 26(4), pp. 2430–2437.

Kim, I.-H. et al. (2013) 'Plum coatings of lemongrass oil-incorporating carnauba wax-based nanoemulsion', *Journal of Food Science*. John Wiley & Sons, Ltd, 78(10), pp. E1551–E1559. doi: 10.1111/1750-3841.12244.

Kim, I.-H. et al. (2014) 'Grape berry coatings of lemongrass oil-incorporating nanoemulsion', *LWT—Food Science and Technology*, 58(1), pp. 1–10. doi: https://doi.org/10.1016/j.lwt.2014.03.018.

Kotta, S. et al. (2012) 'Exploring oral nanoemulsions for bioavailability enhancement of poorly water-soluble drugs', *Expert Opinion on Drug Delivery*, 9(5). doi: 10.1517/17425247.2012.668523.

Kwon, S. S. et al. (2015) 'Formation of stable hydrocarbon oil-in-water nanoemulsions by phase inversion composition method at elevated temperature', *Korean Journal of Chemical Engineering*, 32(3), pp. 540–546. doi: 10.1007/s11814-014-0234-9.

Li, W. et al. (2015) 'Influence of surfactant and oil composition on the stability and antibacterial activity of eugenol nanoemulsions', *LWT-Food Science and Technology*. Elsevier, 62(1), pp. 39–47.

Li, Y. and McClements, D. J. (2014) 'Modulating lipid droplet intestinal lipolysis by electrostatic complexation with anionic polysaccharides: Influence of cosurfactants', *Food Hydrocolloids*. Elsevier, 35, pp. 367–374.

Liang, R. et al. (2012) 'Physical and antimicrobial properties of peppermint oil nanoemulsions', *Journal of Agricultural and Food Chemistry*. American Chemical Society, 60(30), pp. 7548–7555. doi: 10.1021/jf301129k.

Liang, R. et al. (2013) 'Effect of relative humidity on the store stability of spray-dried beta-carotene nanoemulsions', *Food Hydrocolloids*. Elsevier, 33(2), pp. 225–233.

Liu, H. et al. (2013) 'Preparation and characterization of glycerol plasticized (high-amylose) starch–chitosan films', *Journal of Food Engineering*, 116(2), pp. 588–597. doi: https://doi.org/10.1016/j.jfoodeng.2012.12.037.

Liu, F. et al. (2016) 'Utilization of interfacial engineering to improve physicochemical stability of β-carotene emulsions: Multilayer coatings formed using protein and protein–polyphenol conjugates', *Food Chemistry*, 205, pp. 129–139. doi: https://doi.org/10.1016/j.foodchem.2016.02.155.

LLana-Ruiz-Cabello, M. et al. (2014) 'Evaluation of the mutagenicity and genotoxic potential of carvacrol and thymol using the Ames Salmonella test and alkaline, Endo III-and FPG-modified comet assays with the human cell line Caco-2', *Food and Chemical Toxicology*. Elsevier, 72, pp. 122–128.

López-López, I. et al. (2010) 'Frozen storage characteristics of low-salt and low-fat beef patties as affected by Wakame addition and replacing pork backfat with olive oil-in-water emulsion', *Food Research International*. Elsevier, 43(5), pp. 1244–1254.

Mahdi Jafari, S., He, Y. and Bhandari, B. (2006) 'Nano-emulsion production by sonication and microfluidization—a comparison', *International Journal of Food Properties*. Taylor & Francis, 9(3), pp. 475–485. doi: 10.1080/10942910600596464.

Majeed, H. et al. (2016) 'Bactericidal action mechanism of negatively charged food grade clove oil nanoemulsions', *Food Chemistry*, 197, pp. 75–83. doi: https://doi.org/10.1016/j.foodchem.2015.10.015.

Malhotra, B., Keshwani, A. and Kharkwal, H. (2015) 'Antimicrobial food packaging: Potential and pitfalls', *Frontiers in Microbiology*. Frontiers, 6, p. 611.

Martins, J. T. et al. (2015) 'Edible bio-based nanostructures: Delivery, absorption and potential toxicity', *Food Engineering Reviews*, 7(4), pp. 491–513. doi: 10.1007/s12393-015-9116-0.

Mason, T. G. et al. (2006) 'Nanoemulsions: Formation, structure, and physical properties', *Journal of Physics: Condensed Matter*. IOP Publishing, 18(41), pp. R635–R666. doi: 10.1088/0953-8984/18/41/r01.

McClements, D. J., Decker, E. A. and Weiss, J. (2007) 'Emulsion-based delivery systems for lipophilic bioactive components', *Journal of Food Science*. Wiley Online Library, 72(8), pp. R109–R124.

McClements, D. J. and Rao, J. (2011) 'Food-grade nanoemulsions: Formulation, fabrication, properties, performance, biological fate, and potential toxicity', *Critical Reviews in Food Science and Nutrition*. Taylor & Francis, 51(4), pp. 285–330.

Medeiros, A. K. de O. C. et al. (2019) 'Nanoencapsulation improved water solubility and color stability of carotenoids extracted from Cantaloupe melon (Cucumis melo L.)', *Food Chemistry*. Elsevier, 270, pp. 562–572.

Moghimi, R., Aliahmadi, A., et al. (2016) 'Investigations of the effectiveness of nano-emulsions from sage oil as antibacterial agents on some food borne pathogens', *LWT—Food Science and Technology*, 71, pp. 69–76. doi: https://doi.org/10.1016/j.lwt.2016.03.018.

Moghimi, R., Ghaderi, L., et al. (2016) 'Superior antibacterial activity of nanoemulsion of Thymus daenensis essential oil against E. coli', *Food chemistry*. Elsevier, 194, pp. 410–415.

Mustafa, M. A. et al. (2014) 'Ultrasound-assisted chitosan–surfactant nanostructure assemblies: Towards maintaining postharvest quality of tomatoes', *Food and Bioprocess Technology*, 7(7), pp. 2102–2111. doi: 10.1007/s11947-013-1173-x.

Nazzaro, F. et al. (2013) 'Effect of essential oils on pathogenic bacteria', *Pharmaceuticals*. Multidisciplinary Digital Publishing Institute, 6(12), pp. 1451–1474.

Negahdaripour, M. (2020) 'The battle against COVID-19: Where do we stand now?', *Iranian Journal of Medical Sciences*. Shiraz University of Medical Sciences, 45(2), pp. 81–82. doi: 10.30476/IJMS.2020.46357.

Nielsen, F. S., Petersen, K. B. and Müllertz, A. (2008) 'Bioavailability of probucol from lipid and surfactant based formulations in minipigs: Influence of droplet size and dietary state', *European Journal of Pharmaceutics and Biopharmaceutics*. Elsevier, 69(2), pp. 553–562.

Otoni, C. G., Moura, M. R. de, et al. (2014) 'Antimicrobial and physical-mechanical properties of pectin/papaya puree/cinnamaldehyde nanoemulsion edible composite films', *Food Hydrocolloids*, 41, pp. 188–194. doi: https://doi.org/10.1016/j.foodhyd.2014.04.013.

Otoni, C. G., Pontes, S. F. O., et al. (2014) 'Edible films from methylcellulose and nano-emulsions of clove bud (Syzygium aromaticum) and Oregano (Origanum vulgare) essential oils as shelf life extenders for sliced bread', *Journal of Agricultural and Food Chemistry*. American Chemical Society, 62(22), pp. 5214–5219. doi: 10.1021/jf501055f.

Ozaki, A. et al. (2010) 'Emulsification of coenzyme Q10 using gum Arabic increases bioavailability in rats and human and improves food-processing suitability', *Journal of Nutritional Science and Vitaminology*. Center for Academic Publications Japan, 56(1), pp. 41–47.

Ozturk, B. et al. (2015a) 'Formation and stabilization of nanoemulsion-based vitamin E delivery systems using natural biopolymers: Whey protein isolate and gum Arabic', *Food Chemistry*, 188, pp. 256–263. doi: https://doi.org/10.1016/j.foodchem.2015.05.005.

Ozturk, B. et al. (2015b) 'Nanoemulsion delivery systems for oil-soluble vitamins: Influence of carrier oil type on lipid digestion and vitamin D3 bioaccessibility', *Food Chemistry*, 187, pp. 499–506. doi: https://doi.org/10.1016/j.foodchem.2015.04.065.

Pan, H. et al. (2014a) 'Preparation of highly stable concentrated W/O nanoemulsions by PIC method at elevated temperature', *Colloids and Surfaces A: Physicochemical and Engineering Aspects*, 447, pp. 97–102. doi: https://doi.org/10.1016/j.colsurfa.2014.01.063.

Pan, K. et al. (2014b) 'Thymol nanoencapsulated by sodium caseinate: Physical and antilisterial properties', *Journal of Agricultural and Food Chemistry*. American Chemical Society, 62(7), pp. 1649–1657. doi: 10.1021/jf4055402.

Parthasarathi, S., Muthukumar, S. P. and Anandharamakrishnan, C. (2016) 'The influence of droplet size on the stability, in vivo digestion, and oral bioavailability of vitamin E emulsions', *Food & Function*. The Royal Society of Chemistry, 7(5), pp. 2294–2302. doi: 10.1039/C5FO01517K.

Patrignani, F. et al. (2020) 'Combined use of natural antimicrobial based nanoemulsions and ultra high pressure homogenization to increase safety and shelf-life of apple juice', *Food Control*, 111, p. 107051. doi: https://doi.org/10.1016/j.foodcont.2019.107051.

Pinheiro, A. C., Coimbra, M. A. and Vicente, A. A. (2016) 'In vitro behaviour of curcumin nanoemulsions stabilized by biopolymer emulsifiers—effect of interfacial composition', *Food Hydrocolloids*, 52, pp. 460–467. doi: https://doi.org/10.1016/j.foodhyd.2015.07.025.

Ricaurte, L. et al. (2016) 'Production of high-oleic palm oil nanoemulsions by high-shear homogenization (microfluidization)', *Innovative Food Science & Emerging Technologies*, 35, pp. 75–85. doi: https://doi.org/10.1016/j.ifset.2016.04.004.

Ruengvisesh, S. et al. (2015) 'Inhibition of bacterial pathogens in medium and on spinach leaf surfaces using plant-derived antimicrobials loaded in surfactant micelles', *Journal of food science*. Wiley Online Library, 80(11), pp. M2522–M2529.

Saberi, A. H., Fang, Y. and McClements, D. J. (2013) 'Fabrication of vitamin E-enriched nanoemulsions by spontaneous emulsification: Effect of propylene glycol and ethanol on formation, stability, and properties', *Food Research International*, 54(1), pp. 812–820. doi: https://doi.org/10.1016/j.foodres.2013.08.028.

Salminen, H. and Weiss, J. (2014) 'Electrostatic adsorption and stability of whey protein-pectin complexes on emulsion interfaces', *Food Hydrocolloids*, 35, pp. 410–419. doi: https://doi.org/10.1016/j.foodhyd.2013.06.020.

Salvia-Trujillo, L. et al. (2013) 'Influence of particle size on lipid digestion and β-carotene bioaccessibility in emulsions and nanoemulsions', *Food Chemistry*, 141(2), pp. 1472–1480. doi: https://doi.org/10.1016/j.foodchem.2013.03.050.

Salvia-Trujillo, L. et al. (2014a) 'Formulation of antimicrobial edible nanoemulsions with pseudo-ternary phase experimental design', *Food and Bioprocess Technology*. Springer, 7(10), pp. 3022–3032.

Salvia-Trujillo, L. et al. (2014b) 'Impact of microfluidization or ultrasound processing on the antimicrobial activity against Escherichia coli of lemongrass oil-loaded nanoemulsions', *Food Control*. Elsevier, 37, pp. 292–297.

Salvia-Trujillo, L. et al. (2015) 'Use of antimicrobial nanoemulsions as edible coatings: Impact on safety and quality attributes of fresh-cut Fuji apples', *Postharvest Biology and Technology*, 105, pp. 8–16. doi: https://doi.org/10.1016/j.postharvbio.2015.03.009.

Scholz, P. and Keck, C. M. (2015) 'Nanoemulsions produced by rotor–stator high speed stirring', *International Journal of Pharmaceutics*, 482(1), pp. 110–117. doi: https://doi.org/10.1016/j.ijpharm.2014.12.040.

Sessa, M. et al. (2014) 'Bioavailability of encapsulated resveratrol into nanoemulsion-based delivery systems', *Food Chemistry*, 147, pp. 42–50. doi: https://doi.org/10.1016/j.foodchem.2013.09.088.

Sessa, M., Ferrari, G. and Donsi, F. (2015) 'Novel edible coating containing essential oil nanoemulsions to prolong the shelf life of vegetable products', *Chemical Engineering Transactions*, 43, pp. 55–60. doi: 10.3303/CET1543010.

Severino, R. et al. (2014) 'Antimicrobial effects of different combined non-thermal treatments against Listeria monocytogenes in broccoli florets', *Journal of Food Engineering*, 124, pp. 1–10. doi: https://doi.org/10.1016/j.jfoodeng.2013.09.026.

Severino, R. et al. (2015) 'Antimicrobial effects of modified chitosan based coating containing nanoemulsion of essential oils, modified atmosphere packaging and gamma irradiation against Escherichia coli O157:H7 and Salmonella Typhimurium on green beans', *Food Control*, 50, pp. 215–222. doi: https://doi.org/10.1016/j. foodcont.2014.08.029.

Shen, Z. et al. (2005) 'Characterization of fish oil-in-water emulsions using light-scattering, nuclear magnetic resonance, and gas chromatography-headspace analyses', *Journal of the American Oil Chemists' Society*. Springer, 82(11), p. 797.

Silva, H. D., Cerqueira, M. A. and Vicente, A. A. (2015) 'Influence of surfactant and processing conditions in the stability of oil-in-water nanoemulsions', *Journal of food engineering*. Elsevier, 167, pp. 89–98.

Su, D. and Zhong, Q. (2016) 'Lemon oil nanoemulsions fabricated with sodium casein-ate and Tween 20 using phase inversion temperature method', *Journal of Food Engineering*, 171, pp. 214–221. doi: https://doi.org/10.1016/j.jfoodeng.2015.10.040.

Tadros, T. et al. (2004) 'Formation and stability of nano-emulsions', *Advances in Colloid and Interface Science*. Elsevier, 108, pp. 303–318.

Tajkarimi, M. M., Ibrahim, S. A. and Cliver, D. O. (2010) 'Antimicrobial herb and spice compounds in food', *Food Control*, 21(9), pp. 1199–1218. doi: https://doi. org/10.1016/j.foodcont.2010.02.003.

Tan, C. P. and Nakajima, M. (2005) 'Effect of polyglycerol esters of fatty acids on physi-cochemical properties and stability of β-carotene nanodispersions prepared by emulsification/evaporation method', *Journal of the Science of Food and Agriculture*. Wiley Online Library, 85(1), pp. 121–126.

Tanaka, M. et al. (2003) 'Evidence for interpenetration of core triglycerides into surface phospholipid monolayers in lipid emulsions', *Langmuir*. ACS Publications, 19(13), pp. 5192–5196.

Tastan, Ö. et al. (2016) 'Understanding the effect of formulation on functionality of modified chitosan films containing carvacrol nanoemulsions', *Food Hydrocolloids*, 61, pp. 756–771. doi: https://doi.org/10.1016/j.foodhyd.2016.06.036.

Topuz, O. K. et al. (2016) 'Physical and antimicrobial properties of anise oil loaded nano-emulsions on the survival of foodborne pathogens', *Food Chemistry*, 203, pp. 117–123. doi: https://doi.org/10.1016/j.foodchem.2016.02.051.

Velikov, K. P. and Pelan, E. (2008) 'Colloidal delivery systems for micronutrients and nutraceuticals', *Soft Matter*. Royal Society of Chemistry, 4(10), pp. 1964–1980.

Vinogradov, S. V, Bronich, T. K. and Kabanov, A. V (2002) 'Nanosized cationic hydro-gels for drug delivery: Preparation, properties and interactions with cells', *Advanced drug delivery reviews*. Elsevier, 54(1), pp. 135–147.

Vishwanathan, R., Wilson, T. A. and Nicolosi, R. J. (2009) 'Bioavailability of a nano-emulsion of lutein is greater than a lutein supplement', *Nano Biomed Eng*, 1(1), pp. 38–49.

Wei, Z. and Gao, Y. (2016) 'Physicochemical properties of β-carotene bilayer emulsions coated by milk proteins and chitosan–EGCG conjugates', *Food Hydrocolloids*. Elsevier, 52, pp. 590–599.

Wilking, J. N. and Mason, T. G. (2007) 'Irreversible shear-induced vitrification of drop-lets into elastic nanoemulsions by extreme rupturing', *Physical Review E*. APS, 75(4), p. 41407.

Xianquan, S. et al. (2005) 'Stability of lycopene during food processing and storage', *Journal of Medicinal Food*. Mary Ann Liebert, Inc. 2 Madison Avenue Larchmont, NY 10538 USA, 8(4), pp. 413–422.

Yang, X. et al. (2011) 'Inhibition of citral degradation by oil-in-water nanoemulsions combined with antioxidants', *Journal of Agricultural and Food Chemistry*. ACS Publications, 59(11), pp. 6113–6119.

Yuan, Y. et al. (2008) 'Characterization and stability evaluation of β-carotene nanoemulsions prepared by high pressure homogenization under various emulsifying conditions', *Food Research International*. Elsevier, 41(1), pp. 61–68.

Zeeb, B., Thongkaew, C. and Weiss, J. (2014) 'Theoretical and practical considerations in electrostatic depositioning of charged polymers', *Journal of Applied Polymer Science*. Wiley Online Library, 131(7), pp. 40099.

Zhang, C. et al. (2015) 'Influence of anionic alginate and cationic chitosan on physicochemical stability and carotenoids bioaccessibility of soy protein isolate-stabilized emulsions', *Food Research International*, 77, pp. 419–425. doi: https://doi.org/10.1016/j.foodres.2015.09.020.

Zhao, Q., Ho, C.-T. and Huang, Q. (2013) 'Effect of ubiquinol-10 on citral stability and off-flavor formation in oil-in-water (O/W) nanoemulsions', *Journal of Agricultural and Food Chemistry*. ACS Publications, 61(31), pp. 7462–7469.

CHAPTER 12

Application of Essential Oil Nanoemulsions in Food Preservation

Mohammad Aslam,[1] Georgeos Deeb,[2]
Mohd. Aamir Mirza,[3] Javed Ahmad,[4]
Leo M.L. Nollet[5]

[1]Faculty of Pharmacy, Al Hawash Private University, Homs, Syria
[2]Faculty of Science, Damascus University, Damascus, Syria
[3]Department of Pharmaceutics, School of Pharmaceutical Education & Research,
Jamia Hamdard, New Delhi, India
[4]College of Pharmacy, Najran University, Najran, Kingdom of Saudi Arabia
[5]University College, Ghent, Belgium

CONTENTS

12.1 INTRODUCTION

Nowadays the whole world shrinks into a small platform due to political and economic ties between the countries. Due to globalization and advanced technology everything is available from one hook and corner of the world to the other. Foods are the essential part to our daily lives; advanced and organized packaging of food materials are required to reduce the food waste. This has given rise to a novel challenge of food packaging to protect the nutritional and organoleptic value of packaged foods.

Food degradation occurs mainly due to microbial growth and chemical degradation. Microbial growth in food not only reduces the quality of food but also causes food borne diseases (centers for disease control and prevention 2015). While chemically food degrades through the process of oxidation and peroxidation. Lipid is the important part of the food and the food with high content of lipid is more prone to the chemical degradations.

In oxidation process the unsaturated fatty acids present in the food react with the atmospheric oxygen present in the surroundings of the food, leading to production of peroxide radicals. These peroxide radicals start the peroxidation process and hasten the oxidation of lipids present in the food [1–3]. This chemical process accelerates in the presence of environmental conditions such as humidity, high temperature and radiations [2]. These chemical reactions decrease the food nutritional values and alter the organoleptic properties of foods by changing the flavor, odor taste and the texture of food. The final products of these chemical degradations are one of the major reasons leading to cellular mutation, cancer, atherosclerosis, and cardiac problems [2, 4, 5].

With the evolution of human's civilization, humans are using some conventional ways to preserve the foods. The concepts of conventional food preservation methods are still in use in different parts of the world especially in Indian civilization (India, Bangladesh, and Pakistan) and Levantine civilization (Syria, Lebanon, and Iraq) in the forms of salt brine and various condiments as food preservatives. The salt brine mixed with different condiments is renowned for its antimicrobial and antioxidant assets. World health organization suggested to reduce the daily intake of salts due to the increasing risk of cardiovascular diseases.

Antioxidants are the molecules which bind to the free oxygen radicals and can be used to prevent the food degradation by oxidation and peroxidation chemical degradation. Generally used antioxidants in food containing lipids are BHT (butyl hydroxytoluene) and BHA (butyl hydroxyanisole) [6–8]. While many synthetic antibacterial are also used to control the food spoilage by pathogenic microorganisms. These artificial preservatives such as BHT, BHA, sulfite benzoic acid, potassium sorbate, sodium nitrite, and sodium acetate have shown to cause various behavioral diseases such as neuronal toxicity and tumorigenicity in numerous research studies [9]. BHT and BHA have demonstrated to cause the attention deficit hyperactivity disorder [10]. N-nitroso compounds are found to be related with the different types of cancers including ovarian [11], renal [12], colorectal, and pancreatic [13]. As far as these adverse effects of abovementioned synthetic antioxidants, antibacterial and chemical preservatives are concerned, nowadays consumers are switching to natural alternatives for food additives [14].

12.2 OVERVIEW OF ESSENTIAL OILS AND ITS USE AS ALTERNATIVES TO SYNTHETIC PRESERVATIVES

The main phytochemical bioactive compounds are the secondary metabolic products or secondary metabolites. The secondary metabolites are produced from specific precursors after following their biosynthetic pathways [15]. They are generally classified as nitrogen containing compounds, phenolics, polyacetates and terpenoids. Terpenes also called isoprenoids, are one of the largest classes of natural chemicals formed by head to tail arrangement of two or more isoprene molecules. Indeed, the structural diversity associated with at least 40,000 compounds makes the class of terpenoids exemplary amongst all plant chemicals. The evolutionary success of this class of compounds is based on the simplicity of constructing different size molecules. They are important constituents of essential oils [16–18].

Essential oils are the terpenoid aromatic substances produced by plants that can be very well used by several food industries for different purposes including food preservation and imparting flavor as well as aroma to the food products [19]. Generally essential oils are composed of 20–60 bioactive compounds at variable concentrations (commonly ranging from 20 to 70%, and some available in trace amounts). The entities which are present in higher proportions (terpenes, terpenoids and aromatic compounds) are responsible for biological/pharmacological activities [20]. Minor entities/components can also play significant role in exhibition of various biologically functional activities through agonistic and antagonistic effects [21]. So far the activity of essential oils with respect to chemical nature is attributed to aldehydes or phenols as major entities (e.g., Cinnamaldehyde, Citral, Carvacrol, Eugenol, or Thymol) followed by terpene alcohols, ketones or esters (β-myrcene, α-thujone, or geranyl acetate) [22, 23].

The use of salt as alternative for synthetic preservatives is not advisable nowadays, while the use of different condiments is still in choice as alternatives for synthetic preservatives. The active ingredient in these condiments is an essential oil, which is responsible for food preservation action. The pathogens which enter the packaged food with environmental oxygen are the real cause of scourging of food borne illness and this needs to be addressed. The essential oils having antibacterial and antioxidant properties are of great interest for researchers. The successful incorporation of essential oils into raw uncut pieces, minced meat, patties, sausages, fish fillets, and broth has been demonstrated to be advantageous in various research studies. While the importance of modified atmosphere packaging (MAP) in enhancing food safety is being widely recognized, its combination with essential oil is resulting in far more superior results. Essential oils have been widely used in food packaging and found to exert enhancement in shelf-life of the packaged food products.

12.3 APPLICATION OF ESSENTIAL OILS IN DIFFERENT FOOD INDUSTRIES

12.3.1 Meat Industry

Meat and meat products packed using chemical preservatives are leading to sensory deterioration mainly due to the chemical changes that occur in lipids and proteins. These changes are called as post-mortem changes leading to high risks of food borne illness [24]. Therefore, the uses of natural preservatives such as essential oils are being encouraged in preserving meat products.

The hydrophobic extracts are extracted from various parts from the plant such as roots, leaves, stems, barks, fruits, flowers, and seeds. These extracts possess desirable properties of antioxidant, food pathogen inhibition, shelf-life enhancing, and organoleptic characteristic enhancing attributes for preserving meat and meat products [25]. Essential oils having such desirable properties for food preservation and recognized as safe (GRAS) by regulating authorities, can be used in preservation of meat products. Example of some of these natural preservatives for meat and meat like products include

TABLE 12.1 Application of Essential Oils as Preservatives in Meat

S. No.	Essential Oils	Application	Outcome	Reference
1.	Coriander, Clove, Oregano, and Thyme oils	Used to control pathogens and autochthonous spoilage flora in meat	As they caused a marked initial reduction in the viable cell numbers	[26]
2.	Thyme and Cinnamon Oils	Ham	Significantly decrease the L. monocytogenes	[27]
3.	Rosemary/thyme Oils	Mortadella	Shelf-life of mortadella has been extended	[28]
4.	Oregano Oil	Bologna sausage	The shelf-life of bologna sausages was extended	[29]
5.	Garlic and Oregano Oils	Against Salmonella spp., L. monocytogenes and S. aureus	Improved safety of dry-cured sausages	[30]
6.	Clove Oil (concentrations of 0.5% and 1%)	Restricted the growth of L. monocytogenes in meat	Restricted the growth of L. monocytogenes in meat at both 30°C and 7°C	[31]

oregano, mint, rosemary, orange, sage, cinnamon, lavender, bay, clove, cumin, fennel, coriander, peppermint, thyme, balm, tea tree, coriander, laurel, summer savory, eucalyptus, verbena, and hyssop (as elaborated in Table 12.1).

12.3.2 Seafood Products

Seafood usually including mollusks (oysters, clams, and mussels), finfish, marine mammals, fish eggs (roe), and crustaceans (shrimp, crab, and lobster) are highly perishable commodities due to various factors. Worldwide seafood consumption is growing exponentially. To fulfill the requirements of a growing market, the seafood industry has progressively been modernizing by different ways to keep their products fresh and safe. The growth of various bacteria depends on environmental conditions and the microbiological quality of the water. The temperature of water and its salt content, methods of catching and storage conditions are also the important factors [32]. Fresh fish are more liable to spoilage because of its high moisture and nutrients contents. Seafood spoilage agents are mainly bacteria of *Vibrio* spp., *Listeria* spp., *Aeromonas* spp., and many more which can exploit specific metabolites of fish tissues during storage. These bacteria represent an important food safety concern both for fresh and slightly preserved farmed seafood.

Seafood spoilage is any change that makes the product unacceptable for human consumption and it commences just after fatality due to autolysis, microbial invasion and autoxidation [33].

The preservation of overly sensitive seafood products by natural means is an important issue that needs consideration.

Therefore, seafood requires efficient preservation methods. Besides traditional preservation methods such as chilling, salting, and drying; the potentially safe preservatives of natural origin are required to extend shelf life of sea food such as essential oils. The various essential oils have been used for preservation of sea food in different research studies. Kostaki et al, have evaluated the combined effect of Modified Atmosphere Packaging (MAP) condition and 0.2% of thyme oil as a natural preservative to enhance the shelf life of fresh filleted sea bass (*Dicentrarchus labrax*) [34]. It was found that the presence of thyme oil improved the sensory quality of sea bass fillets and improved the shelf life with 17 days as compared to 6 days of the control samples. In another study, Corbo et al have investigated the synergistic effect of natural compounds (mixture of three essential oils containing 0.11% of thymol, 0.10% of grape fruit seed extract, and 0.012% of lemon extracts) on the microbial/sensorial quality decay of packed fish hamburger against the main fish spoilage microorganisms (*Pseudomonas fluorescens*, *Photobacterium phosphoreum* and *Shewanella putrefaciens*) [35]. A significant enhancement in shelf life of the packed fish hamburger was reported and a delay of the sensorial quality decay without compromising the flavor of the fish hamburgers.

12.3.3 Fruits and Vegetables

Fruit preservation and packaging have been practiced since ages to maintain the constant supply of seasonal fruits over lengthened periods round the year. The growing demand of minimally or unprocessed packaged fruits has further aggravated the safety concerns which fueledan extensive research with objectives to develop novel techniques of food processing, preservation, and packaging. There is need for rapid, accurate and early detection of contaminant products/microbes.

Fruit and vegetables are the important part of our recommended diet; the Dietary Guidelines for Americans recommend that daily half plate of fruit and vegetables are essential of for every individual. The vitamins and minerals present in the food are called phytochemicals and are the source of immunity and provide a good source of fiber to the diet and a diverse array of nonessential nutrients.

These fruits before reaching consumers home need up to 5 days in transit postharvest to distribution center and 1–3 days in display in the retail store. The fruits such as tomatoes, apples, melons and peaches (known as climacteric) are harvested before they ripen, and they continue to ripen until they reach their peak color after being detached from the mother plant. There are considerable losses of vitamin C in such process compared to that found if the product had been freshly picked at its peak of maturity [36–38].

Microorganisms may be controlled through the use of heat, cold, dehydration, acid, sugar, salt, smoke, atmospheric composition and radiation. Mild heat treatments in the range of 82–93°Care commonly used to kill bacteria in foods with pH ≥ 4.6, but to ensure spore destruction temperatures of 121°C wet heat for 15 min or longer are required. Highly acidic foods with pH < 4.6 require less heat, and often a treatment of 93°C for 15 min will ensure commercial sterility. Water activity (*a*w) of a food of 0.85 or below requires no thermal process, regardless of the pH of food. Most fruits are highly acidic, except for some fruits which are less acidic such as bananas, figs, mangoes, and some mature stone fruit. Vegetables, on the other hand, are primarily low acidic or alkaline in pH, apart from some 'fruit vegetables' such as tomatoes, whose pH values <4.6.

Another main consideration in choosing the most appropriate method of food preservation is the intended shelf life required for the product. This will dictate to a large extent the method of preservation selected. If the product is meant for consumption within a week or two, fresh-cut or minimal processing may be sufficient, but refrigeration and other means of preventing microbial growth will be required. The main reason for perishing of these cultivated fruits or vegetable stuffs is decaying, due to the fungal infections. Use of synthetic chemicals in post harvesting is able to destroy the fungal and bacterial infection and controls its deterioration, but due to the restriction of the chemicals in food industry alternative ways are always in demand. The use of controlled atmosphere (CA) or modified atmosphere (MAP) with use of different essential oils helps in increasing the shelf life of fruits and vegetables to be preserved. Various essential oils have been used in preservation of various fruit and vegetables as shown in Table 12.2.

12.3.4 Cereal-Based Products

Agriculture based crop products basically consist of barley, wheat corn and rice and they are highly susceptible to fungal infections. The mycotoxins infect various crops to a great extent which is a matter of serious concern. The microorganisms are mainly present in the pericarp of the grains which cover the endosperm. Agricultural crops or cereals are also contaminated from bacteria. The main categories of bacteria which infect cereals come under the species of *Bacillaceae*, *Lactobacillaceae*, *Pseudomonadaceae*, and *Micrococcaceae* [52]. Only few bacteria can survive during the storage, but their spores can survive for a longer period of time. For example, the spore of bacterial species of lactic acid bacteria can survive and can infect the dough made from those cereals [53]. Several research studies have been conducted to get rid of these contaminations and the methods mainly applied include gamma radiation, steam treatment, heat and various fungicides and bactericides. However, at present the use of natural origin preservatives are given preference over all other methods. That's why the antimicrobial effect of essential oil is profoundly explored. The antimicrobial and antifungal activity of these essential oils helps in preservation of cereal based products. In a recent research study thymol mixed with lemon extract and chitosan was tested against mesophilic and psychrotrophic bacteria, total coliforms, *Staphylococcus* spp., yeasts, and molds. The study results have shown improved microbiological stability of refrigerated amaranth-based fresh pasta [54].

12.3.5 Chicken Industry

Due to the wide variety of commercial availability, low-cost productions and higher nutritional values, the production of fresh chicken and their products is prevalent all over the world. The limited shelf life of fresh chicken and their products are the main hurdle in its commercialization. The chicken products are a rich source of proteins; nutrients and presence of moisture make them vulnerable to microbial spoilage [55]. The growth of micro-organisms can cause change in organoleptic characteristic and make the quality of chicken low with shorter shelf life and more susceptibility toward putrefaction [56].

TABLE 12.2   Application of Essential Oils as Preservatives in Fruits and Vegetables

S. No.	Essential Oils (EOs)	Application	Outcome	Reference
1.	Lemongrass and Geraniol	Have been found effective against *E. coli, Salmonella* sp., and *Listeria* spp. inapple, pear, and melon juices	Shelf life of unpasteurized fruit juices is limited by microbial enzymatic spoilage	[39]
2.	Malic acid and EOs extracted from Cinnamon, Palmarosa, and Lemongrass (0.3 and 0.7%) or their major compounds (eugenol, geraniol, and citral)	*Salmonella enteritidis* (10^8 cfu/ml) was used as the target microorganism	To prolong the shelf life of fresh-cut "Piel de Sapo" melon (*Cucumis melo* L.)	[40]
3.	Lemongrass, Oregano oil, and Vanillin	In apples	In an apple puree-alginate edible coating to prolong the shelf-life of fresh-cut Fuji apples	[41]
4.	Vanillin (0.3%w/w)	In general fruits	Preserved sensory quality for at least 21 days at 4°C	[42]
5.	Thyme and Sage EOs	In strawberries	Decreasing in the amount of fungi	[43]
6.	*Echinophora platyloba* EO	All fruits	Antifungal power	[44]
7.	Palmarosa(*Cymbopogon martini*), Red thyme (*Thymus zygis*), Cinnamon leaf (*Cinnamomum zeylanicum*), and Clove buds (*Eugenia caryophyllata*)	Testing numerous extracts from plants and EOs	Evaluated for their antifungal activity *Botrytis cinerea*	[45]
8.	Thyme, Oregano, and Lemon	Strawberry, Tomato, and Cucumber	Reduction of propagation of different fungi	[46]
9.	*Caraway, Cumin, Savory, Thyme and Peppermint*	Strawberries	Check fungus *Botrytis cinerea*	[47]
10.	Fennel, Black caraway, Peppermint, and Thyme	Fruits	Fungicidal effect *on Botrytis cinerea* and *Rhizopus stolonifer* rot fungus	[48]
11.	Eugenol, Thymol, and Carvacrol	Grapes berry decay		[49]
12.	Eugenol, Thymol, or Menthol	"Crimson seedless" table grapes		[49]
13.	Eugenol and menthol EOs	"Crimson seedless" table grapes	Better results during the storage time compared to grapes without such treatments	[50]
14.	Lemongrass, Cinnamon, and Oregano oils,	Pomegranate fruit	Antifungal activity against *Botrytis* sp., *Penicillium* sp., and *Pilidiella granati* pathogens	[51]

The new preservation techniques are being developed to maintain the freshness of chicken and likely products. Essential oils are used for the purpose of chicken preservation to increase the flavor. The essential oils have antimicrobial activity against various pathogenic micro-organisms and therefore it is used for preservation of chicken and related food products [57].

Synthetic antioxidants are common preservatives, but their carcinogenicity is debatable [58], therefore, essential oils are preferred by food technologists as additives for their antioxidant property. The antimicrobial activities of essential oils help to improve the shelf life and taste of fresh chicken and its products.

In chicken meat putrefaction occurs mainly due to the biogenic amines such as putrescine, cadaverine and histamine. The bacteria such as *Pseudomonas* species and endobacteria also are the reason for chicken meat deterioration. Therefore, active packagings with essential oils help in chicken preservation [59].

In a recent study, oregano was used for maintaining the freshness of chicken and it was observed that shelf life of fresh chicken was improved in modified atmospheric packaging [60]. Further, rosemary was used as an active packaging system for chicken meat and it was shown to inhibit the increase of biogenic amines, namely putrescine, cadaverine and histamine, as well as enterobacteria, *Pseudomonas* spp. and *Brochothrix thermosphacta* [59]. These research study findings boost the application of essential oils as preservatives for chicken and similar products.

12.3.6 Dairy Products and Cheese Industry

Contamination of milk and its products can occur either from endogenous source (itself from the animal infection or the disease of mastitis) or exogenous source (during the process of milking via the udders and teats, milking machine, bucket, or milk storage tank, and finally during the storage and transport).

The species of pathogenic bacteria that are present in raw milk and milk products such as cheese are mainly *Listeria monocytogenes*, verocytotoxin-producing *Escherichia coli*, *Staphylococcus aureus*, *Salmonella* sp. and *Campylobacter* sp. [61, 62]. During the ripening of cheese, the physicochemical changes such as increase in pH, makes the conditions more favorable for the growth of pathogenic bacteria. The addition of plants or herbs products influences the growth of the microorganisms in cheese. Many countries use these plant products during production of cheese for their organoleptic properties, or for decorative or preservation purposes. The commonly used products for these purposes include red, black and green peppers, thyme, cloves, cumin, parsley, paprika, onion/garlic, and many more.

The essential oils have been added to milk and cheese products to promote the shelf life, to confer functional properties and to maintain sensory features with very promising conclusions [63].

Essential oils can increase the shelf-life of dairy products, not only by eliminating unwanted microorganisms, but also by decreasing the degree of chemical deterioration during storage and marketing periods.

The antimicrobial effect of essential oils is associated with the composition and characteristics of each food product and specific microorganisms to be eliminated. Additionally, several other factors may affect the antimicrobial effect of essential oils in food products, namely heat treatments, smoking, chemical preservatives, and packaging. The use of essential oils as natural preservatives in food industries meets the current consumer trends of "green", "biological", "natural" and "no chemicals added" labels. The cheese is more susceptible for different types of bacterial contamination. Nowadays, the antimicrobial effect of essential oils is explored with proper packaging for preserving milk and its products as exemplified in Table 12.3.

TABLE 12.3   Application of Essential Oils as Preservatives in Dairy Products

S. No.	Essential Oils	Application	Outcome	Reference
1.	Orange, Lemon, Grapefruit, Madrine, Terpeneless lime, Orange, D-limonene, Terpineol, and Geraniol	Were tested against *Salmonella senftenberg*, *E. coli*, *S. aureus*, and *Pseudomonas* spp.	In different types of milk	[64]
2.	The combination of Thymol dipping and MAP (modified atmosphere packaging)		Prolonged the shelf life by 8 days, without negative effects on the sensory quality and on the growth kinetics of LAB	[65]
3.	Clove, Cinnamon, Bay and Thyme EOs concentrations (0.1%, 0.5% and 1%)	Low-fat and full-fat soft cheese	Anti-microbial effect on *Listeria monocytogenes, Salmonella Enteritidis*	[66]
4.	Oregano and Thyme EOs	Feta cheese	*L. monocytogenes* and *Escherichia coli* O157: H7	[67]
5.	Oregano and Rosemary EOs	Cream cheese	Affected the number of mesophilic microorganisms	[68]
6.	Marjoram and Rosemary EOs	In full cream cheese	Mesophilic bacteria	[69]
7.	Clove EO (concentrations of 0.5% and 1%)	In cheese	Restricted the growth of *L. monocytogenes* in meat at both 30°C and 7°C	[31]
8.	*Zataria multiflora* (thyme-like plant) EO	Gouda cheese	Reduction in the production of histamine and tyramine	[70]

12.4 NANOEMULSION-BASED APPROACH TO IMPROVE THE PROPERTIES OF ESSENTIAL OILS FOR APPLICATION IN FOOD SYSTEMS

The use of essential oils has changed the prospects of preservation of foods, but the direct addition of essential oils into the food is not feasible attributed to low water solubility, high volatility and objectionable odor. These factors lead to a poor bioavailability and the odor of oils alter the aesthetic value of food itself. When the essential oils are mixed with food and food is going through various food elaboration processes, such as heat treatment or air and light exposition, the volatile nature of essential oils cause the degradation with higher rates. That's why several methods for fortification of essential oils are used, so that they can exert their antimicrobial effect efficiently for a longer time. Encapsulation techniques of the essential oils within polymeric particles, liposomes, solid lipid nanoparticles, and nanolipid carriers have been widely studied to increase its stability and efficacy.

12.4.1 Application of Essential Oil Nanoemulsion in Food Systems

Nanoemulsions are a colloidal delivery system in the size range of 10 to 200 nm and are used as carriers of bioactive molecules. Nanoemulsions help in solubilization of lipophilic biomolecules and can be changed into various formulations such as foams, creams, liquids, and sprays. They are helpful in taste masking and have greater surface area providing greater absorption, which causes improved bioavailability of bioactive molecules. In formulations aspect they can be used as substitute to liposomes and vesicles [71] and they require less amount of energy in its preparation. Essential oils have also been used as additives in biodegradable films and coatings for active food packaging [72, 73]. Essential oils can provide films and coatings with antioxidant and/or antimicrobial properties.

The antioxidant activity of oil also depends upon the film's oxygen permeability. The incorporation into edible films can promote the antimicrobial capacity of essential oils. The efficacy of the edible film against microbial growth will depend on the oil's nature and the type of microorganism. Essential oils controlled release from edible films is another aspect that positively affects their effectiveness. Although nanoencapsulation is a promising tool for effective delivery of essential oils into food, still the toxicological aspects of most of the nanocarriers at their molecular target site must be further investigated. The recent advances in nanotechnology have made possible the development of novel carrier agents for the delivery and control release of essential oils in food systems with enhanced chemical, oxidative, and thermal stability [74, 75]. Edible coating of food products is another area where essential oils have proved their potential [76]. For example, nanoemulsions containing lemongrass essential oil as an antimicrobial agent have been used as edible coatings to control the safety and quality parameters of fresh-cut Fuji apples during storage. The performance of essential oils-based products is comparable to that of synthetic molecules and is concentration dependent. In a study it was demonstrated that the encapsulation of carvacrol, limonene, and cinnamaldehyde in the sunflower oil droplets of nanoemulsions has shown concentration reliant antimicrobial activity that was dependent on the concentration of the essential oil in the aqueous phase [77]. Many food scientists have encapsulated essential oils in a nanoemulsion to protect and improve their organoleptic properties and biological activities. This encapsulation into a nanoemulsion system increases their solubility, bioavailability and limits their oxidation process [78]. Li and Chiang prepared a d-limonene nanoemulsion by the ultrasonic-emulsification technique having a particle size of <100 nm and improved the stability against Ostwald ripening at a storage duration of 8 weeks [79]. β-Carotene is an important antioxidant and consumption of a sufficient amount of it can reduce the risk of various chronic diseases and cataracts [80]. However, the lipophilic nature and its low stability against high temperature, oxygen and light limits its use as fortified food. O/W nanoemulsion of β-carotene was prepared by high-pressure homogenization technique. It was observed that bioaccessibility was increased from 3.1 to 35% [81]. Shah et al reported that the eugenol containing essential oil nanoemulsion was more active against *Escherichia coli* O157:H7 and *L. monocytogenes* in bovine milk compared to free essential oil due to their uniform distribution in aqueous-based food products [82]. The limited solubility of free essential oils in aqueous systems makes it challenging to disperse it uniformly in food matrices, ultimately influencing the food quality and antimicrobial efficacy against spoilage. In another study, Donsi et al, investigated the potential of encapsulation of essential oils (terpenes mixture and D-limonene) into nanoemulsion-based delivery systems for integration into fruit juices (orange juice and pear juice), in order to increase their antimicrobial activity and shelf life of the product [83]. The investigator has reported higher antimicrobial activity of the D-limonene containing the essential oil nanoemulsion system with minimal alteration of the organoleptic characteristics of the fruit juice.

12.5 CONCLUSION AND FUTURE DIRECTIONS

The formed nanoemulsion of essential oils are kinetically stable and could have longer shelf-life but the higher volatility of the oils and instability in an acidic environment may cause stability disruption in the gastrointestinal tract milieu [84]. The formed small droplets of a nanoemulsion may go through Ostwald ripening and may lead to the problems of instability such as flocculation, coalescence and sedimentation. Further essential oils processing as nanoemulsion in the food industry is associated with serious concerns of stability, cost and safety of ingredients. These are certain issues with which researchers are dealing at present.

Future perspectives of nanoemulsions are very promising for different applications in food products; especially their incorporation in the formulation of functional foods. An example of future application of nanoemulsions as delivery platform is the encapsulation of probiotics by means of nanoemulsions. Indeed, such bioactive are usually encapsulated in liposomes. Moreover, it will be interesting to investigate *in vivo* performance after the consumption of functional foods fortified with encapsulated bioactive via nanoemulsions.

CONFLICT OF INTEREST

The authors declare no conflict of interest.

REFERENCES

1. Ferrari, C.K.B., 1998. Oxidação lipídica em alimentos e sistemas biológicos: mecanismos gerais e implicações nutricionais e patológicas. *Revista de Nutrição*, *11*(1), pp.3–14.
2. McClements, D.J. and Decker, E.A., 2000. Lipid oxidation in oil-in-water emulsions: Impact of molecular environment on chemical reactions in heterogeneous food systems. *Journal of food science*, *65*(8), pp.1270–1282.

3. Frankel, E.N., 2005. Lipid oxidation, Oily Press Lipid Library,2, pp.1–14.

4. Chanwitheesuk, A., Teerawutgulrag, A. and Rakariyatham, N., 2005. Screening of antioxidant activity and antioxidant compounds of some edible plants of Thailand. *Food chemistry*, 92(3), pp.491–497.

5. Márquez-Ruiz, G., Garcia-Martinez, M.C. and Holgado, F., 2008. Changes and effects of dietary oxidized lipids in the gastrointestinal tract. *Lipid insights*, 2, pp.11–19

6. Ultee, A., Kets, E.P., Alberda, M., Hoekstra, F.A. and Smid, E.J., 2000. Adaptation of the food-borne pathogen Bacillus cereus to carvacrol. *Archives of microbiology*, 174(4), pp.233–238.

7. Moleyar, V. and Narasimham, P., 1992. Antibacterial activity of essential oil components. *International journal of food microbiology*, 16(4), pp.337–342.

8. André, C., Castanheira, I., Cruz, J.M., Paseiro, P. and Sanches-Silva, A., 2010. Analytical strategies to evaluate antioxidants in food: A review. *Trends in food science & technology*, 21(5), pp.229–246.

9. Parke, D.V. and Lewis, D.F.V., 1992. Safety aspects of food preservatives. *Food Additives & Contaminants*, 9(5), pp.561–577.

10. Feingold, B.F., 1982. The role of diet in behaviour. *Ecology of disease*, 1(2-3), pp.153–165.

11. Aschebrook-Kilfoy, B., Heltshe, S.L., Nuckols, J.R., Sabra, M.M., Shuldiner, A.R., Mitchell, B.D., Airola, M., Holford, T.R., Zhang, Y. and Ward, M.H., 2012. Modeled nitrate levels in well water supplies and prevalence of abnormal thyroid conditions among the Old Order Amish in Pennsylvania. *Environmental health*, 11(1), p.6.

12. Dellavalle, C.T., Daniel, C.R., Aschebrook-Kilfoy, B., Hollenbeck, A.R., Cross, A.J., Sinha, R. and Ward, M.H., 2013. Dietary intake of nitrate and nitrite and risk of renal cell carcinoma in the NIH-AARP Diet and Health Study. *British journal of cancer*, 108(1), pp.205–212.

13. DellaValle, C.T., Xiao, Q., Yang, G., Shu, X.O., Aschebrook-Kilfoy, B., Zheng, W., Lan Li, H., Ji, B.T., Rothman, N., Chow, W.H. and Gao, Y.T., 2014. Dietary nitrate and nitrite intake and risk of colorectal cancer in the Shanghai Women's Health Study. *International journal of cancer*, 134(12), pp.2917–2926.

14. Yap, P.S.X., Yiap, B.C., Ping, H.C. and Lim, S.H.E., 2014. Essential oils, a new horizon in combating bacterial antibiotic resistance. *The open microbiology journal*, 8, p.6.

15. Buchanan, B.B., Gruissem, W. and Jones, R.L. eds., 2015. *Biochemistry and molecular biology of plants*. John Wiley & Sons.

16. Rohmer, M., 1999. The discovery of a mevalonate-independent pathway for isoprenoid biosynthesis in bacteria, algae and higher plants. *Natural product reports*, 16(5), pp.565–574.

17. Boucher, Y. and Doolittle, W.F., 2000. The role of lateral gene transfer in the evolution of isoprenoid biosynthesis pathways. *Molecular microbiology*, 37(4), pp.703–716.

18. Lange, B.M., Rujan, T., Martin, W. and Croteau, R., 2000. Isoprenoid biosynthesis: The evolution of two ancient and distinct pathways across genomes. *Proceedings of the national academy of sciences*, 97(24), pp.13172–13177.

19. Pavela, R., 2015. Essential oils for the development of eco-friendly mosquito larvicides: A review. *Industrial crops and products*, 76, pp.174–187.

20. Bakkali, F., 2008. Averbeck, S., Averbeck, D., and Idaomar, M. Biological effects of essential oils-a review. *Food and chemical toxicology*, 46, pp.446–475.
21. Bassolé, I.H.N. and Juliani, H.R., 2012. Essential oils in combination and their antimicrobial properties. *Molecules*, 17(4), pp.3989–4006.
22. Dorman, H.D. and Deans, S.G., 2000. Antimicrobial agents from plants: Antibacterial activity of plant volatile oils. *Journal of applied microbiology*, 88(2), pp.308–316.
23. de Barros, J.C., da Conceição, M.L., Neto, N.J.G., da Costa, A.C.V., Júnior, J.P.S., Junior, I.D.B. and de Souza, E.L., 2009. Interference of Origanum vulgare L. essential oil on the growth and some physiological characteristics of Staphylococcus aureus strains isolated from foods. *LWT—Food science and technology*, 42(6), pp.1139–1143.
24. Casaburi, A., Nasi, A., Ferrocino, I., Di Monaco, R., Mauriello, G., Villani, F. and Ercolini, D., 2011. Spoilage-related activity of Carnobacterium maltaromaticum strains in air-stored and vacuum-packed meat. *Applied and environmental microbiology*, 77(20), pp.7382–7393.
25. Del Nobile, M.A., Lucera, A., Costa, C. and Conte, A., 2012. Food applications of natural antimicrobial compounds. *Frontiers in microbiology*, 3, p.287.
26. Speranza, B. and Corbo, M.R., 2010. Essential oils for preserving perishable foods: Possibilities and limitations. *Application of alternative food-preservation technologies to enhance food safety and stability*, 23, pp.35–37.
27. Dussault, D., Vu, K.D. and Lacroix, M., 2014. In vitro evaluation of antimicrobial activities of various commercial essential oils, oleoresin and pure compounds against food pathogens and application in ham. *Meat science*, 96(1), pp.514–520.
28. Viuda-Martos, M., Ruiz-Navajas, Y., Fernández-López, J. and Pérez-Álvarez, J.A., 2010. Effect of added citrus fibre and spice essential oils on quality characteristics and shelf-life of mortadella. *Meat science*, 85(3), pp.568–576.
29. Viuda-Martos, M., Ruiz-Navajas, Y., Fernández-López, J. and Pérez-Álvarez, J.A., 2010. Effect of orange dietary fibre, oregano essential oil and packaging conditions on shelf-life of bologna sausages. *Food control*, 21(4), pp.436–443.
30. García-Díez, J., Alheiro, J., Pinto, A.L., Soares, L., Falco, V., Fraqueza, M.J. and Patarata, L., 2016. Behaviour of food-borne pathogens on dry cured sausage manufactured with herbs and spices essential oils and their sensorial acceptability. *Food control*, 59, pp.262–270.
31. Menon, K.V. and Garg, S.R., 2001. Inhibitory effect of clove oil on Listeria monocytogenes in meat and cheese. *Food microbiology*, 18(6), pp.647–650.
32. Giuffrida, A., 2003. Application of risk management to the production chain of intensively reared fish. *Veterinary research communications*, 27(1), pp.491–496.
33. Gokoglu, N. and Yerlikaya, P., 2015. *Seafood chilling, refrigeration and freezing: Science and technology*. John Wiley & Sons.
34. Kostaki, M., Giatrakou, V., Savvaidis, I.N., Kontominas, M.G., 2009. Combined effect of MAP and thyme essential oil on the microbiological, chemical and sensory attributes of organically aquacultured sea bass (Dicentrarchus labrax) fillets. *Food microbiology*, 26(5), pp.475–482.
35. Corbo, M.R., Speranza, B., Filippone, A., Granatiero, S., Conte, A., Sinigaglia, M. and Del Nobile, M.A., 2008. Study on the synergic effect of natural compounds on the microbial quality decay of packed fish hamburger. *International journal of food microbiology*, 127(3), pp.261–267.

36. Ahmad, A., Mukherjee, P., Senapati, S., Mandal, D., Khan, M.I., Kumar, R. and Sastry, M., 2003. Extracellular biosynthesis of silver nanoparticles using the fungus Fusarium oxysporum. *Colloids and surfaces B: Biointerfaces*, 28(4), pp.313–318.

37. Shrivastava, S., Bera, T., Roy, A., Singh, G., Ramachandrarao, P. and Dash, D., 2007. Characterization of enhanced antibacterial effects of novel silver nanoparticles. *Nanotechnology*, 18(22), p.225103.

38. Bradford, M.M., 1976. A rapid and sensitive method for the quantitation of microgram quantities of protein utilizing the principle of protein-dye binding. *Analytical biochemistry*, 72(1-2), pp.248–254.

39. Raybaudi-Massilia, R.M., Mosqueda-Melgar, J. and Martin-Belloso, O., 2006. Antimicrobial activity of essential oils on Salmonella enteritidis, Escherichia coli, and Listeria innocua in fruit juices. *Journal of food protection*, 69(7), pp.1579–1586.

40. Raybaudi-Massilia, R.M., Mosqueda-Melgar, J., Sobrino-López, A., Soliva-Fortuny, R., Martín-Belloso, O.2009. Use of malic acid and other quality stabilizing compounds to assure the safety of fresh-cut "Fuji" apples by inactivation of Listeria monocytogenes, Salmonella enteritidis and Escherichia coli O157: H7. *Journal of food safety*, 29(2), pp.236–252.

41. Rojas-Graü, M.A., Raybaudi-Massilia, R.M., Soliva-Fortuny, R.C., Avena-Bustillos, R.J., McHugh, T.H. and Martín-Belloso, O., 2007. Apple puree-alginate edible coating as carrier of antimicrobial agents to prolong shelf-life of fresh-cut apples. *Postharvest biology and technology*, 45(2), pp.254–264.

42. Ayala-Zavala, J.F., González-Aguilar, G.A. and Del-Toro-Sánchez, L., 2009. Enhancing safety and aroma appealing of fresh-cut fruits and vegetables using the antimicrobial and aromatic power of essential oils. *Journal of food science*, 74(7), pp.R84–R91.

43. Campos, T., Barreto, S., Queirós, R., Ricardo-Rodrigues, S., Félix, M.R., Laranjo, M. and Agulheiro-Santos, A.C., 2016. Conservação de morangos com utilização de óleos essenciais. *AGROTEC*, 18, pp.90–96.

44. Moghaddam, M., Taheri, P., Pirbalouti, A.G. and Mehdizadeh, L., 2015. Chemical composition and antifungal activity of essential oil from the seed of Echinophora platyloba DC. against phytopathogens fungi by two different screening methods. *LWT—food science and technology*, 61(2), pp.536–542.

45. Wilson, C.L., Solar, J.M., El Ghaouth, A. and Wisniewski, M.E., 1997. Rapid evaluation of plant extracts and essential oils for antifungal activity against Botrytis cinerea. *Plant disease*, 81(2), pp.204–210.

46. Vitoratos, A., Bilalis, D., Karkanis, A. and Efthimiadou, A., 2013. Antifungal activity of plant essential oils against Botrytis cinerea, Penicillium italicum and Penicillium digitatum. *Notulae botanicae horti agrobotanici Cluj-Napoca*, 41(1), pp.86–92.

47. Beikzadeh, N., Afzali, H. 2020. Impact of six essential oils on strawberry gray mold. *University of Yasouj journals system plant pathology science*, 9(1), pp.129–140.

48. Mohammadi, S., Aroiee, H., Aminifard, M.H., Tehranifar, A. and Jahanbakhsh, V., 2014. Effect of fungicidal essential oils against Botrytis cinerea and Rhizopus stolonifer rot fungus in vitro conditions. *Archives of phytopathology and plant protection*, 47(13), pp.1603–1610.

49. Valero, D., Valverde, J.M., Martínez-Romero, D., Guillén, F., Castillo, S. and Serrano, M., 2006. The combination of modified atmosphere packaging with eugenol or thymol to maintain quality, safety and functional properties of table grapes. *Postharvest biology and technology*, 41(3), pp.317–327.

50. Ricardo-Rodrigues, S., Tim-Tim, S., Véstia, J., Agostinho, J., Derreado, A., Rato, A.E., et al. Avaliação das potencialidades da aplicação de Atmosfera Modificada em combinação com eugenol na conservação de uva de mesa 'Crimson Seedless'. IX Congresso Ibérico de AgroEngenharia; 4–6 September 2017; Bragança, Portugal, 2017.

51. Munhuweyi, K., Caleb, O.J., Lennox, C.L., van Reenen, A.J. and Opara, U.L., 2017. In vitro and in vivo antifungal activity of chitosan-essential oils against pomegranate fruit pathogens. *Postharvest biology and technology*, *129*, pp.9–22.

52. Laca, A., Mousia, Z., Díaz, M., Webb, C. and Pandiella, S.S., 2006. Distribution of microbial contamination within cereal grains. *Journal of food engineering*, *72*(4), pp.332–338.

53. Sabillón, L. and Bianchini, A., 2016. From field to table: A review on the microbiological quality and safety of wheat-based products. *Cereal chemistry*, *93*(2), pp.105–115.

54. Del Nobile, M.A., Di Benedetto, N., Suriano, N., Conte, A., Lamacchia, C., Corbo, M.R. and Sinigaglia, M., 2009. Use of natural compounds to improve the microbial stability of Amaranth-based homemade fresh pasta. *Food microbiology*, *26*(2), pp.151–156.

55. Aymerich, T., Picouet, P.A. and Monfort, J.M., 2008. Decontamination technologies for meat products. *Meat science*, *78*(1-2), pp.114–129.

56. Petrou, S., Tsiraki, M., Giatrakou, V. and Savvaidis, I.N., 2012. Chitosan dipping or oregano oil treatments, singly or combined on modified atmosphere packaged chicken breast meat. *International journal of food microbiology*, *156*(3), pp.264–271.

57. De Martino, L., De Feo, V. and Nazzaro, F., 2009. Chemical composition and in vitro antimicrobial and mutagenic activities of seven Lamiaceae essential oils. *Molecules*, *14*(10), pp.4213–4230.

58. Chen, C., Pearson, A.M. and Gray, J.I., 1992. Effects of synthetic antioxidants (BHA, BHT and PG) on the mutagenicity of IQ-like compounds. *Food chemistry*, *43*(3), pp.177–183.

59. Sirocchi, V., Caprioli, G., Cecchini, C., Coman, M.M., Cresci, A., Maggi, F., Papa, F., Ricciutelli, M., Vittori, S. and Sagratini, G., 2013. Biogenic amines as freshness index of meat wrapped in a new active packaging system formulated with essential oils of Rosmarinus officinalis. *International journal of food sciences and nutrition*, *64*(8), pp.921–928.

60. Chouliara, E., Karatapanis, A., Savvaidis, I.N. and Kontominas, M.G., 2007. Combined effect of oregano essential oil and modified atmosphere packaging on shelf-life extension of fresh chicken breast meat, stored at 4°C. *Food microbiology*, *24*(6), pp.607–617.

61. Hammad, A.M., Hassan, H.A. and Shimamoto, T., 2015. Prevalence, antibiotic resistance and virulence of Enterococcus spp. in Egyptian fresh raw milk cheese. *Food control*, *50*, pp.815–820.

62. Verraes, C., Vlaemynck, G., Van Weyenberg, S., De Zutter, L., Daube, G., Sindic, M., Uyttendaele, M. and Herman, L., 2015. A review of the microbiological hazards of dairy products made from raw milk. *International dairy journal*, *50*, pp.32–44.

63. Carocho, M., Barreira, J.C., Antonio, A.L., Bento, A., Morales, P. and Ferreira, I.C., 2015. The incorporation of plant materials in "Serra da Estrela" cheese improves antioxidant activity without changing the fatty acid profile and visual appearance. *European journal of lipid science and technology*, *117*(10), pp.1607–1614.

64. Fisher, K. and Phillips, C., 2008. Potential antimicrobial uses of essential oils in food: is citrus the answer?. *Trends in food science & technology*, 19(3), pp.156–164.

65. Bevilacqua, A., Corbo, M.R. and Sinigaglia, M., 2007. Combined effects of modified atmosphere packaging and thymol for prolonging the shelf life of caprese salad. *Journal of food protection*, 70(3), pp.722–728.

66. Smith-Palmer, A., Stewart, J. and Fyfe, L., 2001. The potential application of plant essential oils as natural food preservatives in soft cheese. *Food microbiology*, 18(4), pp.463–470.

67. Govaris, A., Botsoglou, E., Sergelidis, D. and Chatzopoulou, P.S., 2011. Antibacterial activity of oregano and thyme essential oils against Listeria monocytogenes and Escherichia coli O157: H7 in feta cheese packaged under modified atmosphere. *LWT—food science and technology*, 44(4), pp.1240–1244.

68. Olmedo, R.H., Nepote, V. and Grosso, N.R., 2013. Preservation of sensory and chemical properties in flavoured cheese prepared with cream cheese base using oregano and rosemary essential oils. *LWT—food science and technology*, 53(2), pp.409–417.

69. Asensio, C.M., Grosso, N.R. and Juliani, H.R., 2015. Quality preservation of organic cottage cheese using oregano essential oils. *LWT—food science and technology*, 60(2), pp.664–671.

70. Laranjo, M., Fernandez-Leon, A.M., Potes, M.E., Agulheiro-Santos, A.C. and Elias, M., 2017. Use of essential oils in food preservation. Antimicrobial Research: Novel Bioknowledge and Educational Programs. pp.177–188.

71. Bouchemal, K., Briançon, S., Perrier, E. and Fessi, H., 2004. Nano-emulsion formulation using spontaneous emulsification: Solvent, oil and surfactant optimisation. *International journal of pharmaceutics*, 280(1–2), pp.241–251.

72. Atarés, L. and Chiralt, A., 2016. Essential oils as additives in biodegradable films and coatings for active food packaging. *Trends in food science & technology*, 48, pp.51–62.

73. Ribeiro-Santos, R., Andrade, M., de Melo, N.R. and Sanches-Silva, A., 2017. Use of essential oils in active food packaging: Recent advances and future trends. *Trends in food science & technology*, 61, pp.132–140.

74. Prakash, B., Kujur, A., Yadav, A., Kumar, A., Singh, P.P. and Dubey, N.K., 2018. Nanoencapsulation: An efficient technology to boost the antimicrobial potential of plant essential oils in food system. *Food control*, 89, pp.1–11.

75. Aureli, P., Costantini, A. and Zolea, S., 1992. Antimicrobial activity of some plant essential oils against Listeria monocytogenes. *Journal of food protection*, 55(5), pp.344–348.

76. Salvia-Trujillo, L., Rojas-Graü, M.A., Soliva-Fortuny, R. and Martín-Belloso, O., 2015. Use of antimicrobial nanoemulsions as edible coatings: Impact on safety and quality attributes of fresh-cut Fuji apples. *Postharvest biology and technology*, 105, pp.8–16.

77. Donsì, F., Annunziata, M., Vincensi, M. and Ferrari, G., 2012. Design of nano-emulsion-based delivery systems of natural antimicrobials: Effect of the emulsifier. *Journal of biotechnology*, 159(4), pp.342–350.

78. Mahfoudhi, N., Ksouri, R. and Hamdi, S., 2016. Nanoemulsions as potential delivery systems for bioactive compounds in food systems: Preparation, characterization, and applications in food industry. In *Emulsions, nanotechnology in the agri-food industry*, 3 (pp. 365–403). Academic Press.

79. Li, P.H. and Chiang, B.H., 2012. Process optimization and stability of D-limonene-in-water nanoemulsions prepared by ultrasonic emulsification using response surface methodology. *Ultrasonics sonochemistry*, 19(1), pp.192–197.

80. Failla, M.L., Huo, T. and Thakkar, S.K., 2008. In vitro screening of relative bio-accessibility of carotenoids from foods. *Asia Pacific journal of clinical nutrition*, *17* (S1), pp. 200–203

81. Liang, R., Shoemaker, C.F., Yang, X., Zhong, F. and Huang, Q., 2013. Stability and bioaccessibility of β-carotene in nanoemulsions stabilized by modified starches. *Journal of agricultural and food chemistry*, *61*(6), pp.1249–1257.

82. Shah, B., Davidson, P.M., Zhong, Q., 2013. Nanodispersed eugenol has improved antimicrobial activity against Escherichia coli O157: H7 and Listeria monocytogenes in bovine milk. *International journal of food microbiology*,*161*(1), pp.53–59.

83. Donsì, F., Annunziata, M., Sessa, M., Ferrari, G., 2011. Nanoencapsulation of essential oils to enhance their antimicrobial activity in foods. *LWT—food science and technology*,*44*(9), pp.1908–1914.

84. Klinkesorn, U. and McClements, D.J., 2009. Influence of chitosan on stability and lipase digestibility of lecithin-stabilized tuna oil-in-water emulsions. *Food chemistry*, *114*(4), pp.1308–1315.

CHAPTER 13

Nanoemulsions to Preserve/ Process Bioactive and Nutritional Food Compounds: Contemporary Research and Applications

Shadab Md,[1] Md. Ali Mujtaba,[2]
Wei Meng Lim,[3] Nabil A. Alhakamy,[1,4,3]
Farrukh Zeeshan,[3] Khaled M. Hosny[1,5]

[1]Department of Pharmaceutics, Faculty of Pharmacy,
King Abdulaziz University, Jeddah, Kingdom of Saudi Arabia
[2]Department of Pharmaceutics, Faculty of Pharmacy,
Northern Border University, Rafha, Kingdom of Saudi Arabia
[3]Department of Pharmaceutical Technology, School of Pharmacy,
International Medical University, Kuala Lumpur, Malaysia
[4]Center of Excellence for Drug Research & Pharmaceutical
Industries, King Abdulaziz University, Jeddah, Kingdom of Saudi Arabia
[5]Department of Pharmaceutics and Industrial Pharmacy,
Faculty of Pharmacy, Beni-Suef University, Beni-Suef, Egypt

CONTENTS

13.1 INTRODUCTION

Bioactive compounds, found mainly in fruits and vegetables, are phytochemicals, such as flavonoids, anthocyanins, tannins, betalains, carotenoids, plant sterols, and glucosinolates. They provide health benefits such as reducing oxidative stress, reducing metabolic disorder, and anti-inflammatory effect (Walia, Gupta and Sharma, 2019). The examples of bioactive compounds, sources and their functions are shown in Table 13.1. The consumption of foods with a high amount of bioactive compounds rich in vitamins and phytonutrients provides a positive effect on human wellbeing. The health benefits include the prevention of various chronic diseases such as hypertension, coronary heart disease, stroke, cancer, eye diseases, osteoporosis, and inflammation related diseases. These bioactive compounds are produced as secondary metabolites in the plant; these metabolites are unique to the individual species which contributes to their defense, attraction and signaling. They are highly accumulated in the skin and the leave of the fruits and vegetables. When consumed by human or animal, these bioactive compounds exert their pharmacological effects, it can be either a positive or negative impact depending on the nature of the substances (Bernhoft, 2010). The composition and level of bioactive compounds varies in different fruits and vegetables because of genetic factors and environment condition, such as light, plant maturity and post harvesting treatment (Deepa et al., 2007). In the food and nutraceutical industry, bioactive compounds are added as additives and food supplements. They can enhance flavor and color of the food, provide nutritional benefits, or act as natural preservatives (Siriwardhana et al., 2013). However, bioactive compounds have low aqueous solubility, low bioavailability and they can be unstable during the food processing (Jin et al., 2016). In order to promote its usage, modification using nanoemulsion provides the solution to overcome the limitations.

13.2 CHALLENGES ASSOCIATED WITH BIOACTIVE FOOD COMPOUNDS

The addition of bioactive compounds in food products are able to improve the nutrition value of the foods and able to enhance health benefits. However, the addition of bioactive compounds is difficult due to their properties. The bioactive compounds have low water solubility and they are sensitive to pH and temperature changes (Jin et al., 2016). These changes during the production processes could promote bioactive compounds oxidation and degradation. In addition, the stability of the bioactive compounds during the storage is important because this will affect its functionality and the shelf life of the food products (Jin et al., 2016). Certain bioactive compounds do possess unpleasant taste or smell, hence masking the taste using flavoring agents are necessary (Vázquez et al., 2018). Fortunately, the invention of nanoemulsion as a carrier system offers a

TABLE 13.1   Examples of Bioactive Compounds, Sources and Their Functions

Category	Bioactive Compounds	Source	Function	References
Lipids	ω-3 PUFA	Fish, chia seeds, flaxseeds, walnuts, kidney beans, soybean oil, avocado	Health benefits associated with cancer, cardiovascular diseases and inflammation	Gammone et al. (2019)
Carotenoids	Beta-carotene	Carrot, pumpkin, kale, apricot, tomato, papaya, spinach, broccoli	Beta-carotene is a precursor of vitamin A. Natural colorant and anti-oxidant	Yuan et al. (2008)
Vitamins	Vitamin A (retinoid)	Egg, cod liver oil, carrot, spinach, lettuce, broccoli, fish, animal liver	Anti-oxidant and vitamin for normal vision and the immune system	Huang et al. (2018)
Vitamins	Vitamin E (tocopherols)	Olive oil, coconut oil, corn, almond, peanuts, palm oil	Anti-oxidant and vitamin for cardiovascular diseases and cell membrane stability	Rizvi et al. (2014)
Vitamins	Vitamin D	Obtain mainly by sun exposure or food source, such as egg, cheese, fish, mushroom, fortified milk	Vitamin needed to improve calcium absorption	Hossein-Nezhad and Holick (2013)
Flavonols	Quercetin	Apples, berries, onions, shallots, tea, and tomatoes	Potential benefits for anti-oxidant, anti-inflammatory, anti-hypertensive, and anti-platelet activities	Li et al. (2016)
Flavonols	Rutin	Apricot, cherries, grapes, grapefruit, plums, oranges	Potential benefits for anti-oxidant, anti-inflammatory, cardioprotective, neuroprotective, anti-diabetic, and anti-cancer activities	Enogieru et al. (2018)
Flavonols	Resveratrol	Grapes, wine, nuts, berries	Potential benefits for anti-oxidant, anti-cancer, anti-aging activities	Chedea et al. (2017)

(Continued)

TABLE 13.1   (Continued)

Category	Bioactive Compounds	Source	Function	References
Flavones	Apigenin	Parsley, chamomile, celery, thyme, oregano, nuts, basil	Potential benefits for anti-oxidant, anti-bacterial, anti-inflammatory, anti-diabetic, and anti-cancer activities	Salehi et al. (2019)
Flavanones	Naringenin	Oranges, grapefruit, lemons, cherries	Potential benefits for anti-diabetes, anti-oxidant, and anti-inflammatory activities	Alam et al. (2014)
Flavanones	Silymarin	Milk thistle (*Silybum marianum*)	Potential benefits for hepato-protective, anti-oxidant, anti-inflammatory, and anti-cancer activities	Vargas-Mendoza et al. (2014)
Isoflavones	Genistein	Soymilk, soy flour, soy protein isolates	Potential benefits for anti-oxidant, cardio protective activities, and prevention of osteoporosis	Dixon and Ferreira (2002)

solution to overcome these limitations, making the application in food industry possible. The formulation of nanoemulsions can overcome the challenges by encapsulating the bioactive compounds to prevent their degradation caused by temperature, pH, and oxidation. Moreover, the small particle size of nanoemulsions can enhance the absorption of bioactive compounds from the gastro-intestinal tract, thus increasing bioavailability (Pathak, 2017).

13.3 PREPARATION AND CHARACTERIZATION OF NANOEMULSION

Nanoemulsions consist of two immiscible liquids, with one liquid being dispersed into another and stabilized by surfactants. Nanoemulsions can be simple water-in-oil (W/O) or oil-in-water emulsions (O/W), or can be multiple water-in-oil-in-water (W/O/W) or oil-in-water-in-oil (O/W/O) emulsions (Singh et al., 2017). The appearance of nanoemulsions is different from that of conventional emulsions because the particle size is smaller (Espitia, Fuenmayor and Otoni, 2019). A comparison of emulsions, micro-emulsions and nanoemulsions is presented in Table 13.2. Nanoemulsions are gaining popularity because they have several advantages. Their high optical clarity is preferred in beverages and cosmeceutical products. Moreover, the small droplet size provides better physical stability against gravitational separation and aggregation (Oca Avalos, Candal and Herrera, 2017).

There are two processes for the preparation of nanoemulsions, these being high energy or low energy emulsification processes. The high energy emulsification process requires the use of mechanical devices to perform homogenization and cavitation to the

TABLE 13.2 Comparison of Emulsions, Microemulsions and Nanoemulsions (Kale and Deore, 2016; Aswathanarayan and Vittal, 2019)

Parameters	Emulsion	Microemulsion	Nanoemulsion
Particle size	1–20 mm	1–100 nm	20–500 nm
Preparation method	High and low energy methods	Low-energy methods	High and low energy methods
Kinetic stability	Stable	Unstable	Stable
Thermodynamic stability	Metastable	Stable	Metastable
Interfacial tension	High	Ultra-low	Ultra-low (<10 dyn cm^{-1})
Polydispersity	High (>40%)	Low (10%)	Low (<10–20%)
Phases	Biphasic	Monophasic	Monophasic
Viscosity	High	Low	Low
Optical isotropy	Anisotropic	Isotropic	Isotropic
Effect on temperature and pH	Stable to temperature and pH changes	Affected by changes in composition, temperature, and pH	Stable to temperature and pH changes

oil and water phases. Examples of this method are high pressure homogenization, micro-fluidization and ultrasonication (Wang et al., 2016). High pressure homogenization uses a pump to push the emulsion through a small orifice using pressures in the range of 10–100 megapascals (Håkansson, 2019). The combination of intense turbulence, hydraulic shear and cavitation can break down the emulsion to nano-size (Kumar et al., 2019). This process involves two stages: at stage one, the emulsion is dispersed into small droplets to increase the surface area, while at stage two, the size reduction and stabilization with surfactants produces nanosize droplets (Aswathanarayan and Vittal, 2019). The principle of micro-fluidization is very similar to high pressure homogenization, except the emulsion is pumped through a Y or T junction. These branches reconnect downstream, causing the two streams of emulsion to impinge and interact with each other at high speed and high pressure. This collision will further break down the emulsion into tiny particles (Maali and Mosavian, 2013). For ultrasonication, the sonicator probe produces ultrasonic agitation by sound waves. When in contact with the emulsion, the cavitation, turbulence, and interfacial waves break down the emulsion to tiny particles. However, this method is more suitable for laboratory scale production, because ultrasonication is unsuitable for large volume production (Kentish et al., 2008). The low energy method produces nanoemulsions as a result of the physicochemical interaction of the ingredients and surfactants. Here, the formulation of nanoemulsions is simple and instantaneous, occurring just by mixing of the ingredients (Anton and Vandamme, 2009). The common low energy methods used are phase inversion temperature, phase inversion composition and solvent diffusion. The phase inversion temperature (PIT) method involves the alteration of the hydrophilicity or lipophilicity of non-ionic surfactants at different temperatures. For example, at low temperature, the surfactants can form oil-in-water emulsion, while as the temperature increases, the surfactants can form water-in-oil emulsion. The temperature at which there is a conversion of oil-in-water to water-in-oil emulsion or vice versa is known as the phase inversion temperature (Ren et al., 2019).

For phase inversion composition (PIC), the temperature is maintained constant but the emulsion composition is changed. Changing the composition of the ingredients such as adding sodium chloride or diluting with water can change the electrical charge in the mixture. This will result in the conversion of an oil-in-water emulsion to a water-in-oil emulsion or vice versa (Maestro et al., 2008). The solvent diffusion method comprises mixing an aqueous phase containing a hydrophilic surfactant with lipophilic compounds in a water-miscible organic solvent. The rapid diffusion of the organic solvent in the aqueous phase can form a nanoemulsion. The solvent is then removed by evaporation (Mori and Anarjan, 2018).

Nanoemulsions can be characterized by their physical property, stability, morphology and rheology based on different applications. Size distribution is measured by using a dynamic light scattering technique; it is based on the fluctuations in the intensity of light scattering by droplets due to Brownian motion (Espitia, Fuenmayor and Otoni, 2019). Zeta potential is a measurement of the surface charge of the particles; it is the interactive forces between particles at the particle surfaces that contributes to the stability of nanoemulsions (Laxmi et al., 2015). Transmission election microscope (TEM) has a very high resolution of 0.2 nm and is commonly used to study the morphology of nanoemulsions. However, the sample must be able to resist high vacuum conditions during the analysis (Silva, Cerqueira and Vicente, 2012). Another option is to use scanning electron microscopy (SEM) which provides three-dimensional images of the droplets. Although SEM has greater resolution and magnification, the operation cost is higher than that of TEM (Luykx et al., 2008). Stability testing using centrifugation assay, freeze-thaw cycle, accelerated stability testing or real time stability testing is used to determine the stability of emulsions over a period of time to ensure that the nanoemulsions can remain stable without coalescence, creaming and sedimentation (Safaya and Rotliwala, 2019).

13.4 DIFFERENT NANOEMULSION-BASED TYPES OF BIOACTIVE FOOD COMPOUNDS

13.4.1 Bioactive Lipids

Conventionally, lipids are regarded as a commendable source of energy for living organisms and are primary components of cell membranes. Nonetheless, the discovery of platelet-activating factor in 1979 as the first biologically active phospholipid revolutionized lipid science. Since then, almost every single lipid class has been assigned a specific biological activity. For instance, lipids as hydrophobic molecules are regarded as chemical messengers among cells. In addition, polar lipids possessing both hydrophilic as well as hydrophobic portions are considered as modulators of the activities of membrane proteins. On the other hand, glycosphingolipids containing carbohydrates impart pivotal roles in immunity (Vázquez et al. 2018). Consequently, numerous biological activities of lipid biomolecules from various nutritional origins are thoroughly reviewed. Extensive research is being conducted on biological lipids to explore their potential benefits with special emphasis on those having robust antioxidant activities (Roby, 2017). Bioactive lipids can be classified into six groups. First, neutral lipids/fatty acids including essential fatty acids and ratio omega-6/omega-3, saturated fatty acids, medium- and short-chain fatty acids, monounsaturated fatty acids, trans-fatty acids and conjugated fatty acids, branched-chain fatty acids, and edible cold-pressed oils. The second class comprises polar lipids, while the third class includes unsaponifiable lipids including terpenes (essential oils and carotenoids such as vitamin A and E), sterols (vitamin D) and polyphenols. The remaining classes comprise glycerol-based ether lipids, isoprenoids and, lastly, phenolic lipids (Bracco, 1994; Magnusson and Haraldsson, 2011; Beller, Lee, and Katz, 2015; Castro-Gomez et al. 2015; Hadacek, 2017).

Omega-3 fatty acids are important dietary ingredients (Vázquez et al. 2018) and cholesterol is pivotal for the synthesis of vitamin D, bile salts and certain hormones. Moreover, several positive effects have been ascribed to seafood-based unsaturated lipids (Vázquez et al. 2018). Lipids are capable of self-organization and can form intriguing structures that regulate interaction among numerous molecules. In addition, lipids form lipid bilayers and membranes upon interactions with aqueous medium. Lipid bilayers comprise the backbone of all biological membranes which are ubiquitous in all cells (Mouritsen, 2005). Dietary bioactive lipids are defined as lipids with distinctive health functions which could be attributed to their physicochemical properties. Lipids provide health benefits primarily by two mechanisms: firstly, altering the composition of fatty acids found in various tissues and secondly by boosting cell signaling pathways (Escribá et al. 2008). Although the most important bioactive lipids are polyunsaturated fatty acids (PUFAs), some health benefits are also ascribed to the consumption of short- to medium-chain fatty acids (Aluko, 2012). The health beneficiary effects of PUFAs are influenced by their isomer configurations and the dominant bioactive form is the cis-isomer which promotes membrane fluidity upon penetration into cells. Enhanced membrane fluidity boosts intercellular communication and helps in the maintenance of homeostasis. The noticeable beneficial effects of biological lipids to the ruminant animals include inhibition of protein degradation and methane production (Cobellis, Trabalza-Marinucci, and Yu, 2016; Patra et al., 2017; Soltan et al., 2018), regulation of ruminal fermentation (Calsamiglia et al., 2007; Mirzaei-Alamouti et al., 2016; Kazemi-Bonchenari et al., 2018), depletion in the growth of pathogens in the intestines, amplification of antioxidant activities and augmentation of immunity (Chowdhury et al., 2018; Kumar et al., 2018) and enhancement in quantity and quality of meat and milk in ruminant animals (Hausmann et al., 2018; Smeti et al., 2018).

Numerous biological lipids, which will now be referred to as "lipophilic actives" have demonstrated health benefits apart from their conventional nutritional role (Wildman 2007), suggesting their potential in the promotion of human health. As a result, their use in food and related industries is increasing (Velikov and Pelan, 2008; McClements, 2009; Sagalowicz and Leser, 2010). These lipophilic actives are employed as nutraceuticals (therapeutic/preventive agents) and cosmetics. However, their formulation is difficult owing to low aqueous solubility, low chemical stability, high oil-water partition coefficient, high melting point and their existence in crystalline state at room temperature (McClements, 2009; 2010a; 2010b). These issues make them incompatible with the aqueous portion of food and beverages, and results in their degradation, either during storage or during gastrointestinal (GI) transit and absorption. These limitations impede their commercial potential, thereby indicating the importance of robust formulation strategies in order to mitigate these issues. The various formulation strategies adopted to address the limitations of bioactive lipids include: (1) mixing of bioactive lipids with an oily phase (generally edible neutral oils) prior to formulating them into conventional emulsion-based products (McClements, 2015), (2) modification of the chemical structure of bioactive lipids primarily by esterification (Lauridsen, Hedemann and Jensen, 2001), and (3) incorporation of bioactive lipids into a nano or micro-particle-based delivery system (McClements, Decker and Park, 2008). Extensive research has been aimed at formulation of bioactive lipid compounds using lipid-based delivery systems (LBDS)

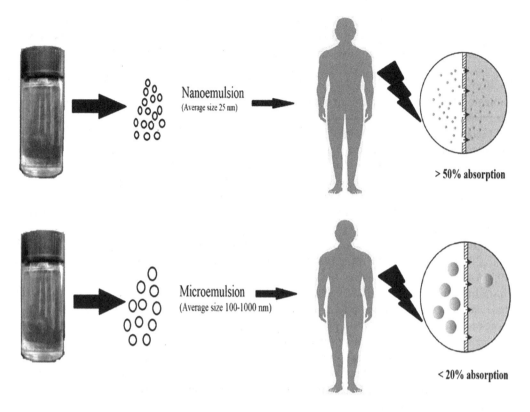

FIGURE 13.1 Differences between nanoemulsion and microemulsion in terms of droplet size and extent of absorption.

(Kalepu, Manthina and Padavala, 2013; Shrestha, Bala and Arora, 2014). The various LBDS reported for administration these bioactive lipid agents include solid lipid nanoparticles particles (SLPs), liposome and phospholipid complexes, nanostructured lipid carriers (NLCs) and emulsion-based systems comprising of nanoemulsions (NEs), microemulsions (MEs) and self-emulsifying delivery systems (SEDSs) (Mouhid et al., 2017). Figure 13.1 depicts the differences between nanoemulsions and microemulsions in terms of droplet size and extent of absorption. The different types of bioactive lipids formulated using Nano emulsion are presented in Table 13.3.

13.4.1.1 Neutral Bioactive Lipids

Neutral bioactive lipids are typically obtained from fish, such as salmon, mackerel, herring, albacore tuna, and trout, walnuts, sunflower seeds, flax seeds or flax oil, corn oi, soybean oil, and safflower oil (Akanbi and Barrow, 2017). Encapsulation of neutral bioactive lipids such as essential polyunsaturated fatty acids (PUFAs), especially omega-3 oils, e.g., eicosapentaenoic acid (EPA), docosahexaenoic acid (DHA), α-linoleic acid and α-linolenic acid in nanoemulsions can mitigate autoxidation, increase compatibility with various food products, mask astringent or bitter tastes and improve functionality (Lane et al., 2016; Lohith Kumar and Sarkar, 2017). Thermodynamically stable EPA and DHA nanoemulsions were prepared using the emulsion phase inversion method and reported to be physically stable under various storage conditions over 20 days. Additionally, the

TABLE 13.3 Types of Bioactive Lipids Commonly Formulated Using Nanoemulsion/
Microemulsion

Bioactive Lipid	Formulation Challenge	Formulation Approach	Merits of Formulation
1. Neutral lipids			
• ω-3 PUFA (α-linolenic acid, eicosapentaenoic acid and docosahexaenoic acid) • Conjugated linoleic acid	• Low water solubility (aqueous matrix incompatibility) • Susceptible to lipid oxidation (problematic long-term storage, rancid off-flavors, and potentially toxic reaction products)	Spray-dried emulsions, multilayer emulsions, multiple emulsions	• Easier incorporation into aqueous medium • Easier handling and storage • Prevention of lipid oxidation
2. Unsaponifiable lipids			
• Essential oils	• Low water solubility • Some water solubility (instability by Ostwald ripening) • Low and variable bioavailability	O/W emulsions, MEs, NEs, multilayer emulsions and multiple emulsions	• Mask undesirable off-flavors
Vitamin A (retinoid) Vitamin E (tocopherols)	• Low water solubility • High melting point • Poor chemical stability • Low and variable bioavailability	O/W emulsions	• Easier incorporation into aqueous medium and handling • Easier handling and storage • Prevention of chemical degradation • Increased efficacy
3. Sterols			
Vitamin D	• Low water solubility • Poor chemical stability • Low and variable bioavailability	O/W emulsions	• Easier incorporation into aqueous medium and handling • Easier handling and storage • Prevention of chemical degradation • Increased efficacy

authors reported stability in the retention rate of DHA/EPA at >60% (Zhang et al., 2019). In another study, an increased uptake rate of PUFAs in a nanoemulsion formulation was reported in the rat small intestine, compared with administration using macroemulsions; this formulation was also more effective in impairing lipopolysaccharide-mediated nitric oxide synthesis (Dey et al., 2019). In yet another study, O/W nanoemulsions of DHA were prepared using the microfluidization technique with sodium caseinate, soy lecithin or Tween 40 as emulsifiers. In this study, Tween 40 was found to be the best agent, with the nanoemulsions showing robust stability and improved lipid digestibility (Karthik and Anandharamakrishnan, 2016). Fortification of food using omega-3 is challenging because of oxidation problem, as well as unpleasant fishy smell (Ismail et al., 2016). To overcome this problem, Karthik and Anandharamakrishnan formulated docosahexaenoic acid (DHA) oil-in-water nanoemulsion using micro-fluidization method. Comparing the stability using different emulsifiers such as Tween 40, sodium caseinate and soya lecithin, Tween 40 nanoemulsion has lower mean diameter of 148 nm, and yield lower peroxidation. Therefore, nanoemulsion is able to increase the stability of omega-3 (Karthik and Anandharamakrishnan, 2016).

13.4.1.2 Essential Oil

Essential oils are aromatic volatile liquids and semi-liquids and are obtained from the seeds, flowers, leaves, buds, fruits, bark, resins, and roots of certain plants, such as oranges, lemon, clove, eucalyptus, lavender, and peppermint (Henley et al., 2007). Essential oils possess antimicrobial activity which is beneficial against numerous strains of fungi, Gram-positive bacteria, and Gram-negative bacteria (Burt, 2004). Essential oils can be used as antibacterial, antifungal, antioxidant and flavoring agents in food industry but their usage is limited due to certain problems such as volatility and hydrophobicity which could mitigate their applications (Bhavaniramya et al., 2019). Nonetheless, the encapsulation of essential oils into nanoemulsions could negate these issues (Liu et al., 2018). A study reported the nanoemulsion formulation of thyme oil and eugenol (2-methoxy-4-(2-propenyl)-phenol) using direct homogenization of these essential oils in the aqueous media containing sodium caseinate (SC) and lecithin (Xue et al., 2015). The authors reported that nanoemulsification of thyme oil and eugenol enhanced their antimicrobial potential compared to those of the free essential oils (Xue et al., 2015). Salvia-Trujillo et al. formulated nanoemulsion based edible coating with lemon grass essential oil to use as coating to preserve the quality of fruits and vegetables during their storage. Lemon grass nanoemulsion exhibits greater bactericidal effect of *Escherichia coli* compare to conventional emulsion (Salvia-Trujillo et al., 2015). In another study, eugenol-loaded nanoemulsions were prepared using spontaneous emulsions method and were stabilized by zein and SC (Wang and Zhang, 2017). These nanoemulsions exhibited narrow size distribution with remarkable stability upon storage for 30 days at ambient temperatures (22°C) following freeze-drying or spray-drying. In another research work, an O/W nanoemulsion comprising of two essential oils namely clove and lemongrass oil was fabricated in order to investigate the antifungal properties. The researchers reported an enhancement in the antifungal potential of clove and lemongrass oil upon their incorporation into nanoemulsions (Sharma et al., 2018). A stable nanoemulsion of *Thymus daenensis* oil with droplet size of 171.88 ± 1.57 nm was developed using Tween 80 and lecithin as emulsifiers (Moghimi et al., 2018). The authors reported superior antibacterial activity to pure oil against *Acinetobacter baumannii*. Moreover, the nanoemulsions demonstrated considerable anti-biofilm activity following the incubation for 24 h at a sub-lethal dose (Moghimi et al., 2018). In yet another reported study, nanoemulsions comprising of two citrus based essential oils (bergamot oil and sweet orange oil) and triacylglycerol oils (medium chain triglyceride and corn oil) at various proportions were prepared (Yang et al., 2018). The results indicated a better stability of nanoemulsions containing citrus oils in combination with triacylglycerol oils compared to that of nano-emulsions of pure citrus oil (Yang et al., 2018). Nanoemulsions containing cinnamaldehyde were prepared using medium chain triglyceride (Tian et al., 2016). The authors reported an enhanced stability and prolonged antibacterial efficacy against *Escherichia coli* of these nanoemulsion compared with pure cinnamaldehyde (Tian et al., 2016).

13.4.1.3 Vitamins

Vitamins are essential micronutrients and play important role in the maintenance of human health. Vitamins can be water-soluble (hydrophilic) and fat-soluble (lipophilic). Vitamins B and C are hydrophilic whilst vitamins A, E, D, and K are lipophilic. The food sources of Vitamins B are soymilk, watermelon, milk, yogurt, cheese, whole and enriched grains, cereals, meat, poultry, fish, fortified and whole grains, mushrooms, avocados, legumes, tofu, bananas, and eggs. Whilst Vitamin C is obtained from citrus fruits, potatoes, broccoli, bell peppers, spinach, strawberries, tomatoes and Brussels sprouts. Vitamin A is obtained from beef, liver, eggs, shrimp, fish, fortified milk, sweet potatoes, carrots, pumpkins, spinach and mangoes. Vitamin D is found in fortified milk, cereals and fatty fish whereas vegetables oils, leafy green vegetables, whole grains and nuts are the main sources of Vitamin E, whereas vitamin K is found in cabbage, eggs, milk, spinach, broccoli, and kale (Rosenfeld, 1997). Lipophilic vitamins possess marginal physical and chemical stability and are prone to oxidation, especially vitamins A and E (Öztürk, 2017). Moreover, light can have a marked effect on the structure of vitamin K. On the other hand, vitamin D has been reported to be stable during processing and storage (Katouzian et al., 2017). The stability of O/W nanoemulsions of vitamin A was evaluated upon exposure to ultraviolet (UV) light and storage at various temperatures (4°C, 25°C, and 40°C). These authors reported an enhancement in the residual vitamin A levels upon increasing the oil concentrations to 10 wt% (v/v) in the emulsion compared to that in the bulk oil which was ascribed to enhanced emulsion turbidity (Park, Mun and Kim, 2019). In another study using vitamin A palmitate, it was shown that the retention ration of vitamin A palmitate in the nanoemulsion surpassed 93% following a three-month storage period at room temperature (Ji et al., 2015). In a study involving a nanoemulsion formulation of vitamin D3, the in vitro bioavailability of vitamin D3 was reported to be increased almost four-fold. Moreover, an in vivo animal study showed an increase in the serum 25(OH) D3 concentration to 73% when vitamin D3 was administered as a nanoemulsion (Kadappan et al., 2018). Another study showed that the nature of the carrier oil, such as flaxseed oil, corn oil and fish oil, markedly influenced the stability and simulated gastrointestinal performance of vitamins D3 nanoemulsions stabilized using pea protein (Schoener et al., 2019). The lipid digestion and vitamin D3 bioaccessibility were evaluated upon exposure of nanoemulsions to a simulated gastrointestinal tract. These authors reported rapid digestion of all three lipids employed and vitamin bioaccessibility was ranked: corn oil > flaxseed oil ≈ fish oil. These findings indicated corn oil is more appropriate for the encapsulation and delivery of vitamin D3 than flaxseed or fish oil (Schoener et al., 2019). In addition, vitamin E loaded nanoemulsions were formulated using corn oil as a carrier along with Quillaja saponins as an emulsifier (Lv et al., 2018). The authors reported a relatively high vitamin E bioaccessibility for the optimized nanoemulsions (53.9%) (Lv et al., 2018). In another study, nanoemulsions of vitamin E resulted in up to a 12-fold higher cellular uptake compared with that of microsized vitamin E (Moradi and Anarjan, 2019). Nanoemulsions were shown to be capable of being used commercially for the delivery of vitamin K1 (Campani et al., 2016).

13.4.2 Carotenoids

Carotenoids are lipophilic pigments found in certain plants, algae and photosynthetic bacteria and produce bright red, yellow and orange colors. Carotenoids have antioxidant properties and have over 600 types rendering health benefits to humans such as mitigation of carcinoma and protection of the eyes (Lohith Kumar and Sarkar, 2017). Foods rich in beta-carotene and other carotenoids include: apricots, asparagus, beef liver, beets, broccoli, cantaloupe, carrots, corn, guava, kale, mangoes, mustard and collard greens, nectarines, peaches, pink grapefruit, pumpkin, squash (yellow and winter), sweet potato, tangerines, tomatoes, and watermelon (Nováková and Moran, 2012). Similar to findings with vitamins, formulating hydrophobic carotenoids as nanoemulsions was shown to protect them from degradation and enhance their bioavailability (Santos and Meireles, 2010; Palafox-Carlos, Ayala-Zavala and González-Aguilar, 2011). β-carotene is used as antioxidant and natural coloring agent in the food industry, but it is easily degraded by heat, light and oxygen. Yuan et al. formulated β-carotene nanoemulsion using medium chain triglyceride oil containing beta carotene as oil phase and Tween 20 as emulsifier. The nanoemulsion formed by high pressure homogenizer method has particle size distribution of 132 nm to 184 nm. Stability study at 25°C reported 25% β-carotene lost after four weeks (Yuan et al., 2008). β-carotene (BC) loaded o/w nanoemulsions were fabricated using whey protein isolate (WPI) and whey protein isolate-dextran (WPI-DT) as emulsifying agents (Fan et al., 2017). The highest β-carotene retention rate following a one-month storage period at 25°C and 50°C was found using nanoemulsions prepared using WPI-DT (5 kDa) conjugate; however, the release and lipolysis of β-carotene from nanoemulsions prepared using WPI-dextran (70 kDa) was profoundly impaired (Fan et al., 2017). Another study showed β-carotene retention to be significantly better for lactalbumin-catechin conjugate-stabilized nanoemulsions than for those stabilized using lactalbumin (Yi et al., 2014). This was ascribed to the enhanced binding and radical-scavenging capability of the lactalbumin-catechin conjugate (Yi et al., 2014). The inclusion of tea polyphenols in the aqueous phase of nanoemulsions of β-carotene enhanced the stability and oral bioavailability of β-carotene (Meng et al., 2019); moreover, this formulation resulted in better recovery rates of β-carotene at phases I and II of the digestion process *in vitro* and enhanced gastrointestinal absorption and increased conversion of β-carotene to vitamin A *in vivo* (Meng et al., 2019). Sotomayor-Gerding et al. prepared nanoemulsions containing carotenoids (astaxanthin or lycopene) using the high-pressure homogenization method (Sotomayor-Gerding et al., 2016). They demonstrated the stability of these nanoemulsions during storage under various environmental conditions and showed that their oxidative stability was increased by Trolox and butylated hydroxytoluene (BHT). Moreover, these nanoemulsions were moderately digested (66%) and were extremely bioaccessible (70–93%) (Sotomayor-Gerding et al., 2016). In another study, astaxanthin was incorporated in nanoemulsions containing various oils such as corn oil, olive oil, and flaxseed oil. The authors reported enhanced bioaccessibility of astaxanthin in oil-containing nanoemulsions compared with nanoemulsions prepared without any lipid. This was attributed to solubilization of the hydrophobic carotenoids by the micelles produced from the carrier oils (Liu et al., 2018). Shen et al. prepared astaxanthin containing nanoemulsions emulsified using WPI. These researchers reported that the cellular uptake of astaxanthin followed the rank order as polymerized whey protein (PWP), WPI–lecithin mixture, PWP–lecithin mixture (5.05 ± 0.1%), lecithin, and Tween 20 (Shen et al., 2019). Citral is one of the main characteristic flavor compounds found in volatile oils of lemon grass, lemon and orange. However, citral is prone to degradation over time. To overcome this problem,

Yang et al. formulated citral O/W nanoemulsion combined with antioxidants such as black tea extract, β-carotene and ascorbic acid to enhance citral stability (Yang et al., 2011).

13.4.3 Curcuminoids

Curcumin is a bioactive ingredient extracted from *Curcuma longa* and is widely used in foods, nutraceutical and pharmaceutical products. Curcumin is a diarylheptanoid, belonging to the group of curcuminoids, which are natural phenols accountable for the distinctive yellow color of turmeric (Manolova et al., 2014). Curcuminoids are accounted for the potential antioxidant characteristics of curcumin leading to inclusion of curcumin in food products and biological preparations as anti-inflammatory, anti-thrombotic, anti-allergenic, anti-microbial, and anti-atherogenic agents (Assadpour and Jafari, 2018). They main problem of curcumin in the formulation is poor water solubility and low bio-availability (Anand et al., 2007). Joung et al. prepared curcumin nanoemulsion by dissolving curcumin in medium chain triglyceride oil and Tween 20 as surfactant. Using high pressure homogenization method, the mean droplet size was 90 to 122 nm, depending on the surfactant concentration. The curcumin nanoemulsion fortified milk showed significant lower lipid oxidation (Joung et al., 2016). Another study reported the incorporation of curcumin into nanoemulsions prepared using sodium caseinate (Kumar et al., 2016). These researchers reported an improved cellular uptake of curcumin which was attributed to the delayed release of curcumin in the intestine. In yet another published study, curcumin-loaded nanoemulsions were developed using three different approaches namely conventional, pH-driven and heat-driven and were compared with three conventional curcumin supplements (Zheng et al., 2018). The authors documented a superior bioaccessibility of curcumin obtained from all three nanoemulsions compared to the conventional curcumin supplements. In another research work, curcumin along with a carrier oil, such as sunflower oil, coconut oil, or linseed oil were developed into a phospholipid-stabilized nanoemulsion. The results indicated better levels of docosahexaenoic acid in serum lipids and tissues as well as enhanced curcumin levels in the heart, brain, liver and serum of rats for the nanoemulsions containing curcumin and linseed oil (Sugasini and Lokesh, 2017).

13.4.4 Bioactive Flavonoids

Flavonoids constitute a class of polyphenol secondary metabolites that are found mostly in vegetables, fruits, grains, teas, and flowers and other plant parts (Kumar and Pandey, 2013). The daily intake of flavonoids varies between 20 mg and 500 mg, which are mainly derived from dietary supplements including tea, tomatoes, onions, apples and red wine. Flavonoids are an important part of both human and animal diets, supporting human wellbeing and helping to minimize disease risk. In recent years, flavonoids have attained great interest because of their extensive biological activities such as anti-cancer, anti-inflammatory, and antioxidant activity (Habbu et al., 2018). Many studies have shown that flavonoids have protective effects against viral and bacterial infections, cancer, cardiovascular diseases and other age-related diseases (Kumar, Gupta and Pandey, 2013; Kumar and Pandey, 2013). Because of their possible role in promoting wellbeing and preventing chronic degenerative diseases, they are called "functional ingredients" or "health promoting biomolecules" (Nijveldt et al., 2001). In both ancient and modern times, they played important roles in effective medical treatments.

Flavonoids are molecules that are extremely lipophilic, and thus difficult to absorb orally. They undergo excessive first-pass metabolism that, upon oral administration, hinders their bioavailability. Although widely distributed and of potential benefit the low bioavailability of flavonoids may have a major impact on their nutritional effects. The low solubility in water of the flavonoids also poses a problem for their therapeutic effectiveness. Emulsion-based delivery systems provide the benefits of securing and delivering poorly water-soluble drugs. These not only protect medicines from oxidation and hydrolysis, but also improve the bioavailability of poorly water-soluble drugs (McClements, 2013). Flavonoids can be split into flavones, flavanol, flavonol, isoflavones, anthocyanins, depending on the molecular structure as shown in Table 13.4.

The use of nanotechnology has been extended to natural products where efforts have been made to improve the bioavailability and therapeutic potential of flavonoid compounds. Nanoemulsions (NEs) can make highly lipophilic substances more bioavailable and have good visibility, transparency and physical stability. Owing to their relatively small droplet size, NEs can be mechanically stable even at very diluted concentrations, or they can be made extremely viscous or gel-like, depending on the final application of food or beverage. For food applications, delivery vehicles should be prepared using food-grade ingredients; this restricts the number of permitted surfactants. Nonionic surfactants such as sorbitan esters (Tweens or Spans) can be used to deliver food ingredients due to their low toxicity, lack of irritability and the ability to easily form NEs (Sanguansri, Oliver, and Leal-Calderon, 2013).

13.4.4.1 Rutin

Rutin (3,3′,4′,5,7-pentahydroxyflavone-3-rhamnoglucoside; RU), is a major flavonoid found in a variety of plants, including orange, grapefruit, tangerine, buckwheat seeds, lemon, and is also known as quercetin 3-rutinoside, or vitamin P (Hosseinzadeh and Nassiri-Asl, 2014). It is an effective antioxidant molecule which has important scavenging effects on oxidizing species including superoxide, OH and peroxyl radicals. It exhibits many pharmacological properties such as anti-inflammatory, vasoactive, anti-allergic, antitumor, antiviral, antibacterial and antiprotozoal activity (Ganeshpurkar and Saluja, 2017). Moreover, it has other potential therapeutic effects such as hypolipidemic, anticarcinogenic and anti-diabetic effects (Sharma et al., 2013). However, its poor aqueous solubility and oral bioavailability hinders its use. There have been many attempts to increase the oral bioavailability of RU by NE techniques.

Macedo et al. developed a stable rutin-loaded nanoemulsion (RU-NE) using the high-pressure homogenization (HPH) technique and determined the release profile of the drug *in vitro*. The mean droplet size, polydispersity index (PDI), zeta potential (ZP) and encapsulation efficiency (EE) was found; 127 nm, 0.168, –3.49 mV, 82%, respectively. The NE was sterically stabilized by soya lecithin and Tween 80 regardless of low zeta potential values. Drug release from NE showed an initial burst release with a sustained release over 24 h. Therefore, this study showed that RU can be successfully encapsulated into a NE with prolonged release over time (Macedo et al., 2014). In another study, RU-NE was developed and incorporated into a gelatin-based film. This RU-NE induces interactions with film matrix, as well as changes in film properties. It was developed to enhance the mechanical properties of gelatin films and produced a significant increase in film resistance to mechanical stress. In addition, these composite films showed higher antioxidant activity than RU against DPPH radical scavenging and reducing power activities. This study showed that RU-NE can act as a potential active packaging system in a gelatin composite film to boost the shelf-life of food products and provide fresh/safe quality products (Dammak et al., 2017).

TABLE 13.4   Nanoemulsion Systems Prepared for Flavonoid Compounds

Flavonoids Groups	Flavonoid	Formulation Challenge	NE Preparation Method	Formulation Outcomes
Flavonols	Quercetin	• Poor aqueous solubility and stability issues	• Spontaneous emulsification technique • High-pressure homogenization (HPH) techniques	• Improved stability • Increased bioavailability • Enhanced skin permeability • Improved therapeutic efficacy
	Rutin	• Poor aqueous solubility	• HPH technique • Solvent-evaporation technique	• Increased antioxidant potential • Act as potential active packaging systems to enhance shelf-life of food products • Prolonged release system
	Resveratrol	• Poor aqueous solubility and stability issues	• Ultrasound cavitation assisted emulsification method • HPH technique	• Increase stability and prevent degradation • Increase therapeutic efficacy • Provided sustained release system • Protection against oxidation
Flavones	Apigenin	• Poor aqueous solubility	• HPH technique	• Enhanced therapeutic outcomes
Flavanones	Naringenin	• Poor aqueous solubility and stability issues	• Aqueous titration method	• Avoids first pass metabolism • Provided targeted delivery • Enhanced therapeutic outcomes
	Silymarin	• Poor aqueous solubility • Poor bioavailability (30%)	• HPH technique • Aqueous titration method	• Increased bioavailability • Enhanced therapeutic efficacy
Isoflavones	Genistein	Poor aqueous solubility Slow absorption through skin	• Spontaneous emulsification • Hybrid homogenization technique	• Enhanced permeation through skin • Provided topical delivery system

13.4.4.2 Quercetin

Quercetin (3,3′,4′,5,7-pentahydroxyflavone; QT) is also an important flavonoid mostly found in vegetables, fruits, red wine and herbal medicines (Liu et al., 2014). It is the most common dietary flavonoid and has several potentially useful effects in the body. Several studies showed that QT has antioxidant, anti-inflammatory, anti-hypertensive, anti-platelet, antiviral, and anticancer activities (Wach, Pyrzyńska, and Biesaga, 2007; Y. Zhang et al., 2008; Ha et al., 2013). However, the low-water solubility and crystalline nature of QT limits its application as a functional food ingredient (Karadag et al., 2013). Furthermore, it is rapidly degraded and discolored under light and heat (Scalia and Mezzena, 2009). The development of appropriate formulations is required to address these properties. A number of encapsulation techniques have been developed to deliver QT in order to improve its aqueous solubility and chemical stability, among them NEs offering the most promising delivery systems for food applications.

QT has been reported to have anti-inflammatory activity similar to that of nonsteroidal anti-inflammatory drugs; however, its low skin permeability hampers its effectiveness by the topical route. In view of its potential therapeutic activity via topical application, Gokhale et al. developed a QT-loaded nanoemulsion (QT-NE) gel for the treatment of rheumatoid arthritis (RA). The NE was spontaneously prepared using oleic acid: arachis oil: Tween 20 and PEG-400 (15:6:6) which could successfully incorporate QT. This NE gel showed sufficient rheological behavior with a good texture profile and an increase in drug permeation compared to that achieved with free QT gel. In contrast to the free QT gel, it also inhibited CFA-induced paw edema in rats over 24 h. Hence this study confirmed the topical application of QT-NE gel as a promising alternative for RA (Gokhale, Mahajan, and Surana, 2019). In another study, Son et al. investigated the hypocholesterolemic effect of a QT-NE in rats fed with high-cholesterol diet. This NE was formulated using O/W NE techniques with complexation and self-assembly using Captex® 355, Tween 80, sodium alginate and soy lecithin. The QT-NE had a droplet size range of 207–289 nm, with a low PDI (<0.47) and was stable during long-term storage at pH between 6.5 and 9.0. The effect of the NE in reducing serum and hepatic cholesterol levels and in increasing the release of bile acid into the feces was greater than that of free QT. Moreover, the QT-NE significantly upregulated the expression of genes involved in cholesterol efflux and mRNA level and CYP7A1 activity linked to the hepatic synthesis of bile acids. (Son et al., 2019).

Karadag et al. prepared QT-NE with nonionic food-grade emulsifiers using HPH. The optimized formulation were prepared with 13% mixed emulsifiers (Tween 80 and Span 20 mixture, 1:1 weight ratio), 17% oil content (limonene oil), and 70 MPa homogenization pressure which showed minimum particle size and the highest stability. This study showed that the loading of QT in NEs had a major impact on particle size and the emulsions stability depending on the oil: emulsifier ratio in the system (Karadag et al., 2013). Jain et al. prepared a QT-loaded self-nanoemulsifying drug delivery system (QT-SNEDDS) which was composed of Capmul MCM (as oily phase), Tween 20 and ethanol (as surfactant and co-surfactant) for enhanced oral bioavailability and improved antioxidant activity. The optimized formulation (Capmul MCM: QT (19:1)/Tween 20/ethanol, 40:40:20 w/w) was selected based on its ability to form a spontaneous NE in SGF. The antioxidant potential of QT-SNEDDS in a DPPH scavenging assay was comparable with that of free QT. A fluorescent dye-loaded SNEDDS formulation when incubated with Caco-2 cells showed fast internalization within 1 h, with 23.75-fold increase in cellular uptake by Caco-2 cells as compared to free QT. Moreover, the QT-SNEDDS showed a 5-fold increase in oral bioavailability as compared to a free QT suspension. This study showed an increase in antioxidant activity with QT-SNEDDS at dose of 100 mg/kg with 65% decrease in tumor growth as compared to free QT (20%) (Jain, Thanki, and Jain, 2014). A QT-NE was also developed using HPH techniques and evaluated for its bioavailability by Pool et al. In this study, QT was successfully incorporated in NEs provided that the QT concentration in the NE did not exceed its saturation level (C_{Sat} = 0.15 mg/mL). In an *in vitro* digestion model, QT incorporated into the NE had a higher bioavailability than QT dissolved in bulk oils or water (Pool et al., 2013). In another study, Ni et al. encapsulated QT into a specially formulated food-grade NE prepared through the HPH technique. The optimum QT-NE shown an average droplet size of 152 ± 6 nm, ZP of –50 ± 2 mV, EE of 93.50 ± 0.35% and had good physical and chemical stability. Moreover, the bioavailability of QT was improved after encapsulation into NE and it showed no significant toxicity to HeLa cells line. Hence this study confirmed that incorporation of QT into an NENEs improved its stability and bioavailability (Ni et al., 2017). Arbain et al. developed QT loaded palm oil-based NE for pulmonary delivery. The optimized NE formulation, prepared using the low and high energy emulsification method, was composed of 1.50 wt.% palm oil: ricinoleic acid (1:1), 1.50 wt.% of lecithin, 1.50 wt.% of Tween 80, 1.50 wt.% of glycerol and 93.90 wt.% of water. This NE had a droplet size of 110.3 nm, PDI value of 0.290 and zeta potential of –37.7 mV and showed good stability when stored at 4°C for 90 days. Evaluation of *in vitro* delivery of the QT-NE showed that the aerosol output, rate and median mass aerodynamic diameter were 99.31%, 0.19 g/min and 4.25 μm, respectively. Based on the above result the authors conclude that the developed QT-NE has the potential for pulmonary delivery specifically for lung cancer treatment (Arbain et al., 2018).

13.4.4.3 Apigenin

Apigenin (4′,5,7-trihydroxyflavone; AP) is a flavone which is commonly present in fruits, vegetables, nuts, onions, oranges, and tea. AP has various pharmacological effects such as anti-oxidant, anti-microbial, anti-inflammatory, anti-proliferative, anti-viral, anti-diabetic and anti-tumor (Salehi et al., 2019). However, it poor solubility in water leads to poor absorption from the gastrointestinal tract (Zhang et al., 2012). Improving the solubility and bioavailability of AP for its pharmaceutical and nutraceutical applications is urgently required.

Jangdey et al. developed a carbopol-based AP-loaded NE gel using tamarind gum emulsifier by HPH method and evaluated it *in vitro* against skin cancer using two different cell lines (HaCaT Cells and A431 cells). The droplet size, PDI and ZP of NE were found 183.31 nm, 0.532, and 31.9 mV, respectively. The formulations containing 3% w/w of emulsifying gum demonstrated good physical stability and did not show phase separation or precipitations. A CLSM study using goat skin showed uniform fluorescence intensity with NE gel treatment across the entire depth of skin, suggesting high skin penetrability of the drug and high drug retention in the skin. This study concluded that the AP-loaded NE could be of value in the treatment of skin cancer (Jangdey, Gupta, and Saraf, 2017).

13.4.4.4 Naringenin

Naringenin (5,7,4-trihydroxyflavanone; NGN) is mainly found in grapefruit and other citrus fruits. It is an interesting flavonoid because of various biological properties and its presence in high amounts in human diets (Jayaraman, Veerappan, and Namasivayam, 2009). NGN has shown anti-cancer, anti-mutagenic, anti-atherogenic, anti-inflammatory and anti-fibrogenic activities as well as free-radical scavenging properties (Parashar et al., 2018). The aqueous solubility and log P value of NGN was found 475 mg/l and 2.42, respectively. NGN is weakly basic and its oral bioavailability is almost 15%. Its absorption takes place through both passive diffusion as well as active transport (Zhang et al., 2015). Poor aqueous solubility and low bioavailability of NGN limits its use. Many researchers have tried to increase its therapeutic effectiveness using NE technology.

Md et al. developed NGN-loaded NE (NGN-NE) formulation for oral delivery and determined apoptotic activity in A549 lung cancer cells. NE globule size was significantly affected by the concentration of Capryol 90 and Tween 20, while the ZP was dependent only on the Tween 20 concentration. The stabilized NEs displayed a spherical surface morphology of 85.6 ± 2.1 nm, a PDI of 0.263 ± 0.02, a ZP of –9.6 ± 1.2 mV, and a drug content of 97.34 ± 1.3%. The release analysis showed an initial NGN burst from the NE, followed by a controlled release up to 24 h. The concentration-dependent cytotoxicity of the NE was larger in A549 lung cancer cells than that of free NGN. In addition, the percentage of NE-induced apoptotic cells and cell cycle arrests in the G2/M and pre-G1 phases was significantly greater than free NGN ($p < 0.05$). NEs have also been found to be more effective in reducing expression of Bcl2 compared to NGN alone, while increasing activity of caspase-3 and Bax protein. These findings indicate that the stable NGN-NE may provide an effective approach to the delivery of medicines for treating lung cancer (Md et al., 2020). In another study, Md et al., evaluated the neuroprotective effects of NGN-NE on neurotoxicity caused by β-amyloid in a human neuroblastoma cell line (SH-SY5Y). NE substantially alleviated the direct neurotoxic effects of beta-amyloid (Aβ) on SH-SY5Y cells; this was correlated with a down-regulation of the expression of amyloid precursor protein (APP) and β-secretase (BACE), suggesting reduced amyloidogenesis. In addition, NE reduced phosphorylated tau concentrations in SH-SY5Y cells exposed to Aβ. This study showed that the NGN-NE had a stronger neuroprotective effect than free NGN, and may be a promising agent for Alzheimer's disease treatment (Md et al., 2018). Gaba et al. developed a stable vitamin E-loaded NGN-NE for direct nose-to-brain delivery to increase the therapeutic effectiveness of NGN in the treatment of Parkinson's disease (PD). The NE was prepared using an aqueous titration method with Capryol 90: vitamin E (1:1) as oil phase, Tween 80 as surfactant and Transcutol-HP as co-surfactant. Here, vitamin E showed an added advantage of providing an additional antioxidant effect. The NGN concentration in the brain was found to be higher when it was administered in the NE through the intranasal route (i.n), proving the targeting efficiency of NE. Intra-nasal administration of NGN-NE along with levodopa to rats with PD induced by 6-OHDA was effective in improving grip strength, swimming activity, and muscle coordination. This study showed that i.n administration of NGN-NE prevented first-pass metabolism and enhanced the uptake of NGN into the brain, resulting in increased brain bioavailability of the drug (Gaba et al. 2019).

13.4.4.5 Resveratrol

Resveratrol (3,5,40-trihydroxy-trans-stilbene; RES) is a compound of polyphenols found in a number of dietary compounds, including grapes, wine, nuts, berries, and many other human foods (Berman et al. 2017). It has a wide variety of biological functions, including anti-oxidant anti-aging, anti-inflammatory, anti-viral and fungal effects (Chedea et al., 2017; Nawaz et al., 2017). RES is available in two isomeric varieties, cis and trans, with the trans isomer being transformed to cis. Biologically active only trans-RES has minimal stability under the influence of light, certain pH levels and temperature. Trans-RES, which is poorly soluble in water (0.03 mg/mL) rapidly isomerizes when exposed to UV or visible light (Zupančič, Lavrič, and Kristl, 2015).

Several attempts have been made to increase the therapeutic efficacy of RES using NE technology. Hussein et al. prepared a RES-loaded NE (RES-NE) and investigated its anticholestatic activity against ethinylestradiol (EE)-induced cholestasis in adult female rats. This NE had a droplet size of 49.5 ± 0.05 nm and a ZP of +15.75 mV and the particles were spherical. In a dose of 39.75 mg/kg the RES-NE was effective in suppressing cholestatic indices, oxidative stress biomarkers, and reducing MMP-2 and -9 in rats with EE-induced cholestasis (Hussein et al., 2019). Another study showed that a RES-NE gel could deliver resveratrol to the upper skin layers *in vitro*. This formulation was effective in preventing UV-induced oxidative skin damage in rats (Sharma et al., 2019). Li et al. developed RES and linseed oil co-loaded NE for the purpose of increasing solubility, bioavailability and stability. The NEs were prepared containing 0.2% w/w RES, 5.0%, w/w linseed oil, 8.8%w/w Tween 80, 2.2% w/w Span 80, 1% w/w soybean powdered lecithin using HPH technology. During storage at 4° C or 25° C for 60 days the resulting NEs had good physicochemical stability. The release study showed sustained release of RES from NEs compared with free RES and increased its antioxidant activity by the synergistic effect of linseed oil. This study also showed that the RES-NE could protect linseed oil from oxidation, with non-encapsulated linseed oil being rapidly digested in the intestine. Thus, RES and linseed oil co-loaded NE may have future applications in functional foods (Li et al., 2018). Vitamin E-loaded RES-NE was prepared for brain targeting in the treatment of PD using vitamin E: sefsol (1:1) as oil process, Tween 80 as surfactant and Transcutol P as co-surfactant using a spontaneous emulsification method followed by HPH. This NE formulation had good antioxidant activity and could deliver a large amount of drug to the brain following intranasal administration (Pangeni et al., 2014). All of the above RES-NE were generally formulated as two-phase systems (O/W) that do not contain imaging agents. As such, *in vivo* biodistribution of these NEs may be difficult to study. For such systems, NE efficacy is usually determined through pharmacological testing in cells and animals. Hence it is not easy to directly correlate RES delivery to specific tissues and cells with pharmacological outcomes. To overcome this, Herneisey et al. developed theranostic RES-NE containing a single therapeutic agent (RES) and dual diagnostic agents. NEs developed using this approach allow a versatile treatment strategy for cancer and inflammatory diseases. This also opens a new avenue for theranostic development and approaches for imaging-supported delivery of RES (Herneisey et al. 2016).

13.4.4.6 Silymarin

Silymarin (Sily) is a naturally flavonoid extracted from the seeds of *Silybum marianum*, which has diverse pharmacological properties, including hepatoprotective, antioxidant, anti-inflammatory, anticancer, and cardioprotective activities. It has been approved as a safe hepatoprotective agent, as well as being a drug of choice for several hepatic disorders (Cetinkunar et al., 2015). Sily is a complex mixture of silybin (60%-70%), silychristin (20%), silydianin (10%), and isosilybin (5%). However, its oral bioavailability is only 23–47% due to its poor aqueous solubility (Javed, Kohli and Ali, 2011). Several researchers have attempted to increase its efficacy using NE technology. Nagi et al. prepared Sily-loaded NE (Sily-NE) by HPH method containing Capryol 90, Solutol HS 15 and Transcutol HP as an oil, surfactant and co-surfactant, respectively. NE has been found to increase the apparent permeability coefficient in everted gut sacs. Pharmacokinetic research showed a substantial increase (p < 0.05) in oral bioavailability of Sily formulated as NE compared with an oral suspension (Nagi et al., 2017).

An optimized Sily-NE, prepared using an aqueous titration method, was composed of oil phase (sefsol 218), S_{mix} (Kolliphor RH40 (surfactant) and PEG 400 (cosurfactant) in 2:1 ratio and distilled water as an aqueous phase. Following oral administration to rats, the C_{max}, and AUC were higher with a shorter T_{max} when Sily was given as an NE compared to values obtained following administration of conventional and standard suspensions of Sily. The Sily-NE decreased cell viability and increased strength of ROS and condensation of chromatin in cells of human hepatocellular carcinoma, without damaging normal cells. These findings indicate that NE in the treatment of human hepatocellular carcinoma could be an effective carrier for the oral delivery of Sily (Ahmad et al., 2018). Sily-NE is composed of 5% w/w of sefsol 218 as an oily form, 35% w/w of Smix containing Tween 80 and ethyl alcohol (2:1) as a surfactant and cosurfactant and 60% w/w as an aqueous phase of distilled water showed nano-droplet size, low viscosity and higher drug release. Following oral administration to rats, the AUC and C_{max} of Sily in the NE were found to be higher as compared to simple suspension of Sily. This study also showed that administration in the NE formulation of less than half the dose of Sily provided similar safety against CCl4 mediated toxicity in rats as compared to Sily in solution and suspension form. Thus Sily-NE appeared to have better bioavailability and hepatoprotective activity and easily prepared (Parveen et al., 2011).

13.4.4.7 Genistein

Genistein [4′,5,7-trihydroxyisoflavone; GNT] is one of the most abundant and best-studied soy isoflavones and has received great attention for its potential human health benefits. It is an essential isoflavone in soy and soy food products that people in Asian countries consume regularly (Ronis, 2016). It has potential applications for skin cancer treatment and chemoprevention. However, the clinical use of GNT is hindered because of its poor water solubility (0.9 μg/mL) and low oral bioavailability (Chen et al., 2013). Brownlow et al. investigated the dermal delivery of GNT using NE formulations. The results showed a slow drug release using both liquid and cream formulations of Tocomin® NE loaded with genistein (GNT-Tocomin® NE). GNT-Tocomin® NE also demonstrated good biocompatibility and offered significant protection of cultured subcutaneous L929 fibroblasts against UVB. These studies suggest that Tocomin® NE has the potential to provide effective antioxidant defense and to prevent skin damage caused by UVB-induced skin damage (Brownlow et al., 2015). De Vargas et al. developed GNT-loaded NE and incorporated these into acrylic-acid hydrogels for topical application. The formulations exhibited non-Newtonian pseudoplastic behavior. The skin permeation study showed that high amount of GNT was detected using a NE hydrogel composed of a medium-chain triglycerides oil core. Thus hydrogels containing GNT-loaded NEs could be a promising formulation to deliver isoflavones into the skin (De Vargas et al., 2012).

13.5 CONCLUSION AND FUTURE PERSPECTIVES

Nanoemulsions are very popular in nutraceutical, pharmaceutical, cosmeceutical and food industry due to their unique properties. Numerous *in vivo* and *in vitro* studies have showed consistent findings that they are able to improve solubility, bioavailability, stability and absorption of the bioactive and nutritional food compounds. In addition, the methods of preparation for nanoemulsions are extensively developed, thus it is feasible to scale up in the industry. Despite the advantages, the application in the industry does carry limitations, such as the increase in production cost, investment of expensive instruments, additional characterization and the quality control processes for the nanoemulsions. In addition, optimization of preparation methods is required because the selection of emulsifying agents which are safe for consumption are limited. Majority of the synthetic or semi-synthetic emulsifiers are not suitable for consumption. Moreover, the usage of harmful organic solvents should be avoided, thus making the preparation more difficult. The major concern about nanoemulsions is the safety issue, because the ingestion of large amount of emulsifiers and additives may increase the risk of toxicity because they may interact with the cell membrane. Therefore, further research work is required to understand the safety profile of the nanoemulsions consumption and the authorities should develop a stringent inspection process to grant approval for the utilization in food industry. Overcoming the challenges is not impossible, hence the future perspectives of nanoemulsions application in food industry is very promising.

CONFLICTS OF INTEREST

The authors declare no conflict of interest.

ACKNOWLEDGMENTS

The Deanship of Scientific Research (DSR) at King Abdulaziz University, Jeddah, funded this project, under grant no. (**RG-12-166-38**). The authors, therefore, acknowledge with thanks the DSR for technical and financial support.

The authors are thankful to Brian L. Furman (University of Strathclyde, Glasgow, UK) for critical reading of the manuscript and language editing.

REFERENCES

Ahmad, U., Akhtar, J., Singh, S. P., Badruddeen, B., Ahmad, F. J., Siddiqui, S., & Wahajuddin, W. (2018). Silymarin nanoemulsion against human hepatocellular carcinoma: development and optimization. *Artificial Cells, Nanomedicine and Biotechnology*, 46(2), 231–41.

Akanbi, T. O., & Barrow, C. J. (2017). Candida Antarctica lipase A effectively concentrates DHA from fish and thraustochytrid oils. *Food Chemistry*, 229, 509–16.

Alam, M. A., Subhan, N., Rahman, M. M., Uddin, S. J., Reza, H. M., & Sarker, S.D. (2014). Effect of citrus flavonoids, naringin and naringenin, on metabolic syndrome and their mechanisms of action. *Advances in Nutrition*, 5(4), 404–17.

Aluko, R. (2012). Functional foods and nutraceuticals. New York, Springer.

Anand, P., Kunnumakkara, A. B., Newman, R. A., & Aggarwal. B. B. (2007). Bioavailability of curcumin: Problems and promises. *Molecular Pharmaceutics*, 4(6), 807–18.

Anton, N., & Vandamme, T. F. (2009). The universality of low-energy nano-emulsification. *International Journal of Pharmaceutics*, 377(1–2), 142–7.

Arbain, N. H., Salim, N., Wui, W. T., Basri, M., & Rahman, M. B. A. (2018). Optimization of quercetin loaded palm oil ester based nanoemulsion formulation for pulmonary delivery. *Journal of Oleo Science*, 67(8), 933–40.

Assadpour, E., & Jafari, S. M. (2018). A systematic review on nanoencapsulation of food bioactive ingredients and nutraceuticals by various nanocarriers. *Critical Reviews in Food Science and Nutrition*, 59, 3129–3151.

Aswathanarayan, J. B., & Vittal, R. R. (2019). Nanoemulsions and their potential applications in food industry. *Frontiers in Sustainable Food Systems*, 3, 95.

Beller, H. R., Lee, T. S., & Katz, L. (2015) Natural products as biofuels and bio-based chemicals: fatty acids and isoprenoids. *Natural Product Reports*, 32(10), 1508–26.

Berman, A. Y., Motechin, R. A., Wiesenfeld, M. Y., & Holz, M. K. (2017). The therapeutic potential of resveratrol: a review of clinical trials. *NPJ Precision Oncology*, 1, 35.

Bernhoft, A. (2010). Bioactive compounds in plants – benefits and risks for man and animals. Proceedings from a symposium held at the Norwegian Academy of Science and Letters, Oslo, 13–14 November 2008.

Bhavaniramya, S., Vishnupriya, S., Al-Aboody, M. S., Vijayakumar, R., & Baskaran, D. (2019). Role of essential oils in food safety: antimicrobial and antioxidant applications. *Grain & Oil Science and Technology*, 2(2), 49–55.

Bracco, U. (1994). Effect of triglyceride structure on fat absorption. *American Journal of Clinical Nutrition*, 60(6), 1002S–9S.

Brownlow, B., Nagaraj, V. J., Nayel, A., Joshi, M., & Elbayoumi, T. (2015). Development and in vitro evaluation of vitamin E-enriched nanoemulsion vehicles loaded with genistein for chemoprevention against UVB-induced skin damage. *Journal of Pharmaceutical Sciences*, 104(10), 3510–23.

Burt, S. (2004). Essential oils: their antibacterial properties and potential applications in foods-A review. *International Journal of Food Microbiology*, 94, 223–53.

Calsamiglia, S., Busquet, M., Cardozo, P. W., Castillejos, L., & Ferret, A. (2007). Invited review: essential oils as modifiers of rumen microbial fermentation. *Journal of Dairy Science*, 90, 2580–95.

Campani, V., Biondi, M., Mayol, L., Cilurzo, F., Pitaro, M., & de Rosa, G. (2016). Development of nanoemulsions for topical delivery of vitamin K1. *International Journal of Pharmaceutics*, 511, 170–77.

Castro-Gomez, P., Garcia-Serrano, A., Visioli, F., & Fontecha, J. (2015) Relevance of dietary glycerophospholipids and sphingolipids to human health. *Prostaglandins, Leukotrienes & Essential Fatty Acids (PLEFA)*, 101, 41–51.

Cetinkunar, S., Tokgoz, S., Bilgin, B. C., Erdem, H., Aktimur, R., Can, S., Erol, H. S., Isgoren, A., Sozen, S., & Polat, Y. (2015). The effect of silymarin on hepatic regeneration after partial hepatectomy: is silymarin effective in hepatic regeneration? *International Journal of Clinical and Experimental Medicine*, 8(2), 2578–85.

Chedea, V. S., Vicaş, S. I., Sticozzi, C., Pessina, F., Frosini, M., Maioli, E., & Valacchi, G. (2017). Resveratrol: from diet to topical usage. *Food and Function*, 8, 3879–92.

Chen, F., Peng, J., Lei, D., Liu, J., & Zhao, G. (2013). Optimization of genistein solubilization by κ-carrageenan hydrogel using response surface methodology. *Food Science and Human Wellness*, 2(3–4), 124–31.

Chowdhury, S., Mandal, G. P., Patra, A. K., Kumar, P., Samanta, I., Pradhan, S. & Samanta, A. K. (2018). Different essential oils in diets of broiler chickens: 2. Gut microbes and morphology, immune response, and some blood profile and antioxidant enzymes. *Animal Feed Science and Technology*, 236, 39–47.

Cobellis, G., Trabalza-Marinucci, M., & Yu, Z. (2016). Critical evaluation of essential oils as rumen modifiers in ruminant nutrition: a review. *Science of the Total Environment*, 545–546, 556–68.

Dammak, I., de Carvalho, R. A., Trindade, C. S. F., Lourenço, R. V., & do Amaral Sobral, P. J. (2017). Properties of active gelatin films incorporated with rutin-loaded nanoemulsions. *International Journal of Biological Macromolecules*, 98, 39–49.

De Vargas, B. A., Bidone, J., Oliveira, L. K., Koester, L. S., Bassani, V. L., & Teixeira, H. F. (2012). Development of topical hydrogels containing genistein-loaded nanoemulsions. *Journal of Biomedical Nanotechnology*, 8(2), 330–6.

Deepa, N., Kaur, C., George, B., Singh, B., & Kapoor, H. (2007). Antioxidant constituents in some sweet pepper (Capsicum annuum L.) genotypes during maturity. *LWT-Food Science and Technology*, 40(1), 121–29.

Dey, T. K., Koley, H., Ghosh, M., Dey, S., & Dhar, P. (2019). Effects of nano-sizing on lipid bioaccessibility and ex vivo bioavailability from EPA-DHA rich oil in water nanoemulsion. *Food Chemistry*, 275, 135–42.

Dixon, R. A., & Ferreira, D. (2002). Genistein. *Phytochemistry*, 60(3), 205–11.

Enogieru, A.B., Haylett, W., Hiss, D. C., Bardien, S., & Ekpo, O. E. (2018). Rutin as a potent antioxidant: Implications for neurodegenerative disorders. Oxidative Medicine and Cellular Longevity. Article ID 6241017.

Escribá, P. V., González-Ros, J. M., Goñi, F. M., Kinnunen, P. K. J., Vigh, L., Sánchez-Magraner, L., Fernández, A. M., Busquets, X., Horváth, I., & Barceló-Coblijn, G. (2008). Membranes: a meeting point for lipids, proteins and therapies. *Journal of Cellular and Molecular Medicine*, 12(3), 829–75.

Espitia, P. J. P., Fuenmayor, C. A., & Otoni, C. G. (2019). Nanoemulsions: synthesis, characterization, and application in bio-based active food packaging. *Comprehensive Reviews in Food Science and Food Safety*, 18(1), 264–85.

Fan, Y., Yi, J., Zhang, Y., Wen, Z., & Zhao, L. (2017). Physicochemical stability and in vitro bioaccessibility of β-carotene nanoemulsions stabilized with whey protein-dextran conjugates. *Food hydrocolloids*, 63, 256–64.

Gaba, B., Khan, T., Haider, M. F., Alam, T., Baboota, S., Parvez, S., & Ali, J. (2019). Vitamin E loaded naringenin nanoemulsion via intranasal delivery for the management of oxidative stress in a 6-OHDA Parkinson's disease model. BioMed Research International, Article ID 2382563.

Gammone, M. A., Riccioni, G., Parrinello, G., & D'orazio, N. (2019). Omega-3 polyunsaturated fatty acids: benefits and endpoints in sport. *Nutrients*, 11(1):46.

Ganeshpurkar, A., & Saluja, A. K. (2017). The pharmacological potential of rutin. *Saudi Pharmaceutical Journal*, 25(2), 149–64.

Gokhale, J. P., Mahajan, H. S., & Surana, S. S. (2019). Quercetin loaded nanoemulsion-based gel for rheumatoid arthritis: in vivo and in vitro studies. *Biomedicine and Pharmacotherapy*, 112, 108622.

Ha, H. K., Kim, J. W., Lee, M. R., & Lee, W. J. (2013). Formation and characterization of quercetin-loaded chitosan oligosaccharide/β-lactoglobulin nanoparticle. *Food Research International*, 52(1), 82–90.

Habbu, P., Hiremath, M., Madagundi, S., Vankudri, R., Patil, B., & Savant, C. (2018). Phytotherapeutics of polyphenolic-loaded drug delivery systems: a review. *Pharmacognosy Reviews*, 12, 7–19.

Hadacek, F. (2017). Phenolic lipids in plants: functional diversity of. In: Wenk M. (eds) Encyclopedia of Lipidomics. Springer, Dordrecht.

Håkansson, A. (2019). Emulsion formation by homogenization: current understanding and future perspectives. *Annual Review of Food Science and Technology*, 10(1), 239–58.

Hausmann, J., Deiner, C., Patra, A. K., Immig, I., Starke, A., & Aschenbach, J. R. (2018). Effects of a combination of plant bioactive lipid compounds and biotin compared with monensin on body condition, energy metabolism and milk performance in transition dairy cows. *PloS One*, 13(3), e0193685.

Henley, D. V., Lipson, N., Korach, K. S., & Bloch, C. A. (2007). Prepubertal gynecomastia linked to lavender and tea tree oils. *New England Journal of Medicine*, 356 (5), 479–85.

Herneisey, M., Williams, J., Mirtic, J., Liu, L., Potdar, S., Bagia, C., Cavanaugh, J. E., & Janjic, J. M. (2016). Development and characterization of resveratrol nanoemulsions carrying dual-imaging agents. *Therapeutic Delivery*, 7(12), 795–808.

Hossein-Nezhad, A., & Holick, M. F. (2013). Vitamin D for health: a global perspective. *Mayo Clinic Proceedings*, 88(7), 720–55.

Hosseinzadeh, H., & Nassiri-Asl, M. (2014). Review of the protective effects of rutin on the metabolic function as an important dietary flavonoid. *Journal of Endocrinological Investigation*, 37(9), 783–8.

Huang, Z., Liu, Y., Qi, G., Brand, D., & Zheng, S. (2018). Role of vitamin A in the immune system. *Journal of Clinical Medicine*, 7(9), 258.

Hussein, M. A., Kasser, A. K., Mohamed, A. T., Eraqy, T. H., & Asaad, A. (2019). Resveratrol nanoemulsion: a promising protector against ethinylestradiol-induced hepatic cholestasis in female rats. *Journal of Biomolecular Research and Therapeutics*, 8, 175.

Ismail, A., Bannenberg, G., Rice, H. B., Schutt, E., & Mackay, D. (2016). Oxidation in EPA- and DHA-rich oils: an overview. *Lipid Technology*, 28(3–4), 55–9.

Jain, A. K., Thanki, K., & Jain, S. (2014). Novel self-nanoemulsifying formulation of quercetin: implications of pro-oxidant activity on the anticancer efficacy. *Nanomedicine: Nanotechnology, Biology, and Medicine*, 10(5), 959–69.

Jangdey, M. S., Gupta, A., & Saraf, S. (2017). Fabrication, in-vitro characterization, and enhanced in-vivo evaluation of carbopol-based nanoemulsion gel of apigenin for uv-induced skin carcinoma. *Drug Delivery*, 24(1), 1026–36.

Javed, S., Kohli, K., & Ali, M. (2011). Reassessing bioavailability of silymarin. *Alternative Medicine Review*, 16(3), 239–49.

Jayaraman, J., Veerappan, M., & Namasivayam, N. (2009). Potential beneficial effect of naringenin on lipid peroxidation and antioxidant status in rats with ethanol-induced hepatotoxicity. *Journal of Pharmacy and Pharmacology*, 61(10), 1383–90.

Ji, J., Zhang, J., Chen, J., Wang, Y., Dong, N., Hu, C., Chen, H., Li, G., Pan, X., & Wu, C. (2015). Preparation and stabilization of emulsions stabilized by mixed sodium caseinate and soy protein isolate. *Food Hydrocolloids*, 51, 156–65.

Jin, W., Xu, W., Liang, H,. Li, Y., Liu, S., & Li, B. (2016). Nanoemulsions for food: properties, production, characterization, and applications. *Emulsions*, 3, 1–36.

Joung, H. J., Choi, M. J., Kim, J. T., Park, S. H., Park, H. J., & Shin, G. H. (2016). Development of food-grade curcumin nanoemulsion and its potential application to food beverage system: antioxidant property and in vitro digestion. *Journal of Food Science*, 81(3), N745–53.

Kadappan, A. S., Guo, C., Gumus, C. E., Bessey, A., Wood, R. J., McClements, D. J. & Liu, Z. (2018). The efficacy of nanoemulsion-based delivery to improve vitamin D absorption: comparison of in vitro and in vivo studies. *Molecular Nutrition & Food Research*, 62(4). Article ID 1700836.

Kale, S. N., & Deore, S. L. (2016). Emulsion micro emulsion and nano emulsion: a review. *Systematic Reviews in Pharmacy*, 8(1), 39–47.

Kalepu, S., Manthina, M., & Padavala, V. (2013). Oral lipid-based drug delivery systems–an overview. *Acta Pharmaceutica Sinica B*, 3(6), 361–72.

Karadag, A., Yang, X., Ozcelik, B., & Huang, Q. (2013). Optimization of preparation conditions for quercetin nanoemulsions using response surface methodology. *Journal of Agricultural and Food Chemistry*, 61(9), 2130–9.

Karthik, P., & Anandharamakrishnan, C. (2016). Enhancing omega-3 fatty acids nanoemulsion stability and in-vitro digestibility through emulsifiers. *Journal of Food Engineering*, 187, 92–105.

Katouzian, I., Esfanjani, A. F., Jafari, S. M., & Akhavan, S. (2017). Formulation and application of a new generation of lipid nano-carriers for the food bioactive ingredients. *Trends in Food Science and Technology*, 68, 14–25.

Kazemi-Bonchenari, M., Falahati, R., Poorhamdollah, M., Heidari, S. R., & Pezeshki, A. (2018). Essential oils improved weight gain, growth and feed efficiency of young dairy calves fed 18 or 20% crude protein starter diets. *Journal of Animal Physiology and Animal Nutrition*, 102, 652–61.

Kentish, S., Wooster, T. J., Ashokkumar, M., Balachandran, S., Mawson, & R., Simons, L. (2008). The use of ultrasonics for nanoemulsion preparation. *Innovative Food Science and Emerging Technologies*, 9(2), 1466–8564.

Kumar, M., Bishnoi, R. S., Shukla, A. K., & Jain, C. P. (2019). Techniques for formulation of nanoemulsion drug delivery system: a review. *Preventive Nutrition and Food Science*, 24(3), 225–34.

Kumar, S., Gupta, A., & Pandey, & A. K. (2013). Calotropis procera root extract has the capability to combat free radical mediated damage. *International Scholarly Research Notices*. Article ID 691372.

Kumar, D. D., Mann, B., Pothuraju, R., Sharma, R., Bajaj, R., & Minaxi. (2016). Formulation and characterization of nanoencapsulated curcumin using sodium caseinate and its incorporation in ice cream. *Food & Function*, 7, 417–24.

Kumar, S., & Pandey, & A. K. (2013). Chemistry and biological activities of flavonoids: an overview. *The Scientific World Journal*. Article ID 162750.

Kumar, P., Patra, A. K., Mandal, G. P., & Debnath, B. C. (2018). Carcass characteristics, chemical and fatty acid composition and oxidative stability of meat from broiler chickens fed black cumin (Nigella sativa) seeds. *Journal of Animal Physiology and Animal Nutrition*, 102, 769–77.

Lane, K. E., Li, W., Smith, C. J., & Derbyshire, E. J. (2016). The development of vegetarian omega-3 oil in water nanoemulsions suitable for integration into functional food products. *Journal of Functional Foods*, 23, 306–14.

Lauridsen, C., Hedemann, M. S., & Jensen, S. K. (2001). Hydrolysis of tocopheryl and retinyl esters by porcine carboxyl ester hydrolase is affected by their carboxylate moiety and bile acids. *Journal of Nutritional Biochemistry*, 12(4), 219–24.

Laxmi, M., Bhardwaj, A., Mehta, S., & Mehta, A. (2015). Development and characterization of nanoemulsion as carrier for the enhancement of bioavailability of artemether. *Artificial Cells, Nanomedicine, and Biotechnology*, 43(5), 334–44.

Li, T., Huang, J., Wang, Q., Xia, N., & Xia, Q. (2018). Resveratrol and linseed oil co-delivered in O/W nanoemulsions: preparation and characterization. *Integrated Ferroelectrics*, 190 (1), 101–11.

Li, Y., Yao, J., Han, C., Yang, J., Chaudhry, M. T., Wang, S., Liu, H., & Yin, Y. (2016). Quercetin, inflammation and immunity. *Nutrients*, 8(3), 167.

Liu, L., Tang, Y., Gao, C., Li, Y., Chen, S., Xiong, T., Li, J., Du, M., Gong, Z., Chen, H., & Yao, P. (2014). Characterization and biodistribution in vivo of quercetin-loaded cationic nanostructured lipid carriers. *Colloids and Surfaces B: Biointerfaces*, 115:125–31.

Liu, X., Zhang, R., McClements, D. J., Li, F., Liu, H., Cao, Y., & Xiao, H. (2018). Nanoemulsion-based delivery systems for nutraceuticals: influence of long-chain triglyceride (LCT) type on in vitro digestion and astaxanthin bioaccessibility. *Food Biophysics*, 13, 412–21.

Lohith Kumar, D. H., & Sarkar, P. (2017). Encapsulation of bioactive compounds using nanoemulsions. *Environmental Chemistry Letters*, 16, 59–70.

Luykx, D. M. A. M., Peters, R. J. B., Van Ruth, S. M., & Bouwmeester, H. (2008). A review of analytical methods for the identification and characterization of nano delivery systems in food. *Journal of Agricultural and Food Chemistry*, 56 (18), 8231–47.

Lv, S., Gu, J., Zhang, R., Zhang, Y., Tan, H., & McClements, D. J. (2018). Vitamin E encapsulation in plant-based nanoemulsions fabricated using dual-channel microfluidization: formation, stability, and bioaccessibility. *Journal of Agricultural and Food Chemistry*, 66, 10532–42.

Maali, A., & Mosavian, M. T. H. (2013). Preparation and application of nanoemulsions in the last decade (2000-2010). *Journal of Dispersion Science and Technology*, 34(1), 92–105.

Macedo, A. S., Quelhas, S., Silva, A. M., & Souto, E. B. (2014). Nanoemulsions for delivery of flavonoids: Formulation and in vitro release of rutin as model drug. *Pharmaceutical Development and Technology*, 19(6), 677–80.

Maestro, A., Solè, I., González, C., Solans, C., & Gutiérrez, J. M. (2008). Influence of the phase behavior on the properties of ionic nanoemulsions prepared by the phase inversion composition method. *Journal of Colloid and Interface Science*, 327(2), 433–9.

Magnusson, C. D., & Haraldsson, G. G. (2011). Ether lipids. *Chemistry and Physics of Lipids*, 164 (5), 315–40.

Manolova, Y., Deneva, V., Antonov, L., Drakalska, E., Momekova, D., & Lambov, N. (2014). The effect of the water on the curcumin tautomerism: a quantitative approach. *Spectrochimica Acta. Part A, Molecular and Biomolecular Spectroscopy*, 132, 815–20.

McClements, D. J. (2010a). Design of nano-laminated coatings to control bioavailability of lipophilic food components. *Journal of Food Science*, 75(1), R30–R42.

McClements, D. J. (2010b). Emulsion design to improve the delivery of functional lipophilic components. *Annual Review of Food Science and Technology*, 1, 241–69.

McClements, D. J. (2013). Nanoemulsion-based oral delivery systems for lipophilic bioactive components: nutraceuticals and pharmaceuticals. *Therapeutic Delivery*, 4(7), 841–57.

McClements, D. J. (2015). Nanoparticle- and microparticle-based delivery systems. Boca Raton, CRC Press.

McClements, D. J., Decker, E. A., & Park, Y. (2008). Controlling lipid bioavailability through physicochemical and structural approaches. *Critical Reviews in Food Science and Nutrition*, 49(1), 48–67.

McClements, D. J., Decker, E. A., Park, Y., & Weiss, J. (2009). Structural design principles for delivery of bioactive components in nutraceuticals and functional foods. *Critical Reviews in Food Science and Nutrition*, 49(6), 577–606.

Md, S., Alhakamy, N. A., Aldawsari, H. M., Husain, M., Kotta, S., Abdullah, S. T., Fahmy, U. A., Alfaleh, M. A., & Asfour, H. Z. (2020). Formulation design, statistical optimization, and in vitro evaluation of a naringenin nanoemulsion to enhance apoptotic activity in a549 lung cancer cells. *Pharmaceuticals*, 13(7), 152.

Md, S., Gan, S. Y., Haw, Y. H., Ho, C. L., Wong, S., & Choudhury, H. (2018). In vitro neuroprotective effects of naringenin nanoemulsion against β-amyloid toxicity through the regulation of amyloidogenesis and tau phosphorylation. *International Journal of Biological Macromolecules*, 118, 1211–9.

Meng, Q., Long, P., Zhou, J., Ho, C. T., Zou, X., Chen, B., & Zhang, L. (2019). Improved absorption of beta-carotene by encapsulation in an oil-in-water nanoemulsion containing tea polyphenols in the aqueous phase. *Food Research International*, 116, 731–73.

Mirzaei-Alamouti, H., Moradi, S., Shahalizadeh, Z., Razavian, M., Amanlou, H., Harkinezhad, T., Shahalizadeh, Z. (2016). Both monensin and plant extract alter ruminal fermentation in sheep but only monensin affects the expression of genes involved in acid-base transport of the ruminal epithelium. *Animal Feed Science and Technology*, 219, 132–43.

Moghimi, R., Aliahmadi, A., Rafati, H., Abtahi, H. R., Amini, S., & Feizabadi, M. M. (2018). Antibacterial and anti-biofilm activity of nanoemulsion of Thymus daenensis oil against multi-drug resistant Acinetobacter baumannii. *Journal of Molecular Liquids*, 265, 765–70.

Moradi, S., & Anarjan, N. (2019). Preparation and characterization of alpha-tocopherol nanocapsules based on gum Arabic-stabilized nanoemulsions. *Food Science and Biotechnology*, 28, 413–21.

Mori, Z, & Anarjan, N. (2018). Preparation and characterization of nanoemulsion based β-carotene hydrogels. *Journal of Food Science and Technology*, 55(12), 5014–24.

Mouhid, L., Corzo-Martínez, M., Torres, C., Vázquez, L., Reglero, G., Fornari, T., & Ramírez de Molina, A. (2017). Improving in vivo efficacy of bioactive molecules: an overview of potentially antitumor phytochemicals and currently available lipid-based delivery systems. *Journal of Oncology*. Article ID 7351976.

Mouritsen, O. G. (Ed.). (2005). Prologue: Lipidomics – a science beyond stamp collection. In: Life – as a matter of fat: The emerging science of lipidomics. Springer, Berlin/Heidelberg, 1–5.

Nagi, A., Iqbal, B., Kumar, S., Sharma, S., Ali, J., & Baboota, S. (2017). Quality by design based silymarin nanoemulsion for enhancement of oral bioavailability. *Journal of Drug Delivery Science and Technology*, 40, 35–44.

Nawaz, W., Zhou, Z., Deng, S., Ma, X., Ma, X., Li, C., & Shu, X. (2017). Therapeutic versatility of resveratrol derivatives. *Nutrients*, 9(11), 1188.

Ni, S., Hu, C., Sun, R., Zhao, G., & Xia, Q. (2017). Nanoemulsions-based delivery systems for encapsulation of quercetin: preparation, characterization, and cytotoxicity studies. *Journal of Food Process Engineering*, 40, e12374.

Nijveldt, R. J., Van Nood, E., Van Hoorn, D. E. C., Boelens, P. G., Van Norren, K., & Van Leeuwen, P. A. M. (2001). Flavonoids: a review of probable mechanisms of action and potential applications. *American Journal of Clinical Nutrition*, 74 (4), 418–25.

Nováková, E., & Moran, N. A. (2012). Diversification of genes for carotenoid biosynthesis in aphids following an ancient transfer from a fungus. *Molecular Biology and Evolution*, 29(1), 313–23.

Oca Avalos, J. M. M., Candal, R. J., & Herrera, M. L. (2017). Nanoemulsions: stability and physical properties. *Current Opinion in Food Science*, (16), 1–6.

Öztürk, B. (2017). Nanoemulsions for food fortification with lipophilic vitamins: production challenges, stability, and bioavailability. *European Journal of Lipid Science and Technology*, 119, 1500539.

Palafox-Carlos, H., Ayala-Zavala, J. F., & González-Aguilar, G. A. (2011). The role of dietary fiber in the bioaccessibility and bioavailability of fruit and vegetable antioxidants. *Journal of Food Science*, 76, R6–R15.

Park, H., Mun, S., & Kim, Y. R. (2019). UV and storage stability of retinol contained in oil-in-water nanoemulsions. *Food Chemistry*, 272, 404–10.

Pangeni, R., Sharma, S., Mustafa, G., Ali, J., & Baboota, S. (2014). Vitamin E loaded resveratrol nanoemulsion for brain targeting for the treatment of Parkinson's disease by reducing oxidative stress. *Nanotechnology*, 25(48), 485102.

Parashar, P., Rathor, M., Dwivedi, M., & Saraf, S. A. (2018). Hyaluronic acid decorated naringenin nanoparticles: Appraisal of chemopreventive and curative potential for lung cancer. *Pharmaceutics*, 10(1), 33.

Parveen, R., Baboota, S., Ali, J., Ahuja, A., Vasudev, S. S., & Ahmad, S. (2011). Oil based nanocarrier for improved oral delivery of silymarin: In vitro and in vivo studies. *International Journal of Pharmaceutics*, 413(1-2), 245–53.

Pathak, M. (2017). Nanoemulsions and their stability for enhancing functional properties of food ingredients. In: Oprea, A. E. and Grumezescu, A. M. (eds) *Nanotechnology Applications in Food: Flavor, Stability, Nutrition and Safety*. Academic Press, Cambridge, 87–106.

Patra, A. K., Park, T., Kim, M., & Yu, Z. (2017). Rumen methanogens and mitigation of methane emission by anti-methanogenic compounds and substances. *Journal of Animal Science and Biotechnology*, 8, 13.

Pool, H., Mendoza, S., Xiao, H., & McClements, D. J. (2013). Encapsulation and release of hydrophobic bioactive components in nanoemulsion-based delivery systems: impact of physical form on quercetin bioaccessibility. *Food and Function*, 4(1), 162–74.

Ren, G., Sun, Z., Wang, Z., Zheng, X., Xu, Z., & Sun, D. (2019). Nanoemulsion formation by the phase inversion temperature method using polyoxypropylene surfactants. *Journal of Colloid and Interface Science*, 540, 177–84.

Rizvi, S., Raza, S. T., Ahmed, F., Ahmad, A., Abbas, S., & Mahdi, F. (2014). The role of vitamin E in human health and some diseases. *Sultan Qaboos University Medical Journal*, 14(2), e157–e65.

Roby, M. H. H. (2017). Synthesis and characterization of phenolic lipids. In: Soto-Hernandez, M., Palma-Tenango, M., Garcia-Mateos, M. R. (eds) Ch. 4 – Phenolic compounds – natural sources, importance and applications. InTech Open, Rijeka.

Ronis, M. J. (2016). Effects of soy containing diet and isoflavones on cytochrome P450 enzyme expression and activity. *Drug metabolism reviews*, 48(3), 331–41.

Rosenfeld, L. (1997). Vitamine – vitamin. The early years of discovery. *Clinical Chemistry*, 43(4), 680–5.

Safaya, M., & Rotliwala, Y. C. (2019). Nanoemulsions: a review on low energy formulation methods, characterization, applications and optimization technique. *Materials Today: Proceedings*, 27(1), 454–9.

Sagalowicz, L., & Leser, M. E. (2010). Delivery systems for liquid food products. *Current Opinion in Colloid & Interface Science*, 15(1), 61–72.

Salehi, B., Venditti, A., Sharifi-Rad, M., Kręgiel, D., Sharifi-Rad, J., Durazzo, A., Lucarini, M., Santini, A., Souto, E. B., Novellino, E., Antolak, H., Azzini, E., Setzer, W. N., & Martins, N. (2019). The therapeutic potential of Apigenin. *International Journal of Molecular Sciences*, 20(6), 1305.

Salvia-Trujillo, L., Rojas-Graü, M. A., Soliva-Fortuny, R., & Martín-Belloso, O. (2015). Use of antimicrobial nanoemulsions as edible coatings: impact on safety and quality attributes of fresh-cut Fuji apples. *Postharvest Biology and Technology*, 105, 8–16.

Sanguansri, L., Oliver, C. M. & Leal-Calderon, F. (2013). Nanoemulsion technology for delivery of nutraceuticals and functional-food ingredients. In: Bagchi, D., Bagchi, M., Moriyama, H., and Shahidi, F. (eds) Bio-nanotechnology: A revolution in food, biomedical and health sciences. Blackwell Publishing Ltd., Oxford, UK.

Santos, D. T., & Meireles, M. A. A. (2010). Carotenoid pigments encapsulation: fundamentals, techniques and recent trends. *The Open Chemical Engineering Journal*, 4, 42–50.

Scalia, S., & Mezzena, M. (2009). Incorporation of quercetin in lipid microparticles: effect on photo- and chemical-stability. *Journal of Pharmaceutical and Biomedical Analysis*, 49(1), 90–4.

Schoener, A. L., Zhang, R., Lv, S., Weiss, J., & McClements, D. J. (2019). Fabrication of plant-based vitamin D3-fortified nanoemulsions: influence of carrier oil type on vitamin bioaccessibility. *Food & Function*, 10, 1826–35.

Sharma, S., Ali, A., Ali, J., Sahni, J. K., & Baboota, S. (2013). Rutin: therapeutic potential and recent advances in drug delivery. *Expert Opinion on Investigational Drugs*, 22(8), 1063–79.

Sharma, B., Iqbal, B., Kumar, S., Ali, J., & Baboota, S. (2019). Resveratrol-loaded nanoemulsion gel system to ameliorate UV-induced oxidative skin damage: from in vitro to in vivo investigation of antioxidant activity enhancement. *Archives of Dermatological Research*, 311(10), 773–93.

Sharma, A., Sharma, N. K., Srivastava, A., Kataria, A., Dubey, S., Sharma, S., & Kundu, B. (2018). Clove and lemongrass oil based non-ionic nanoemulsion for suppressing the growth of plant pathogenic Fusarium oxysporum f.sp. lycopersici. *Industrial Crops and Products*, 123, 353–62.

Shen, X., Fang, T., Zheng, J., & Guo, M. (2019). Physicochemical Properties and Cellular Uptake of Astaxanthin-Loaded Emulsions. *Molecules*, 24, 727.

Shrestha, H., Bala, R., & Arora, S. (2014). Lipid-based drug delivery systems. *Journal of Pharmaceutics*. Article ID 801820.

Silva, H. D., Cerqueira, M. Â., & Vicente, A. A. (2012). Nanoemulsions for food applications: development and characterization. *Food and Bioprocess Technology*, 5, 854–67.

Singh, Y., Meher, J. G., Raval, K., Khan, F. A., Chaurasia, M., Jain, N. K., & Chourasia, M. K. (2017). Nanoemulsion: concepts, development and applications in drug delivery. *Journal of Controlled Release*, 252, 28–49.

Siriwardhana, N., Kalupahana, N. S., Cekanova, M., LeMieux, M., Greer, B., & Moustaid-Moussa, N. (2013). Modulation of adipose tissue inflammation by bioactive food compounds. *Journal of Nutritional Biochemistry*, 24(4), 613–23.

Smeti, S., Hajji, H., Mekki, I., Mahouachi, M., & Atti, N. (2018). Effects of dose and administration form of rosemary essential oils on meat quality and fatty acid profile of lamb. *Small Ruminant Research*, 158, 62–68.

Soltan, Y. A., Natel, A. S., Araujo, R. C., Morsy, A. S., & Abdalla, A. L. (2018). Progressive adaptation of sheep to a microencapsulated blend of essential oils: ruminal fermentation, methane emission, nutrient digestibility, and microbial protein synthesis. *Animal Feed Science and Technology*, 237, 8–18.

Son, H. Y., Lee, M. S., Chang, E., Kim, S. Y., Kang, B., Ko, H., Kim, I. H., Zhong, Q., Jo, Y. H., Kim, C. T., & Kim, Y. (2019). Formulation and characterization of quercetin-loaded oil in water nanoemulsion and evaluation of hypocholesterolemic activity in rats. *Nutrients*, 11(2), 244.

Sotomayor-Gerding, D., Oomah, B. D., Acevedo, F., Morales, E., Bustamante, M., Shene, C., & Rubilar, M. (2016). High carotenoid bioaccessibility through linseed oil nanoemulsions with enhanced physical and oxidative stability. *Food Chemistry*, 199, 463–70.

Sugasini, D., & Lokesh, B. R. (2017). Curcumin and linseed oil co-delivered in phospholipid nanoemulsions enhances the levels of docosahexaenoic acid in serum and tissue lipids of rats. *Prostaglandins, Leukotrienes & Essential Fatty Acids*, 119, 45–52.

Tian, W.L., Lei, L. L., Zhang, Q., & Li, Y. (2016). Physical stability and antimicrobial activity of encapsulated cinnamaldehyde by self-emulsifying nanoemulsion. *Journal of Food Process Engineering*, 39, 462–71.

Vargas-Mendoza, N., Madrigal-Santillán, E., Morales-González, A., Esquivel-Soto, J., Esquivel-Chirino, C., García-Luna Y González-Rubio, M., Gayosso-de-Lucio, J. A., & Morales-González, J. A. (2014). Hepatoprotective effect of silymarin. *World Journal of Hepatology*, 6(3), 144–9.

Vázquez L., Corzo-Martínez M., Arranz-Martínez P., Barroso E., Reglero G., & Torres C. (2018). Bioactive lipids. In: Mérillon J. M., & Ramawat K. (eds) Bioactive molecules in food. Reference series in phytochemistry. Springer, Cham, 1–61.

Velikov, K. P., & Pelan, E. (2008). Colloidal delivery systems for micronutrients and nutraceuticals. *Soft Matter*, 4(10), 1964–80.

Wach, A., Pyrzyńska, K., & Biesaga, M. (2007). Quercetin content in some food and herbal samples. *Food Chemistry*, 100(2), 699–704.

Walia, A., Gupta, A. K. & Sharma, V. (2019). Role of bioactive compounds in human health. *Acta Scientific Medical Sciences*, 3(9), 25–33.

Wang, Z., Neves, M. A., Isoda, H., & Nakajima, M. (2016). Preparation and characterization of micro/nanoemulsions containing functional food components. *Japan Journal of Food Engineering*, 16(4), 263–76.

Wang, L., & Zhang, Y. (2017). Eugenol nanoemulsion stabilized with zein and sodium caseinate by self-assembly. *Journal of Agricultural and Food Chemistry*, 65, 2990–8.

Wildman, R. E. (2007). Nutraceuticals and functional foods. In: Wildman, R. E. (ed) Handbook of nutraceuticals and functional foods. CRC press, Boca Raton.

Xue, J., Davidson, P. M., & Zhong, Q. (2015). Antimicrobial activity of thyme oil co-nanoemulsified with sodium caseinate and lecithin. *International Journal of Food Microbiology*, 210, 1–8.

Yang, X., Tian, H., Ho, C. T., & Huang, Q. (2011). Inhibition of citral degradation by oil-in-water nanoemulsions combined with antioxidants. *Journal of Agricultural and Food Chemistry*, 59(11), 6113–9.

Yang, Y., Zhao, C., Tian, G., Lu, C., Li, C., Bao, Y., Tang, Z., McClements, D. J., Xiao, H., & Zheng, J. (2018). Characterization of physical properties and electronic sensory analyses of citrus oil-based nanoemulsions. *Food Research International*, 109, 149–58.

Yi, J., Zhang, Y., Liang, R., Zhong, F., & Ma, J. (2014). Beta-carotene chemical stability in nanoemulsions was improved by stabilized with beta-lactoglobulin–catechin conjugates through free radical method. *Journal of Agricultural and Food Chemistry*, 63, 297–303.

Yuan, Y., Gao, Y., Zhao, J., & Mao, L. (2008). Characterization and stability evaluation of β-carotene nanoemulsions prepared by high pressure homogenization under various emulsifying conditions. *Food Research International*, 41(1), 61–8.

Zhang, J., Liu, D., Huang, Y., Gao, Y., & Qian, S. (2012). Biopharmaceutics classification and intestinal absorption study of apigenin. *International Journal of Pharmaceutics*, 436(1-2), 311–7.

Zhang, L., Song, L., Zhang, P., Liu, T., Zhou, L., Yang, G., Lin, R., & Zhang, J. (2015). Solubilities of naringin and naringenin in different solvents and dissociation constants of naringenin. *Journal of Chemical and Engineering Data*, 60(3), 932–40.

Zhang, Y., Yang, Y., Tang, K., Hu, X., & Zou, G. (2008). Physicochemical characterization and antioxidant activity of quercetin-loaded chitosan nanoparticles. *Journal of Applied Polymer Science*, 107(2), 891–7.

Zhang, L., Zhang, F., Fan, Z., Liu, B., Liu, C., & Meng, X. (2019). DHA and EPA nanoemulsions prepared by the low-energy emulsification method: process factors influencing droplet size and physicochemical stability. *Food Research International*, 121, 359–66.

Zheng, B., Peng, S., Zhang, X., & McClements, D. J. (2018). Impact of delivery system type on curcumin bioaccessibility: comparison of curcumin-loaded nanoemulsions with commercial curcumin supplements. *Journal of Agricultural and Food Chemistry*, 66, 10816–26.

Zupančič, Š., Lavrič, Z., & Kristl, J. (2015). Stability and solubility of trans-resveratrol are strongly influenced by pH and temperature. *European Journal of Pharmaceutics and Biopharmaceutics*, 93, 197–204.

CHAPTER 14

Nano-Food Safety and Regulatory Perspectives

Faraat Ali

Department of Inspection and Enforcement, Laboratory Services,
Botswana Medicines Regulatory Authority, Gaborone, Botswana

CONTENTS

14.1 INTRODUCTION

Nanotechnology now has tremendous potential that can be used in food applications to strengthen consistency and physicochemical characteristics, enhance consistency and life span (shelf life), expand the nutritional quality of food products, enhance the bioavailability of nutritional supplements and impart safety at the contact surface through the use of new packaging materials (Jafari and Esfanjani, 2017a). Because of its relatively broad surface area to mass ratio, the particular designable physical and chemical properties of nanomaterials (NMs) have numerous functionalities in applications.

Food organic NMs produced from protein, and lipids, such as nanoemulsions (NEs), nanotubes (NTs), nanoparticles (NPs), nanocarriers (NCs), nanocapsules (NCPs), and inorganic NPs, along with silver, gold, clay, silicon dioxide, and carbon NTs are the numerous components of the nanoscale created by different pathways and used for specific food industry applications. The most extensive nanotechnology applications in food are the production of NCs for food, reinforcement and functional packaging, and the health promotion of sustained delivery of nutraceutical products (Acosta, 2009).

A growing number of food materials containing nanoscale constituents and vehicles have already been placed in the market (Bouwmeester et al., 2014; Chaudhry et al., 2008). Various food products, packaging ingredients, food supplements, contact devices, and pediatric foods containing engineered NPs have been documented in reported literature (Sohal et al., 2018). European Union inventories acknowledged NPs containing foods were declared as well. The consistently rising market value of the food industry's nano-technology products is projected at about US\$14.8 billion by 2020 (Naseer et al., 2018).

Novel food formulation developed by using the tools of nanotechnology offers several values to consumers and industry. Moreover, when used ineffectively in food products, various NMs varying from organic to metal inorganic substances can familiarize some inconsistencies and unforeseeable effects.

The possible hazards to health and the environment resulting from the highest levels of exposure and extensive use result in significant community and eco-friendly concerns (Amenta et al., 2015). Such aspects must be considered when developing laws and regulations for general safety and use of food products based on nanotechnology. The extensive use of nanotechnology-based materials results in a high potential for NPs being exposed across various routes. There have been three main paths to NMs human exposure (Jafari and Esfanjani, 2017a). Mostly in case of contact with cosmetics and medications containing NPs, these are dermal penetration occurring through the skin; inhalation occurring through volatile mixtures containing NPs; and absorption taking place through the absorption of NPs enhanced food content.

These ultrafine particles will enter the body in all three conditions, pass through the cell barriers and penetrate tissues and organs, and hence could be deposited in different parts of the body. The body's reaction and potential consequences resulting from exposure will depend greatly on the concept and physiochemical properties of the NPs (Wani et al., 2018). Organic NPs focused on carbohydrates, proteins, lipids, and vitamins are deemed digestible, but inorganic NPs based on metal chemicals like silver, titanium, and silica are assumed indigestible. The extant research contains many scientific papers in vitro and in vivo that track possible health implications due to a prolonged degradation of nano ingredients in agri-food products.

Countries across the world, pay strong attention to regulating the effective and safe production and supervision of NMs for use in the food industry (Jafari and McClements, 2017b). Regulations, legislation, exemptions, recommendations, and a guidance document presented by legal food controlling authorities are important tools to be considered in the assessment of potential risks and safety guidelines of nanotechnology used in food regulation (Amenta et al., 2015). Standard regulatory guidelines and procedures indicate the identification of NMs by their size, surface textures, chemical characteristics, and stability, which are very important in assessing their possible interactions and persistence within the body, thereby providing data for risk assessment (Figure 14.1).

Specific analytical methodologies accompanied by theoretical techniques are commonly used for the detection and identification of NPs in foods to determine their potential threats and hazards. Also, *in vitro* and *in vivo* studies provide very significant evidence to establish potential responses to NPs in the body. Inconsistencies and inadequate awareness about the effects of NPs embedded in foods contributing to health and safety concerns could be removed globally through defined regulations and legislation. This would enhance the transparency in the manufacture and distribution of nanofood materials, thereby promoting public acceptance. Possible routes of exposure, toxicological effects, and regulations, and recommendations for risk assessment and safety evaluation are discussed in the proposed chapter risk assessment.

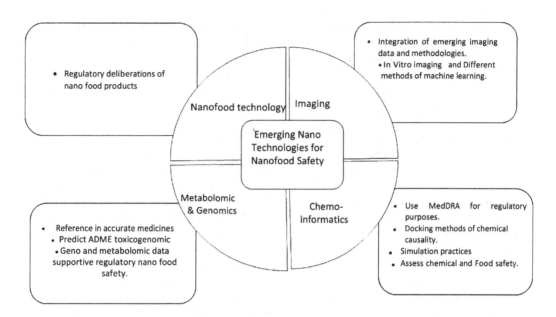

FIGURE 14.1   Represented techniques and methods for the standardization of nanofood products and hazards assessment.

Courtesy W. Slikker et al., 2018

14.2 WORLDWIDE STATUS OF NANOFOOD MARKET

Nanofood is a cultivated, produced, processed, or packaged product that uses nano-technology or that harbors NMs to strengthen its characteristics (Joseph and Morrison, 2006). Nanotechnology enables the beneficial modification of food materials to enhance physicochemical characteristics, extend shelf life, assess sustainable safety and quality, and improve health. The demands and preferences of consumers have directed producers to produce novel foods with attractive features. Because of these successful impacts, many food manufacturers are very keen to produce food by using this innovative technology.

Worldwide, many major firms have been investigating nanotechnology applications for food production, packaging, and processing, including Heinz, Nestle, Unilever, and Kraft (Nano 2030, 2014). There are also many food products available in the market which contain organic and inorganic NPs. Several European Union inventories have reported lists of items claiming to contain NPs already in the market (ANEC; BEUC; Inventory of Consumer Goods, 2019; Nanodatabase, 2019; Nanoproducts, 2013; Nanotech-data, 2013). Such databases show industrial NPs food items that are found in the food itself and embedded into packaging products. Bouwmeester et al. (2014) confirmed that 140 food-related items were identified with the assertion of harboring numerous NPs based on online database inventories. Different products (farming, dietary supplements, probi-otics, packaging NMs), categories, and producers are also separately mentioned in the product list. More than 500 food-related products have been reported to be placed on the market recently (Nanodatabase, 2019; Nanotechnology Project, 2019).

Nanofoods are composed of organic NPs such as nanoliposomes (NLs), micelles, and NCs for controlled nutraceutical delivery as well as inorganic materials, such as titanium, silica, zinc, clay, and silver (Bazana et al., 2019; Park et al., 2006). Because of their physical and chemical properties, those NPs are used for specific applications. For example, clay NPs are among the most industrially used NMs with their excellent barrier properties in packaging materials (De Paiva et al., 2008). NPs of silver and titanium dioxide are incorporated into the packaging material for food preservation due to their antimicrobial potential (Bouwmeester et al., 2014).

According to Bumbudsanpharoke and Ko, the food and beverage industry is showing a constantly increasing trend in the use of NMs for processing and packaging specific purposes (Bumbudsanpharoke and Ko, 2015). In specific the food and beverage industry has preferred packaging materials that are functionalized with NPs. According to market analysis estimates, the global sales volume of NMs in the food industry in 2013 was around $6.6 billion, increased by 13.0% per annum, and is estimated to reach about $19.90 billion by 2020 (Naseer et al., 2018; Persistence Market Research, 2014). In 2020, the European Institute for Health and Consumer Protection also announced the projected size of the demand for food packaging using NMs to be about $20 billion (Bumbudsanpharoke and Ko, 2015).

Manufacturing companies, consumers, and producers are all beneficiaries involved in the nano market for the agri-food sector. Such new products or preservatives, in addition to their beneficial service in food processing, packaging, and nutrient delivery, can also pose considerable risks to the environment and human health. Distinctive physicochemical characteristics, a large surface area compared to the tiny size and common use allow them accessibility to uncertain hazardous implications for overdose exposures. All of these issues will trigger public nanofood safety concerns and affect public priorities. Additionally, limited regulatory methodologies for safety assessment may prevent investors from engaging with the nanotechnology business sector. Safety confirmation of nano-enabled food products to be performed through global legislation by taking into account excellently-defined NMs, dose limits for exposure, potential hazards, and therefore informing the community would be effective in reducing community suspicions and supporting the industry. Risk and hazard assessment will be discussed in the following sections, by considering various critical aspects.

14.3 RISK ASSESSMENT OF NANOSTRUCTURE FOOD

Nanotechnological developments in the agri-food sector is growing tremendously. A huge concern in the facility provided by this application in detection of pesticides, microbe monitoring, quality food maintenance, and monitoring, increasing shelf life, and enhancing nutritional content has led to the introduction of several nanofood products into the market (Dasgupta et al., 2015). The optimal characteristics can be used with all food-based organic particles or other inorganic particles. It can result in unregulated exposure to those NPs by humans, animals, and the environment. Some can cross biological barriers inside the human body and can penetrate vessels, tissues, and cells (Bajpai et al., 2018; Su and Li, 2004). With excess exposure or release, inorganic NPs such as carbon nanotubes, silver, silica, titanium, and zinc NPs may present unpredictable and hazardous safety issues for health and the environment. Some evidence suggests NPs being recorded in organs such as the liver, lungs, spleen, heart, and brain (Kreyling et al., 2002; Oberdörster et al., 2002).

Due to the unique characteristics of the different NMs used, the implementation of nanotechnology in the food industry can lead to potential risks. They are relatively small in size and it can be engineered at the cellular level, likely exhibiting various innovative physicochemical and biological characterization. Given that this topic is specifically linked to human nutrition, health issues and debates are currently being addressed about how to determine the protection of NPs foods. Relevant authorities in Europe and America have developed a risk assessment roadmap and suggested safety evaluation guidance for the use of NPs in foodstuffs (EFSA, 2011a, b, 2018; FDA, 2011, 2012).

Potential risks and toxicity problems arising from inhalation and ingestion of such tiny NPs will lead to issues and these should be explained and removed. Due to the increasing exposure of such NMs, specifically inorganic chemicals can be accumulated in tissues, organs, and this can be correlated with various health problems. Thus, manufacturers and suppliers should properly follow the risk assessment framework defined by international agencies (FDA, ESFA) (Bajpai et al., 2018; Shi et al., 2013; Yu et al., 2012). Production of organic and sustainable food should be strictly protected while making people aware of NMs and nanofoods, health problems, and health sustainability.

NMs risk assessment guidance is provided (OECD, 2013; ESFA, 2011, 2018). However, there are methods of research for hazard evaluation, classification, optimization, and verification of methods that need to be further developed to better identify possible risks (Amenta et al., 2015; OECD, 2013; SCENIHR, 2007). According to the European food safety document (2011a, b) presenting "Risk Assessment Guidelines for the Application of Nanoscience and Nanotechnology in the Food and Feed Chain," physicochemical characterization and identification of NMs hazards were achieved, and potential risks were identified because of the application of nanotechnology in foods.

For the nanotoxicity assessment, multiple parameters are proposed as follows: exposure assessment, toxicity assessment, distribution commitment, persistence and transformation capabilities, recycled content, and NMs sustainability (Chau et al., 2007; Dreher, 2004). Possible exposure mechanisms and effective research methods are highlighted to define and describe the NMs risks. As mentioned previously, post-ingestion or inhalation transformation of NPs can alter their surface properties and stability (Amenta et al., 2015). Consequently, NPs that are transported through the gastrointestinal tract or circulatory system produce different characteristics. *In vitro* studies performed through simulated body conditions need to be checked thoroughly during *in vivo* experiments to track NMs reaction to changing conditions.

Because of their organic and inorganic composition, the associated risk threshold of NPs blended in foods can vary. Organic NPs based on carbohydrates, proteins, and lipids are often associated with enhanced absorption of such NPs, probably resulting in allergic reactions. Additionally, inorganic NPs may have more severe effects, such as organ deposition and chronic heart disease.

14.3.1 Analytical Characterization of NMs in Foods

Physical and chemical properties such as particle shape, size, accumulation and cluster condition, stability, molecular configuration, surface charging, penetrability, and structural characteristics are deemed important for evaluating the toxic potential and hazardous quality of NMs (Oberdorster et al., 2005). To fulfill the regulatory requirements an array of procedures is crucial. Size and morphology investigation, chemical properties, surface composition, operational features, and accessibility offer data to assess suitability for safe ingesting of NMs holding foods. Regulatory definitions, particularly related to size implement the usage of NPs with the lowest limit of detection (LLOD). Whenever the particle size decreases, so does the sensitivity decrease (Bouwmeester et al., 2014).

Globally, regulatory agencies including WHO, EC, EFSA, the FDA, and EPA have implemented standards for the safety evaluation of NPs. Method development and validation of analytical procedures for evaluating the compliance of NMs with the regulations are referenced. Robustness, ruggedness, repeatability, and reproducibility of the analytical methods used are critical for an adequate safety assessment of NPs.

Nanostructures in foods are not so easy to detect as they are complicated matrix multiplication and interruptions are indeed very feasible. Consequently, sample treatment is essential before the analysis. The sample with the least treatment must be separated from the sample matrix first. The targeted nanostructure must be isolated from other elements of the matrix before detection.

Specific analytical techniques, especially spectroscopy and chromatography, are commonly used for detection, isolation, quantification, and detailed analysis. High-resolution microscopy (HRM) such as non-field optical random mapping is also an additional tool for the investigation of NMs. Using these approaches, organic NPs, particularly carbohydrates, proteins and lipids, and inorganics, namely metal NPs, used in products can be detected both quantitatively and qualitatively (Jafari and McClements, 2017b).

Electron Microscopy (EM) is widely used to examine macroscopically properties such as size, shape, texture, and external coarseness of the NMs in foods. Energy dispersive X-ray spectroscopy (EDX), combined with EM, also allows the determination of the NPs elemental percentage composition (Peters et al., 2012). A dynamic light scattering (DLS) is the most commonly used spectrophotometric technique for evaluating hydrodynamic diameter size and size distribution (Filipe et al., 2010; Graveland-Bikker et al., 2004). Many instrumental methods for measuring NPs in foods are spectrometry and chromatography. Combined with different detectors, hydrodynamic chromatography (HDC) are efficient in separating NPs since they provide size-based assessment analysis (Peters et al., 2011; Von der Kammer et al., 2011). Such methodology allows for significantly excellent sensitivity, specificity, and reproducibility, with a lower limit of detection (DL). Restricted treatment of a sample leads to extremely trusted data before review. ICP-OES and ICP-MS are also recommended for use per an ng/L range detection and quantitation edge (EC, 2011) when detecting NPs.

14.3.2 Routes of Exposure to Nano Food Products

Food nano-ingredients can reach the body through tissues and organs, due to their smaller size. Three essential methods of exposure to these NPs are dermal touch, inhalation, and ingestion. Of course, there may be several food ingredients in the nanoscale, but they are not NMs specifically designed. Also, when related foods are consumed their intake amounts can be deemed acceptable. The consumption of a particular nanoingredient-enriched food product, however, might still result in exposures, an allergic reaction, and toxic effects. Because there is a limited understanding of possible nanostructure exposure risks, management of that should be better controlled. Relevant authorities such as the EC and FDA are publishing guidelines and standards for NMs handling and use. The three main mechanisms of exposure referenced above are discussed in the following sections.

14.4 REGULATIONS IN USING NMS FOR FOOD PRODUCTS

Worldwide initiatives are underway to govern the manufacturing and safe use of NPs through legislation, recommendations, and guidance in agriculture and the food processing industry. Production of food materials using NPs should meet regulatory or guideline procedures. A potential health implication with the embedded NPs in food requires production control procedures to ensure food safety standards are met. Regulatory structures with the capacity to manage risks are important for the application of nanotechnology approaches in the food industry.

Special criteria for the test series and specifications were stated based on EU regulations for nanotechnology-related food applications (ESFA, 2011). Throughout the United States, the FDA and EPA have introduced legislation for NMs used for food processing and food packaging.

NMs are under examination for health threats and hazards. Consumers need to be shielded from harmful exposure threats by legal legislation since they are not yet fully explained. As per EFSA, products to be used in marketing schemes should not potentially affect human health and should not be deceptive to the customer (De Jong et al., 2005). The European and FDA regulations state the maximum permitted levels of NPs in foods (Chaudhry et al., 2008).

Some other problem relating to NMs control is the economic criteria in the use of nanotechnology for commercial products. Manufacturing of appropriate good packaging material using NMs, for example, involves additional costs. It has a relatively small profit margin and thus it is difficult to meet regulatory criteria (Mihindukulasuriya and Lim, 2014). The EU-derived labeling requirements should be suggested for NMs found in food content (European Parliament and Council, 2011, 2012). Once used in food applications, inorganic NPs need to be given more consideration when considering the potential toxicity. Carbon black, titanium nitrate (TiN), and silicon dioxide nanocomposites are approved for use in packaged foods as per EU legislation. Moreover, NPs of silver, aluminum, zinc oxide, and clay were not accepted (Bumbudsanpharoke and Ko, 2017).

Regulatory authorities in the USA and Europe have established NMs management strategies and guidelines in foods (EFSA, 2011a, b; FDA, 2014a, b). Little scientific data and ambiguity are difficulties in handling FDA approval of food ingredients for NPs (McClements et al., 2016). The FDA has identified the required evaluation and adequate testing of new NMs. Measuring physical and chemical properties prior, throughout, then after exposure to dose-response toxicity tests can aid in assessing NMs transformation and characteristics. As per EFSA, to assess the survival of these NPs in the gastrointestinal tract, the physical and chemical properties of NPs and their retention throughout the food matrix and in the body after ingestion must be assessed. The major NPs-related concerns are the effects of these fine particles on gastrointestinal system dynamics and reactions, varying absorption profiles through the digestive tract, unregulated antimicrobial activity, and translocation through organs and tissues (McClements et al., 2016).

Regulatory guidelines are concentrated on specific concerns and the fundamental aspects. These are mentioned in the subsection below.

14.4.1 Regulatory Prospects of NMs in Food

Regulatory guidelines establish safe pathways for producers, importers, and consumers to safeguard the quality of nanofood products in the market (Amenta et al., 2015). Health risks and health issues must be addressed separately during the production, refining, packaging, and use processes of NPs-integrated food products. Unfortunately, the legislative aspects of NPs in food-related areas do not even have specific legislation applicable globally (Bajpai et al., 2018). Also, most countries have no safety assessment regulations for nanoproducts at all. The critical aspects restricting the market potential of nanoproducts developed by rapidly evolving technology are inadequate awareness of risk assessment for human health and climate, and lack of regulations for proper handling (Bazana et al., 2019).

Until authorization for use, food products and additives produced by nanotechnology must undergo safety review (Rasmussen et al., 2019). The FDA has adopted recommendations for the safety assessment (Pathakoti et al., 2017). Amenta et al. (2015) addressed extensively EU and US laws governing the safe handling of NMs from foods. NMs need a description to define and differentiate them from the others and EU legislation and policies have provided guidelines for a NMs description (European Commission (EC), 2011). The EC description using the range from 1 to 100 nm and serves as a reference broadly applied for defining the particle size only by legislative units. This refers to external an aspect of individual NPs in the form of aggregates or agglomerates standing unbound or existing (Amenta et al., 2015; Rauscher et al., 2014; Roebben et al., 2014). For evaluating the size/size distribution of NMs, it is recommended that correct validated methods be combined (Linsinger et al., 2012).

EC guidelines specify the use of NMs as novel foods or new food ingredients in food production, particularly newly developed, creative foods (Amenta et al., 2015). The "Novel Food Regulations," an updated version of the previous regulations, regulate the production and processing of vitamins, minerals, and other materials containing engineered NPs (EC, 2013). The EC Regulations also introduce synthetic inorganic compounds such as silica (E551) and titanium dioxide (TiO2; E171), applied to food to enhance such technical features of foods (Amenta et al., 2015; European Parliament and Council, 2002a, 2008b, c, d, e). Nanoscale types of vitamins and minerals need safety evaluations in comparison to their macroscale types due to potential differences in nutritional value and bioavailability (EC, 2013). Permission is needed before the new products and food ingredients are sold based on a risk evaluation to ensure healthy ingredients are used in products. The evaluation will also meet new safety assessment standards due to the changing features during the manufacturing or processing of NMs. Concerning inorganic NPs commonly used as food additives, calcium carbonate ($CaCO_3$; E170) and silicon dioxide (SiO_2; E551), titanium dioxide (TiO_2; E171), silver (E174), and gold (E175) were investigated to determine their possible safety and toxicity hazards (Amenta et al., 2015).

Regulations associated with food contact surfaces, such as active smart packaging materials, plastics, ceramics, and adhesives, relate to common health evaluation and authorization concepts and measurements (EC, 2009; European Parliament and Council, 2011).

14.4.2 Recent Legislation Related to Nanofood

Regulatory policies and guidelines for the proper handling and use of NMs were already introduced worldwide throughout the agri-food industry. Several countries adopt different approaches for assessing the safety of nanoparticulated food materials for different applications. They also can check their regulatory processes on applications of nanotechnology throughout the food industry.

14.4.2.1 Regulated Market

In the food regulated market, throughout the United States, the FDA is the major regulatory agency responsible for overseeing recent innovations throughout food applications. The EPA has also announced a NMs research strategy and established typical NMs that are used in the food industry, choosing 6 of them to be evaluated. The OECD (USEPA, 2013) has rendered another effort to investigate the potential effects of NMs like nanosilver and nanotitania as packaged food products for human and ecological use (Amenta et al., 2015). The FDA recently published guidance that provides manufacturers with guidelines on potential effects, safety risks, and health effects arising through the use of NMs in FDA-approved materials. Another report, "Guidance for Industry: Assessing the Properties of Important Process Changes in Manufacturing, including developing technologies, on Health and Regulatory Status" the FDA developed in 2014 (USFDA, 2014b) for food compositions and substances, with food products that are color additives. The FDA proposes comprehensively evaluating the toxicological characteristics of food products by referring to the CFR 21 (USFDA, 2011).

European Commission has introduced deliberate action plans and officialdoms for applications in nanoscience and nanotechnology. "Risk evaluation guidelines for the application of nanoscience and nanotechnology in the food and feed chain" released by EFSA reported the possible risks of NPs used in industrial goods. The need for appropriate and validated testing methodologies and protocol to provide a consistent identification and characterization of NMs has helped researchers perform additional testing and evaluation to remove qualms (EC, 2010a, b; FAO/WHO, 2013). EU law defines the amount of NPs guided or incorporated into packaging materials to be used in food. The findings of migration analysis and potential health risks should be presented for packaging applications (EC, 2004, 2007). Altogether food constituents in the type of engineered NMs must be indicated on the food product label according to EC (2013). The names of such products in brackets (FAO/WHO, 2013) must be followed by the word "nano."

In Canada, CFIA and PHAC competent agencies are responsible for the regulation of food NMs and safety. Food NMs comes under the title of nanospecific regulations are threatened by present legislation and regulatory landscape, no stringent legislation for the food NMs. The current governmental frameworks seek to reduce food NMs safety risks (Takeuchi et al., 2014).

FSANZ is the authority that regulates the safe management and use of food nano products developed by using new technologies in New Zealand and Australia. This authority outlines legislative methods for NMs safety. Based on the survey results highlighting inadequate criteria for the coding of food packaging combined with NPs, FSANZ agreed to conduct future migration risk assessments (FAO/WHO, 2013; FSANZ, 2014; Tager, 2014).

14.4.2.2 Semi-Regulated Market

In semi regulated countries such as Turkey, Switzerland, and Russia, the safety of NMs is measured with current regulations. NMs with different characteristics relating to structure, size, shape, and functionalization must be recorded before application. These are required to submit by the Swiss Rule of 2010 for registration (Gesamte Rechtsvorschrift für Pflanzenschutzmittelverordnung 2011). "The Russian Federation's Federal Service for the Monitoring of Consumer Rights Safety and Human Well-being Office" is the agency that is responsible for the use of nanotechnology in industry. The Russian Federation 's chief state health officer has developed and accepted a range of principles and methods for risk assessment and a safety assessment of NMs.

So many Asian countries, such as India, Iran, have established nanoscale reference and qualification systems. The OECD Working Party on Manufactured NMs participates with Japan and Korea (Amenta et al., 2015). There is no complete regulation for NMs existing in Japan, so guidelines are done by prevailing legislation. In Japan over the last decade, development into nanotechnology has been emphasized and given priority. Government funding has been given for research projects that concentrate on the carcinogenicity of NMs (Bumbudsanpharoke and Ko, 2015). According to a 2010 study, the development of basic procedures and approaches for the safety evaluation of NMs in the event of health risks emerging was required (FAO/WHO, 2013).

In 2011, South Korea adopted guidelines on the safety administration of NMs. The published document aimed at guaranteeing producers and consumers the protection and advantages of NMs, enhancing public understanding of NMs, and fostering supportable industrial growth (Bumbudsanpharoke and Ko, 2015). Other Korean government startups involving setting up a record for an efficacy and safety evaluation of NMs with a structured method to generating a framework for professional administration for entire shareholders. The South Korea Ministry of Food and Drug Health (MFDS) has introduced new guidance and safety standards on the use of NMs for food packaging applications (Hwang et al., 2012). Some other Asian countries, Indonesia and Malaysia, currently lack clear risk evaluation assessment and safety regulations for NMs (FAO/WHO, 2013; Takeuchi et al., 2014).

14.4.2.3 Non-Regulated Market

In the non-regulated market including African countries such as South Africa, Ghana, Botswana, Nigeria, Tanzania, Egypt, Zimbabwe, Namibia, Kenya, and Ethiopia there is no such regulation for the safety concern of NMs in food.

14.5 CONCLUSION

Nano food materials loaded with organic and inorganic NPs can deliver unique characteristics that benefit producers, suppliers, and consumers. In addition to the favorable technical influences of NMs in the food industry, there are also apprehensions about the unexpected adverse effects, environmental effects, and risks. Potential consequences of exposure, absorption, and deposition to these innovative substances involve extensive inspection of food products based on nanotechnology. Legislative policy for the safe handling and use of nanotechnology, as well as future risk assessments, has been proposed internationally. Up-to-date recognized agencies from the USA and EU have adopted a range of regulatory and guidance documents that are useful for guiding NMs in the industry.

The NMs market today is around hit $21 billion per annum. A variety of innovative agri-food products based on nanotechnology have been offered in the market and are currently showing a growing trend. Therefore, more and more attention is focusing on regulatory mechanisms to regulate the development of NMs and the possible risks. Precautions must be taken to ensure a high degree of protection for human health and the environment while offering beneficial features and functionality and global promotion for the production of novel food items (Amenta et al., 2015). This chapter discussed possible health hazards, exposure paths, and regulatory guidelines and regulations for nanotechnology-based food products.

Nanotechnology's most popular uses are seen in packaged foods containing inorganics and using organic food supplements (Bouwmeester et al., 2014). Nanotechnology provides health benefits through the regulated delivery of bioactive compounds and enhanced bioavailability. Moreover, determining the safety and toxicity risks of such novel materials needs extensive testing, including in vitro and in vivo studies, and encouraging and supporting stable marketing and use by global legislation (Bazana et al., 2019).

For the safety evaluation of nanofoods and straightforwardly maintained advertising, a very precise world regulatory approach should be established. The federal government is required to develop rules for regulating nano-enabled food items, in particular with global cooperation, to develop international security frameworks for detecting NPs in imported foodstuffs (Bumbudsanpharoke, 2017). Every nation should also adopt regulatory guidelines intended to protect health and the environment from unauthorized use of NMs in food items. The absence of information on safety and environmental consequences and possible threats of unknown toxicity levels affect consumer acuity and receipt for the feasting of food products containing NPs.

BIBLIOGRAPHY

Acosta, E. (2009). Bioavailability of nanoparticles in nutrient and nutraceutical delivery. Current Opinion in Colloid & Interface Science, 14, 3–15.

Akhavan, S., Assadpour, E., Katouzian, I., & Jafari, S. M. (2018). Lipid nano scale cargos for the protection and delivery of food bioactive ingredients and nutraceuticals. Trends in Food Science & Technology, 74, 132–146.

Amenta, V., Aschberger, K., Arena, M., Bouwmeester, H., Moniz, F. B., Brandhoff, P., Peters, R. J. (2015). Regulatory aspects of nanotechnology in the agri/feed/food sector in EU and non-EU countries. Regulatory Toxicology and Pharmacology, 73, 463–476.

Ammendolia, M. G., Iosi, F., Maranghi, F., Tassinari, R., Cubadda, F., Aureli, F., ... De Berardis, B. (2017). Short-term oral exposure to low doses of nano-sized TiO2 and potential modulatory effects on intestinal cells. Food and Chemical Toxicology, 102, 63–75.

ANEC. https://www.anec.eu/

Bacchetta, R., et al. (2014). Evidence and uptake routes for zinc oxide nanoparticles through the gastrointestinal barrier in Xenopus laevis. Nanotoxicology, 8, 728–744.

Bajpai, V. K., Kamle, M., Shukla, S., Mahato, D. K., Chandra, P., Hwang, S. K., ... Han, Y. K. (2018). Prospects of using nanotechnology for food preservation, safety, and security. Journal of Food Drug Analysis, 26, 1201–1214.

Ball, G. F. M. (2006). Vitamins in foods: Analysis, bioavailability, and stability. CRC Press.

Bazana, M. T., Codevilla, C. F., & de Menezes, C. R. (2019). Nanoencapsulation of bioactive compounds: Challenges and perspectives. Current Opinion in Food Science, 26, 47–56.

Bettini, S., Boutet-Robinet, E., Cartier, C., Come´ra, C., Gaultier, E., Dupuy, J., ... Houdeau, E. (2017). Food-grade TiO2 impairs intestinal and systemic immune homeostasis, initiates preneoplastic lesions and promotes aberrant crypt development in the rat colon. Nature Scientific Reports, 7, 40373. Available from https://doi.org/10.1038/srep40373.

BEUC. https://www.beuc.eu/. Bouwmeester, H., Brandhoff, P., Marvin, H. J. P., Weigel, S., & Peters, R. J. B. (2014). State of the safety assessment and current use of nanomaterials in food and food production. Trends in Food Science & Technology, 40, 200–210.

Bouwmeester, H., Dekkers, S., Noordam, M., Hagens, W., Bulder, A., de Heer, C., et al., (2007). Health impact of nanotechnologies in food production, Report 2007.014, RIKILT—Institute of Food Safety, NL.

Brown, D. M., Stone, V., Findlay, P., MacNee, W., & Donaldson, K. (2000). Increased inflammation and intracellular calcium caused by ultrafine carbon black is independent of transition metals or other soluble components. Occupational and Environmental Medicine, 57, 685–691.

Bumbudsanpharoke, N., & Ko, S. (2015). Nano-food packaging: An overview of market. Migration Research, and Safety Regulations Journal of Food Science, 80(5), 910–923.

Buzea, C., Blandino, I. I. P., & Robbie, K. (2007). Nanomaterials and nanoparticles: Sources and toxicity. Biointerphases, 2, MR17–MR172.

Cao, X., Han, Y., Li, F., Li, Z., McClements, D. J., He, L., et al. (2019). Impact of protein nanoparticle interactions on gastrointestinal fate of ingested nanoparticles: Not just simple protein corona effects. NanoImpact, 13, 37–43.

Cele, L. M., Ray, S. S., & Coville, N. J. (2009). Nanoscience and nanotechnology in South Africa. South African Journal of Science, 105, 7–8.

Cha, K., et al. (2008). Comparison of acute responses of mice livers to short-term exposure to nano-sized or micro-sized silver particles. Biotechnology Letters, 30, 1893–1899.

Chau, C.-F., Wu, S.-H., & Yen, G.-C. (2007). The development of regulations for food nanotechnology. Trends in Food Science & Technology, 18, 269–280.

Chaudhry, Q., Scotter, M., Blackburn, J., Ross, B., Boxall, A., Castle, L., ... Watkins, R. (2008). Applications and implications of nanotechnologies for the food sector. Food Additives and Contaminants, 25(3), 241–258.

Chen, L. Y., Remondetto, G. E., & Subirade, M. (2006). Food protein-based materials as nutraceutical delivery systems. Trends in Food Science & Technology, 17, 272–283.

Chen, H., Zhao, R., Wang, B., Cai, C., Zheng, L., Wang, H., ... Feng, W. (2017). The effects of orally administered Ag, TiO2 and SiO2 nanoparticles on gut microbiota composition and colitis induction in mice. Nanoimpact, 8, 80–88.

Cho, W. S., Kang, B. C., Lee, J. K., Jeong, J., Che, J. H., & Seok, S. H. (2013). Comparative absorption, distribution, and excretion of titanium dioxide and zinc oxide nanoparticles after repeated oral administration. Particle and Fibre Toxicology, 10(2013), 9.

Cockburn, A., Bradford, R., Buck, N., Constable, A., Edwards, G., Haber, B., et al. (2012). Approaches to the safety assessment of engineered nanomaterials (ENM) in food. Food and Chemical Toxicology, 50, 2224–2242.

Cooper, T. A. (2013). Developments in plastic materials and recycling systems for packaging food, beverages and other fast-moving consumer goods. In N. Farmer (Ed.), Trends in packaging of food, beverages and other fast-moving consumer goods (FMCG): Markets, materials and technologies (p. 79). Philadelphia, PA: Woodhead Publishing Limited.

Cushen, M., Kerry, J., Morris, M., Cruz-Romero, M., & Cummins, E. (2012). Nanotechnologies in the food industry-recent developments, risks and regulation. Trends in Food Science & Technology, 24(1), 30–46.

Cushen, M., Kerry, J., Morris, M., Cruz-Romero, M., & Cummins, E. (2013). Migration and exposure assessment of silver from a PVC nanocomposite. Food Chemistry, 139(1-4), 389–397.

Das, M., Saxena, N., & Dwivedi, P. D. (2008). Emerging trends of nanoparticles application in food technology: Safety paradigms. Nanotoxicology, 3, 10–18.

Dasgupta, N., Ranjan, S., Mundekkad, D., Ramalingam, C., Shanker, R., & Kumar, A. (2015). Nanotechnology in agro-food: From field to plate. Food Research International, 69, 381–400.

De Jong, A. R., Boumans, H., Slaghek, T., Van Veen, J., Rijk, R., & Van Zandvoort, M. (2005). Active and intelligent packaging for food: Is it the future? Food Additives and Contaminants, 22, 975–979.

De Paiva, L. B., Morales, A. R., & Valenzuela Díaz, F. R. (2008). Organoclays: Properties, preparation and applications. Applied Clay Science, 42, 8–24.

Dekkers, S., et al. (2011). Presence and risks of nanosilica in food products. Nanotoxicology, 5, 393–405.

Duan, Y., et al. (2010). Toxicological characteristics of nanoparticulate anatase titanium dioxide in mice. Biomaterials, 31, 894–899.

Dudefoi, W., Moniz, K., Allen-Vercoe, E., Ropers, M.-H., & Walker, V. K. (2017). Impact of food grade and nano-TiO2 particles on a human intestinal. Community. Food and Chemical Toxicology, 106, 242–249.

Duncan, T. V. (2011). Applications of nanotechnology in food packaging and food safety: Barrier materials, antimicrobials and sensors. Journal of Colloid and Interface Science, 363(1), 1–24.

Duran, N., & Marcato, P. D. (2013). Nanobiotechnology perspectives. Role of nanotechnology in the food industry: A review. International Journal of Food Science & Technology, 48(6), 1127–1134.

Dreher, K. L. (2011). Health and environmental impact of nanotechnology: Toxicological assessment of manufactured nanoparticles. Toxicological Sciences, 77, 3–5.

Echegoyen, Y., & Nerin, C. (2013). Nanoparticle release from nanosilver antimicrobial food containers. Food and Chemical Toxicology, 62, 16–22.

European Food Safety Authority (EFSA). (2011a). Scientific opinion on guidance on the risk assessment of the application of nanoscience and nanotechnologies in the food and feed chain. EFSA Journal, 9(5), 36. Available from https://doi.org/10.2903/j.efsa.2011.2140.

European Food Safety Authority (EFSA). (2011b). Guidance on the risk assessment of the application of nanoscience and nanotechnologies in the food and feed chain. EFSA Journal, 9, 36.

European Food Safety Authority (EFSA). (2012). Scientific Opinion on the safety evaluation of the substance, titanium nitride, nanoparticles, for use in food contact materials. EFSA Journal, 10(3), 2641.

European Food Safety Authority (EFSA). (2014). Statement on the safety assessment of the substance silicon dioxide, silanated, FCM Substance No 87 for use in food contact materials. EFSA Journal, 12(6), 3712.

European Food Safety Authority (EFSA). (2016a). Safety assessment of the substance zinc oxide, nanoparticles, for use in food contact materials. EFSDA Journal, 14(4408), 4401–4408.

European Food Safety Authority (EFSA). (2016b). Recent developments in the risk assessment of chemicals in food and their potential impact on the safety assessment of substances used in food contact materials. EFSA Journal, 14(1), 4357.

European Food Safety Authority (EFSA). (2018). Guidance on risk assessment of the application of nanoscience and nanotechnologies in the food and feed chain: Part 1, human and animal health.

Esmaeillou, M., Moharamnejad, M., Hsankhani, R., Tehrani, A. A., & Maadi, H. (2013). Toxicity of ZnO nanoparticles in healthy adult mice. Environmental Toxicology and Pharmacology, 35, 67–71.

European Commission (EC). (2004). Regulation (EC) No. 1935/2004 of the European Parliament and the Council of 27 October 2004 on materials and articles intended to come into contact with food and repealing directives 80/590/EEC and 89/109 EEC. Official Journal of the European Union, 338, 4.

European Commission (EC). (2007). Commission Directive 2007/19/EC of 2 April 2007 amending Directive 2002/72/EC relating to plastic materials and articles intended to come into contact with food and Council Directive 85/572/EEC laying down the list of simulants to be used for testing migration of constituents of plastic materials and articles intended to come into contact with foodstuffs. Official Journal of the European Union, 97, 50.

European Commission (EC) (2009). Commission Regulation (EC) No 450/2009 of 29 May 2009 on active and intelligent materials and articles intended to come into contact with food. The Official Journal of the European Union, 1, L135, 3–11.

European Commission (EC). (2010a). Towards a strategic nanotechnology action plan (SNAP) 2010-2015, http://ec.europa.eu/research/consultations/snap/report_en.pdf. Accessed November 2014.

European Commission (EC). (2010b). Commission Regulation (EC) No 257/2010 of 25 March 2010 setting up a programme for the re-evaluation of approved food additives in accordance with Regulation (EC) No 1333/2008 of the European Parliament and of the Council on food additives. Official Journal of the European Union, L80, 19–27.

European Commission (EC). (2011). Commission Regulation (EU) No 10/2011 of 14 January 2011 on plastic materials and articles intended to come into contact with food. Official Journal of the European Union, L328, 20–29.

European Commission (EC). (2013). Proposal for a regulation of the European parliament and of the council on novel foods. COM (2013) 894 final, http://ec.europa.eu/food/food/biotechnology/novelfood/documents/novel-cloning_com2013-894_final_en.pdf.

European Parliament and Council. (2011). Regulation (EU) No 1169/2011 of the European Parliament and of the Council of 25 October 2011 on the provision of food information to consumers. Official Journal of the European Union, L304, 18–63.

European Parliament and Council. (2012). Regulation (EU) No 528/2012 of the European Parliament and of the Council of 22 May 2012 concerning the making available on the market and use of biocidal products. Official Journal of the European Union, L167, 1–123.

FAO/WHO. (2013). Food and Agriculture Organization of the United Nations/World Health Organization. State of the art on the initiatives and activities relevant to risk assessment and risk management of nanotechnologies in the food and agriculture sectors: FAO/WHO technical paper report (p. 48) Rome, Italy: FAO/WHO.

Faridi Esfanjani, A., Assadpour, E., & Jafari, S. M. (2018). Improving the bioavailability of phenolic compounds by loading them within lipid-based nanocarriers. Trends in Food Science & Technology, 76, 56–66.

FDA. (2011). Considering whether an FDA-regulated product involves the application of nanotechnology. https://www.fda.gov/regulatory-information/search-fda-guidance-documents/considering-whether-fda-regulated-product-involves-application-nano-technology

FDA. (2012). Guidance for industry: Assessing the effects of significant manufacturing process changes, including emerging technologies, on the safety and regulatory status of food ingredients and food contact substances, including food ingredients that are color additives. https://www.fda.gov/regulatory-information/search-fda-guidance-documents/guidance-industry-assessing-effects-significant-manufacturing-process-changes-including-emerging

FDA. (2014a). Guidance for industry: Assessing the effects of significant manufacturing process changes, including emerging technologies, on the safety and regulatory status of food ingredients and food contact substances, including food ingredients that are color additives. https://www.fda.gov/regulatory-information/search-fda-guidance-documents/guidance-industry-assessing-effects-significant-manufacturing-process-changes-including-emerging

FDA. (Ed.) (2014b). Guidance for industry: Considering whether an FDA-regulated product involves the application of nanotechnology. https://www.federalregister.gov/documents/2014/06/27/2014-15033/guidance-for-industry-considering-whether-a-food-and-drug-administration-regulated-product-involves

Foladori, G., & Invernizzi, N. (2013). Inequality gaps in nanotechnology development in Latin America. Journal of Arts and Humanities, 2(3), 35–45.

Foladori, G., & Lau, E. Z. (2014). The regulation of nanotechnologies in Mexico. Nanotechnology Law & Business, 11, 164.

Food Standards Australia New Zealand (FSANZ) (2014). Summary of responses to FSANZ's industry packaging survey. Parliament of Australia Website. Available from: http://www.aph.gov.au/~/media/Committees/clac_ctte/estimates/bud_1415/DoH/answers/SQ14-000586_Att2.pdf.

Frohlich, E., & Roblegg, E. (2012). Models for oral uptake of nanoparticles in consumer products. Toxicology, 291, 10–17.

Fulgoni, V. L., Keast, D. R., Bailey, R. L., & Dwyer, J. (2011). Foods, fortificants, and supplements: Where do Americans get their nutrients? Journal of Nutrition, 141, 1847–1854.

Gaillet, S., & Rouanet, J. M. (2015). Silver nanoparticles: Their potential toxic effects after oral exposure and underlying mechanisms-a review. Food and Chemical Toxicology., 77, 58–63.

Gambaro, A. (2018). Projective techniques to study consumer perception of food. Current Opinion in Food Science, 21, 46–50.

Gesamte Rechtsvorschrift für Pflanzenschutzmittelverordnung (2011). https://www.ris.bka.gv.at/GeltendeFassung.wxe?Abfrage=Bundesnormen&Gesetzesnummer=20007374

Giles, E. L., Kuznesof, S., Clark, B., Hubbard, C., & Frewer, L. J. (2015). Consumer acceptance of and willingness to pay for food nanotechnology: A systematic review. Journal of Nanoparticle Research, 17, 467.

Graveland-Bikker, J. F., Ipsen, R., Otte, J., & de Kruif, C. G. (2004). Influence of calcium on the self-assembly of partially hydrolyzed a-lactalbumin. Langmuir, 20, 6841–6846.

Guo, Z., Martucci, N. J., Moreno-Olivas, F., Tako, E., & Mahler, G. J. (2017). Titanium dioxide nanoparticle ingestion alters nutrient absorption in an in vitro model of the small intestine. Nanoimpact, 5, 70–82.

Hock J., Behra R., Bergamin L., Bourqui-Pittet M., Bosshard C., Epprecht T., Furrer V., et al., (2018). Guidelines on the precautionary matrix for synthetic nanomaterials. Federal Office of Public Health and Federal Office for the Environment, Berne, Version 3.1.

Hansen, S. F., Baun, A., & Alstrup-Jensen, K. (2011). NanoRiskCat—a conceptual decision support tool for nanomaterials. Danish Ministry of the Environment. Environmental Project, No. 1372, 2011.

Hendrickson, O. D., et al. (2016). Toxicity of nanosilver in intragastric studies: Biodistribution and metabolic effects. Toxicology Letters, 241, 184–192.

Heock, J., Epprecht, T., Furrer, E., Gautschi, M., Hofmann, H., Heohener, K., et al. (2013). Guidelines on the precautionary matrix for synthetic nanomaterials. Federal Office of Public Health and Federal Office for the Environment, Version 3.0. Berne: Federal Office of Public Health and Federal Office for the Environment.

Hilty, F. M., et al. (2010). Iron from nanocompounds containing iron and zinc is highly bioavailable in rats without tissue accumulation. Nature Nanotechnology, 5, 374–380.

Hoet, P. H., Brüske-Hohlfeld, I., & Salata, O. V. (2004). Nanoparticles – known and unknown health risks. Journal of Nanobiotechnology, 2(1), 12. Available from https://doi.org/10.1186/1477-3155-2-12.

Hoseinnejad, M., Jafari, S. M., & Katouzian, I. (2018). Inorganic and metal nanoparticles and their antimicrobial activity in food packaging applications. Critical Reviews in Microbiology, 44(2), 161–181.

Hussain, N., Jaitley, V., & Florence, A. T. (2001). Recent advances in the understanding of uptake of microparticulates across the gastrointestinal lymphatics. Advanced Drug Delivery Reviews, 50, 107–142.

Hwang, M., Lee, E. J., Kweon, S. Y., Park, M. S., Jeong, J. Y., Um, J. H., et al. (2012). Risk assessment principle for engineered nanotechnology in food and drug. Toxicological Research, 28(2), 73–79.

Jafari, S. M., & Esfanjani, A. F. (2017a). 14—Instrumental analysis and characterization of nanocapsules. Nanoencapsulation technologies for the food and nutraceutical industries (pp. 524–544).

Jafari, S. M., & McClements, D. J. (2017b). Nanotechnology approaches for increasing nutrient bioavailability. Advances in food and nutrition research. Academic Press UK.

Jafari, S. M., Esfanjani, A. F., Katouzian, I., & Assadpour, E. (2017a). Chapter 10—Release, characterization, and safety of nanoencapsulated food ingredients. Nanoencapsulation of food bioactive ingredients (pp. 401–453). Academic Press UK.

Jafari, S. M., Katouzian, I., & Akhavan, S. (2017b). 15—Safety and regulatory issues of nanocapsules. Nanoencapsulation technologies for the food and nutraceutical industries (pp. 545–590). Academic Press UK.

Jeong, G. N., et al. (2010). Histochemical study of intestinal mucins after administration of silver nanoparticles in Sprague-Dawley rats. Archives of Toxicology., 84, 63–69.

Jones, O. G., & McClements, D. J. (2010). Functional biopolymer particles: Design, fabrication, and applications. Comprehensive Reviews in Food Science and Food Safety, 9, 374–397.

Joseph, T., & Morrison, M. (2006, May). Nanotechnology in agriculture and food. A Nanoforum Report. Institute of Nanotechnology., www.nanoforum.org.

Joye, I. J., Davidov-Pardo, G., & McClements, D. J. (2014). Nanotechnology for increased micronutrient bioavailability. Trends in Food Science and Technology, 40, 168–182.

Kang, T. S., et al. (2015). Cytotoxicity of zinc oxide nanoparticles and silver nanoparticles in human epithelial colorectal adenocarcinoma cells. Food Science and Technology, 60, 1143–1148.

Katouzian, I., & Jafari, S. M. (2016). Nano-encapsulation as a promising approach for targeted delivery and controlled release of vitamins. Trends Food Science and Technology., 53, 34–48.

Katouzian, I., & Jafari, S. M. (2019). Protein nanotubes as state-of-the-art nanocarriers: Synthesis methods, simulation and applications. Journal of Controlled Release, 303, 302–318.

Kay, L., & Shapira, P. (2009). Developing nanotechnology in Latin America. Journal of Nanoparticle Research, 11(2), 259–278.

Khemani, K., Moad, G., Lascaris, E., Li, G., Simon, G., Habsuda, J., …, inventors; Plantic Technologies Ltd., assignee. (2008). Starch nanocomposite materials. USA Patent US20100307951 A1.

Kim, Y. R., Lee, S. Y., Lee, E. J., et al. (2014). Toxicity of colloidal silica nanoparticles administered orally for 90 days in rats. International Journal of Nanomedicine, 9, 67–78.

Kim, H., Liu, X., Kobayashi, T., Kohyama, T., Wen, F. Q., Romberger, D., et al. (2003). Ultrafine carbon black particles inhibit human lung fibroblast-mediated collagen gel 696 Handbook of Food Nanotechnology contraction. American Journal of Respiratory Cell and Molecular Biology, 28, 111–121.

Kim, Y. S., et al. (2008). Twenty-eight-day oral toxicity, genotoxicity, and gender-related tissue distribution of silver nanoparticles in Sprague-Dawley rats. Inhalation Toxicology, 20, 575–583.

Kim, Y. S., et al. (2010). Subchronic oral toxicity of silver nanoparticles. Particle and Fibre Toxicology, 7, 20.

Kim, T. H., et al. (2012). Size-dependent cellular toxicity of silver nanoparticles. Journal of Biomedical Materials Research Part A, 100A, 1033–1043.

Kreilgaard, M. (2002). Influence of microemulsions on cutaneous drug delivery. Advanced Drug Delivery Reviews, 54, S77–S98.

Kreyling, W. G., Semmler, M., Erbe, F., Mayer, P., Takenaka, S., Schulz, H., et al. (2002). Translocation of ultrafine insoluble iridium particles from lung epithelium to extrapulmonary organs is size dependent but very low. Journal of Toxicology and Environmental Health, 65, 1513–1530.

Lai, S. K., O'Hanlon, D. E., Harrold, S., Man, S. T., Wang, Y. Y., & Cone, R. (2007). Rapid transport of large polymeric nanoparticles in fresh undiluted human mucus. Proceedings of the National Academy of Sciences of the United States of America, 104, 1482–1487.

Le Corre, D., Bras, J., & Dufresne, A. (2010). Starch nanoparticles: A review. Biomacromolecules, 11, 1139–1153.

Lee, K. E., Cho, S. H., Lee, H. B., Jeong, S. Y., & Yuk, S. H. (2003). Microencapsulation of lipid nanoparticles containing lipophilic drug. Journal of Microencapsulation, 20, 489–496.

Lee, J. A., Kim, M. K., Paek, H. J., Kim, Y. R., Kim, M. K., Lee, J. K., et al. (2014). Tissue distribution and excretion kinetics of orally administered silica nanoparticles in rats. International Journal of Nanomedicine, 9(Suppl 2), 251–260.

Lefebvre, D. E., Venema, K., Gombau, L., et al. (2015). Utility of models of the gastrointestinal tract for assessment of the digestion and absorption of engineered nanomaterials released from food matrices. Nanotoxicology, 9(4), 523–542. Available from https://doi.org/10.3109/17435390.2014.948091.

Limnach, L. K., Wick, P., Manser, P., Gras, R. N., Bruinink, A., & Stark, W. J. (2007). Exposure of engineered nanoparticles to human lung epithelial cells: Influence of chemical composition and catalytic activity on oxidative stress. Environmental Science & Technology, 41, 4158–4163.

Linsinger, T. P. J., Roebben, G., Gilliland, D., Calzolai, L., Rossi, F., Gibson, N., & Klein, C. (2012). Requirements on measurements for the implementation of the European Commission Definition of the Term "nanomaterial". European Commission Joint Research Centre, Luxembourg., http://publications.jrc.ec.europa.eu/repository/bitstream/111111111/26399/2/irmm_nanomaterials%20%28online%29.pdf.

Livney, Y. D. (2015). Nanostructured delivery systems in food: Latest developments and potential future directions. Current Opinion In Food Science, 3, 125–135.

Loeschner, K., Hadrup, N., Qvortrup, K., Larsen, A., Gao, X., Vogel, U., et al. (2011). Distribution of silver in rats following 28 days of repeated oral exposure to silver nanoparticles or silver acetate. Particle and Fibre Toxicology, 8, 18.

Mackevica, A., Olsson, M. E., & Hansen, S. F. (2016). Silver nanoparticle release from commercially available plastic food containers into food simulants. Journal of Nanoparticle Research, 18, 1–11.

Mahler, G. J., Esch, M. B., Tako, E., Southard, T. L., Archer, S. D., Glahn, R. P., et al. (2012). Oral exposure to polystyrene nanoparticles affects iron absorption. Nature Nanotechnology, 7, 264–271.

Marvin, H. J. P., Bouwmeester, H., Bakker, M., Kroese, E. D., van de Meent, D., Bourgeois, F., et al. (2013). Exploring the development of a decision support system (DSS) to prioritize engineered nanoparticles for risk assessment. Journal of Nanoparticle Research: An Interdisciplinary Forum for Nanoscale Science and Technology, 15(1839). Available from https://doi.org/10.1007/s11051-013-1839-3.

Maynard, A. D. (2006). Nanotechnology: assessing the risks. Nanotoday, 1, 22–33.

McCarron, E., 2016. Nanotechnology and food: Investigating consumers' acceptance of foods produced using nanotechnology. Dublin, Ireland.

McClements, D. J., DeLoid, G., Pyrgiotakis, G., Shatkin, J. A., Xiao, H., & Demokritou, P. (2016). The role of the food matrix and gastrointestinal tract in the assessment of biological properties of ingested engineered nanomaterials (iENMs): State of the science and knowledge gaps. Nanoimpact, 3-4, 47–57.

McClements, D. J., & Rao, J. (2011). Food-grade nanoemulsions: Formulation, fabrication, properties, performance, biological fate, and potential toxicity. Critical Reviews in Food Science and Nutrition, 51, 285–330.

McClements, D. J., & Xiao, H. (2017). Is nano safe in foods? Establishing the factors impacting the gastrointestinal fate and toxicity of organic and inorganic food-grade nanoparticles. NPJ Science of Food, 1–6. Available from https://doi.org/10.1038/s41538-017-0005-1.

McCracken, C., Zane, A., Knight, D. A., Dutta, P. K., & Waldman, W. J. (2013). Minimal intestinal epithelial cell toxicity in response to short- and long-term food-relevant inorganic nanoparticle exposure. Chemical Research in Toxicology, 26, 1514–1525.

Mihindukulasuriya, S. D. F., & Lim, L. T. (2014). Nanotechnology development in food packaging: A review. Trends in Food Science & Technology, 40(2014), 149–167.

Mohanty, A. K., Misra, M., & Nalwa, H. S. (2009). Packaging nanotechnology. Los Angeles, CA: American Scientific Publishers.

Mwilu, S. K., El Badawy, A. M., Bradham, K., Nelson, C., Thomas, D., Scheckel, K. G., et al. (2013). Changes in silver nanoparticles exposed to human synthetic stomach fluid: Effects of particle size and surface chemistry. Science of the Total Environment, 447, 90–98.

Myrick, J. M., Vendra, V. K., & Krishnan, S. (2014). Self-assembled polysaccharide nanostructures for controlled-release applications. Nanotechnology Reviews, 3, 319–346.

Nano 2030. (2014). Manufactured nanomaterials by 2030 (pp. 1–270). INRS. https://en.inrs.fr/news/manufactured-nanomaterials-by-2030.html

Nanodatabase. (2019). The Ecological Council, Danish Consumer Council, DTU Environment. https://nanodb.dk/

Nanoproducts. (2013). http://nanoproducts.de. Accessed September 2018.

Nanotech-data. (2013). http://www.nanodaten.de/site/page_de_garde.html. Accessed September 2018.

Nanotechnology Industries Association (NIA). (2011). Opportunities & risks in nanomedicine — How to build an infrastructure in support of the next generation of health care, 698 Handbook of Food Nanotechnology August 2931, 2011, Sao Paulo. Brussels, Belgium: Nanotechnology Industries Association.

Naseer, B., Srivastava, G., Qadri, O. S., Faridi, S. A., Islam, R. U., & Younis, K. (2018). Importance and health hazards of nanoparticles used in the food industry. Nanotechnology Reviews, 7, 623–641.

Nel, A., Xia, T., Madler, L., & Li, N. (2006). Toxic potential of materials at the nanolevel. Science, 311, 622–627.

Nemmar, A., Hoet, P. H. M., Vanquickenborne, B., Dinsdale, D., Thomeer, M., Hoylaerts, M. F., et al. (2002). Passage of inhaled particles into the blood circulation in humans. Circulation, 105, 411–414.

Oberdorster, G., Maynard, A., Donaldson, K., Castranova, V., Fitzpatrick, J., Ausman, K., et al. (2005). Principles for characterizing the potential human health effects from exposure to nanomaterials: Elements of a screening strategy. Particle and Fibre Toxicology. Available from. Available from http://www.particleandfibretoxicology.com/content/2/1/8.

Oberdörster, G. (2001). Pulmonary effects of inhaled ultrafine particles. International Archives of Occupational and Environmental Health, 74, 1–8.

Oberdörster, G., Ferin, J., & Lehnert, B. E. (1994). Correlation between particle size, in vivo particle persistence, and lung injury. Environmental Health Perspectives, 102(Suppl. 5), 173–179.

Oberdörster, G., Sharp, Z., Atudorei, V., Elder, A., Gelein, R., Lunts, A., et al. (2002). Extrapulmonary translocation of ultrafine carbon particles following whole-body inhalation exposure of rats. Journal of Toxicology and Environmental Health, 65, 1531–1543.

OECD. (2013). Regulatory frameworks for nanotechnology in foods and medical products: summary results of a survey activity. OECD science, technology and industry policy papers. Paris: OECD Publishing, No. 4.

Paik, S. Y., Zalk, D. M., & Swuste, P. (2008). Application of a pilot control banding tool for risk level assessment and control of nanoparticle exposures. The Annals of Occupational Hygiene, 52, 419–428.

Park, E. J., et al. (2010). Repeated-dose toxicity and inflammatory responses in mice by oral administration of silver nanoparticles. Environmental Toxicology and Pharmacology, 30, 162–168.

Park, J. Y., Li, S. F. Y., & Kricka, L. J. (2006). Nanotechnologic nutraceuticals: Nurturing or nefarious? Clinical Chemistry, 52, 331–332.

Pathakoti, K., Manubolu, M., & Hwang, H. M. (2017). Nanostructures: Current uses and future applications in food science. Journal of Food and Drug Analysis, 25, 245–253.

Patil, U. S., et al. (2015). In vitro/in vivo toxity evaluation and quantification of iron oxide nanoparticles. International Journal of Molecular Sciences, 16, 24417–24450.

Persistence Market Research. (2014). Global market study on nano-enabled packaging for food and beverages: Intelligent packaging to witness highest growth by 2020. PersistenceMarket ResearchWebsite. Available from:, http://www.persistencemarketresearch.com/market-research/nano-enabled-packaging-market.asp. Accessed January 2015.

Peters, R., et al. (2012). Presence of nano-sized silica during in vitro digestion of foods containing silica as a food additive. ACS Nano, 6, 2441–2451.

Pico, Y., & Blasco, C. (2012). Nanomaterials in food, Which way forward? In D. Barcelo, & M. Farre (Eds.), Analysis and risk of nanomaterials in environmental and food samples (p. 305). Great Britain: Elsevier.

Rajendran, S., Udenigwe, C. C., & Yada, R. Y. (2016). Nanochemistry of protein-based delivery agents. Frontiers in Chemistry, 4, 31.

Rasmussen, K., Rauscher, H., Kearns, P., Gonzalez, M., & Sintes, J. R. (2019). Developing OECD test guidelines for regulatory testing of nanomaterials to ensure mutual acceptance of test data. Regulatory Toxicology and Pharmacology, 104, 74–83.

Rauscher, H., Roebben, G., Amenta, V., Boix Sanfeliu, A., Calzolai, L., Emons, H., ..., Stamm, H. (2014). Towards a review of the EC recommendation for a definition of the term "nanomaterial": Part 1: Compilation of information concerning the experience with the definition. European Commission Joint Research Centre, https://ec.europa.eu/jrc/sites/default/files/lbna26567enn.pdf.

Rezaei, A., Fathi, M., & Jafari, S. M. (2019). Nanoencapsulation of hydrophobic and low soluble food bioactive compounds within different nanocarriers. Food Hydrocolloids, 88, 146–162.

Rischitor, G., Parracino, M., La Spina, R., Urbán, P., Ojea-Jiménez, I., Bellido, E., ... Colpo, P. (2016). Quantification of the cellular dose and characterization of nanoparticle transport during in vitro testing. Particle and Fibre Toxicology, 13, 47.

Roebben, G., Rauscher, H., Amenta, V., Aschberger, K., Boix Sanfeliu, A., Calzolai, L., ..., Stamm, H. (2014). Towards a review of the EC recommendation for a definition of the term "nanomaterial": Part 2: Assessment of collected information concerning the experience with the definition. European Commission Joint Research Centre., https://ec.europa.eu/jrc/en/publication/eur-scientific-and-technical-research-reports/towardsreview-ecrecommendation-definition-term-nanomaterial-part-2-assessment-collected.

Ropers, M. H. (2019). Nanomaterials and food security: The next challenge for consumers, food industries and policies. Encyclopedia of Food Security and Sustainability, 2, 575–581.

Rostamabadi, H., Falsafi, S. R., & Jafari, S. M. (2019). Starch-based nanocarriers as cutting-edge natural cargos for nutraceutical delivery. Trends in Food Science & Technology, 88, 397–415.

SCENIHR, 2007. Opinion in the appropriateness of the risk assessment methodology in accordance with the technical guidance documents for new and existing substances for assessing the risks of nanomaterials. scientific committee on emerging and newly identified health risks, Brussels from http://ec.europa.eu/health/archive/ph_risk/committees/04_scenihr/docs/scenihr_o_010.pdf

SCENIHR (2009). Scientific committee on emerging and newly identified health risks. risk assessment of products of nanotechnologies. Available from http://ec.europa.eu/health/archive/ph_risk/committees/04_scenihr/docs/scenihr_o_023.pdf.

Sekhon, B. S. (2010). Food nanotechnology—an overview. Nanotechnology, Science and Applications, 3(1), 1.

Sharma, V. K., Siskova, K. M., Zboril, R., & Gardea-Torresdey, J. L. (2014). Organic-coated silver nanoparticles in biological and environmental conditions: Fate, stability and toxicity. Advances in Colloid and Interface Science, 204, 15–34.

Shi, S., Wang, W., Liu, L., Wu, S., Wei, Y., & Li, W. (2013). Effect of chitosan/nano-silica coating on the physicochemical characteristics of longan fruit under ambient temperature. Journal of Food Engineering, 118, 125–131.

Shin, G. H., Kim, J. T., & Park, H. J. (2015). Recent developments in nanoformulations of lipophilic functional foods. Trends in Food Science and Technology, 46, 144–157.

Sia, P. (2017). Nanotechnology among innovation, health and risks. Trends in Food Science and Technology, 237, 1076–1080.

Silvestre, C., Duraccio, D., & Cimmino, S. (2011). Food packaging based on polymer nanomaterials. Progress in Polymer Science, 36(12), 1766–1782.

Sirelkhatim, A., et al. (2015). Review on zinc oxide nanoparticles: Antibacterial activity and toxicity mechanism. Nano-Micro Letters, 7, 219–242.

Sohal, I. S., O'Fallon, K. S., Gaines, P., Demokritou, P., & Bello, D. (2018). Ingested engineered nanomaterials: State of science in nanotoxicity testing and future research needs. Particle and Fibre Toxicology, 15, 29.

Su, S. L., & Li, Y. (2004). Quantum dot biolabeling coupled with immunomagnetic separation for detection of Escherichia coli O157:H7. Analytical Chemistry, 76(16), 4806–4810.

Szakal, C., Roberts, S. M., Westerhoff, P., Bartholomaeus, A., Buck, N., Illuminato, I., et al. (2014). Measurement of nanomaterials in foods: Integrative consideration of challenges and future prospects. ACS Nano, 8, 3128–3135.

Tager, J. (2014). Nanomaterials in food packaging: FSANZ fails consumers again. Chain Reaction, 122, 16–17.

Takeuchi, M. T., Kojima, M., & Luetzow, M. (2014). State of the art on the initiatives and activities relevant to risk assessment and risk management of nanotechnologies in the food and agriculture sectors. Food Research International, 64, 976–981.

Tassinari, R., Cubadda, F., Moracci, G., Aureli, F., D'Amato, M., Valeri, M., ... Maranghi, F. (2014). Oral, short-term exposure to titanium dioxide nanoparticles in Sprague–Dawley rat: Focus on reproductive and endocrine systems and spleen. Nanotoxicology, 8(6), 654–662.

Tinkle, S. S., Antonini, J. M., Rich, B. A., Roberts, J. R., Salmen, R., DePree, K., & Adkins, E. J. (2003). Skin as a route of exposure and sensitization in chronic beryllium disease. Environmental Health Perspectives, 111, 1202–1208.

United States Environmental Protection Agency (USEPA). (2013). Nanomaterials EPA is assessing. The United States Environmental Protection Agency Website. Available from http://www.epa.gov/nanoscience/quickfinder/nanomaterials.htm. Accessed January 2015.

United States Food and Drug Administration (USFDA). (2011). Draft guidance for industry: Dietary supplements: New dietary ingredient notifications and related issues. 2011 July ed., Maryland. Available from http://www.fda.gov/Food/GuidanceRegulation/GuidanceDocumentsRegulatoryInformation/DietarySupplements/ucm257563.htm. Accessed November 2014.

United States Food and Drug Administration (USFDA). (2014a). CFR-Code of federal regulations title 21 – Food and Drugs. Available from https://www.fda.gov/medical-devices/medical-device-databases/code-federal-regulations-title-21-food-and-drugs (accessed January 25, 2021).

United States Food and Drug Administration (USFDA). (2014b). Guidance for industry: Assessing the effects of significant manufacturing process changes, including emerging technologies, on the safety and regulatory status of food ingredients and food contact substances, including food ingredients that are color additives. Available from https://www.fda.gov/regulatory-information/search-fda-guidance-documents/guidance-industry-assessing-effects-significant-manufacturing-process-changes-including-emerging (accessed January 25, 2021).

Van de Waterbeemd, H., Lennernas, H., & Artursson, P. (2003). Drug bioavailability: Estimation of solubility, permeability, absorption and bioavailability. Wiley-VCH.

Van der Zande, M., Vandebriel, R. J., Van Doren, E., Kramer, E., Herrera Rivera, Z., Serrano-Rojero, C. S., et al. (2012). Distribution, elimination, and toxicity of silver Safety and regulatory issues of nanomaterials in foods 701 nanoparticles and silver ions in rats after 28-day oral exposure. ACS Nano, 6, 7427–7442.

Van Duuren-Stuurman, B., Vink, S. R., Verbist, K. J., Heussen, H. G., Brouwer, D. H., Kroese, D. E., et al. (2012). Stoffenmanager Nano version 1.0: A web-based tool for risk prioritization of airborne manufactured nano objects. The Annals of Occupational Hygiene, 56, 525–541.

Van Kesteren, P. C. E., et al. (2015). Novel insights into the risk assessment of the nanomaterial synthetic amorphous silica, additive E551, in food. Nanotoxicology, 9, 442–452.

Von der Kammer, F., Legros, S., Larsen, E. H., Loeschner, K., & Hofmann, T. (2011). Separation and characterization of nanoparticles in complex food and environmental samples by field-flow fractionation. Trends in Analytical Chemistry, 30(3), 425–436.

Wakefield, G., Green, M., Lipscomb, S., & Flutter, B. (2004). Modified titania nanomaterials for sunscreen applications e reducing free radical generation and DNA damage. Materials Science and Technology, 20, 985–988.

Walczak, A. P., Fokkink, R., Peters, R., Tromp, P., Herrera Rivera, Z. E., Rietjens, I. M., et al. (2013). Behaviour of silver nanoparticles and silver ions in an in vitro human gastrointestinal digestion model. Nanotoxicology, 7(7), 1198–1210.

Walters, C., Pool, E., & Somerset, V. (2016). Nanotoxicity in aquatic invertebrates. In M. L. Larramendy (Eds.), Invertebrates experimental models in toxicity screening (pp. 13–34). IntechOpen, London, UK.

Wang, J., et al. (2007). Acute toxicity and biodistribution of different sized titanium dioxide particles in mice after oral administration. Toxicology Letters, 168, 176–185.

Wang, Y., et al. (2013). Susceptibility of young and adult rats to the oral toxicity of titanium dioxide nanoparticles. Small, 9, 1742–1752.

Wang, Y. L., et al. (2014). A combined toxicity study of zinc oxide nanoparticles and vitamin C in food additives. Nanoscale, 6, 15333–15342.

Wani, T. A., Masoodi, F. A., Jafari, S. M., & McClements, D. J. (2018). Chapter 19— Safety of nanoemulsions and their regulatory status. In S.M. Jafri, D.J. McClements (Eds.), Nanoemulsions pp. 613–628. Academic Press, Cambridge, MA.

Weir, A., Westerhoff, P., Fabricius, L., Hristovski, K., & von Goetz, N. (2012). Titanium dioxide nanoparticles in food and personal care products. Environmental Science & Technology, 46, 2242–2250.

Williams, K., et al. (2015). Effects of subchronic exposure of silver nanoparticles on intestinal microbiota and gut-associated immune responses in the ileum of Sprague-Dawley rats. Nanotoxicology, 9, 279–289.

Wu, H. H., Yin, J. J., Wamer, W. G., Zeng, M. Y., & Lo, Y. M. (2014). Reactive oxygen species related activities of nano-iron metal and nano-iron oxides. Journal of Food and Drug Analysis., 22, 86–94.

Yang, Y., et al. (2016). Survey of food-grade silica dioxide nanomaterial occurrence, characterization, human gut impacts and fate across its lifecycle. Science of the Total Environment, 565, 902–912.

Yao, M., McClements, D. J., & Xiao, H. (2015). Improving oral bioavailability of nutraceuticals by engineered nanoparticle-based delivery systems. Current Opinion In Food Science, 2, 14–19.

Yu, Y. W., Zhang, S. Y., Ren, Y. Z., Li, H., Zhang, X. N., & Di, J. H. (2012). Jujube preservation using chitosan film with nano-silicon dioxide. Journal of Food Engineering, 113, 408–414.

Zalk, D. M., Paik, S. Y., & Swuste, P. (2009). Evaluating the control banding nanotool: A qualitative risk assessment method for controlling nanoparticle exposures. Journal of Nanoparticle Research, 11, 1685–1704.

Zhou, Y., Fang, X., Gong, Y., Xiao, A., Xie, Y., Liu, L., & Cao, Y. (2017). The Interactions between ZnO nanoparticles (NPs) and -linolenic acid (LNA) complexed to BSA did not influence the toxicity of ZnO NPs on HepG2 cells. Nanomaterials, 7, 91.

Zimmermann, M. B., & Hilty, F. M. (2011). Nanocompounds of iron and zinc: Their potential in nutrition. Nanoscale, 3, 2390–2398.

Index